HAROLD K. STEEN, EDITOR

FOREST AND WILDLIFE SCIENCE IN AMERICA

A History

Forest History Society

1999

The Forest History Society is a nonprofit, educational institution
dedicated to the advancement of historical understanding of man's
interaction with the forest environment. It was established in 1946.
Interpretations and conclusions in FHS publications are those of the
authors; the institution takes responsibility for the selection of
topics, the competence of the authors, and their freedom of inquiry.

This publication was developed in cooperation with the USDA
Forest Service, Southern Forest Experiment Station, Asheville,
North Carolina, and the USDA Forest Service History Program.

Library of Congress Cataloging-in-Publication Data
Forest and wildlife science in America : a history / Harold K. Steen.
          p.   cm.
Includes bibliographical references.
ISBN 0-89030-057-7 (pbk.)
1. Forests and forestry—United States—History. 2. Forest
ecology—United States—History. 3. Forests and forestry—
Research—United States—History. 4. Forest ecology—
Research—United States—History. I. Steen, Harold K.
SD143.F624 1999
634.9′0973–dc21   98-53507
                                          CIP

Cover and title page illustration: The U. S. Forest Service established
its first experiment station in 1908 on the Coconino National Forest
in Arizona. Here, two employees enjoy collecting soil samples.
(USDA Forest Service photo by Gus A. Pearson, 1910.)
Book design by Teresa Smith Perrien

CONTENTS

Dedication

v

Introduction

vii

*To the memory of
David Bruce, Marion Clawson,
Stanley Gessel, and George Staebler—
mentors and friends*

INTRODUCTION

Just as forestry is an amalgam of disciplines, forest science draws from a broad spectrum of natural and social sciences. Trees are plants and thus have a botanical dimension. Trees produce wood that economists evaluate and that wood technologists study. Trees produce habitat for wildlife and they grow on soil. Trees are a component of the forest ecosystem and fall under the scrutiny of theoretical ecologists and the more applied silviculturalists. Trees have a social value beyond their economic contribution and are of interest to sociologists and the related recreational sciences. And if nature's tree is not good enough, the geneticist strives to do better. The list is much longer, but you get the idea.

Not only is forest science a mix of disciplines, but it is advanced by institutions with different missions. The Forest Service and the Park Service are land management agencies, and their research programs reflect their land base. But the Forest Service by tradition and later by law also investigates all of the forested lands of the United States. The Soil Conservation Service, now the Natural Resource Conservation Service, not having land of its own works to advance the needs of its myriad clients.

University forest science has many parallels to its government counterparts; although not typical it is not unusual for scientists to transfer back and forth during their careers. But the primary mission of universities is education, and the generally short tenure of graduate students—apprentice scientists—influences the nature of many of the studies undertaken. Too, the major infusion of federal money into university forest science since the early 1960s has encouraged more than a few professors to shift their area of interest.

Because of its largely proprietary nature, the contribution of corporate research to forest science is more difficult to measure—corporate scientists publish less than their university and government colleagues. Capturing the benefits of products development is probably easier to accomplish through patents than is biological research, and industry wide there is much more investment in learning how to use wood than how to grow trees.

The sixteen chapters that follow well illustrate the above characterization, with a mix of broad strokes and focused accounts. The authors, too, are a mix, with historians writing about science and scientists trying their hands at history. The realities of space and budget, and the lack of availability of qualified authors, are reflected in the obvious chapters that are missing. But the strengths of this book far outweigh whatever deficiencies the gaps may impose. Needed is yet another compilation on the history of forest science.

Many thanks are due the authors, a cheerful and understanding lot in the face of a much postponed publication.

PART ONE

Institutional Programs

HAROLD K. STEEN[1]

THE FOREST SERVICE
AND THE HISTORY
OF FOREST SCIENCE

For well more than a century, the U.S. Forest Service and its predecessors have displayed an active and productive interest in forest science. In fact, it is fair to say that the agency has set the standards for others to emulate. Of course, university faculty conducted similar research, but it would not be until the 1962 McIntire-Stennis act that predictable and substantial research funds became available on campus for forest science. The private sector traditionally focused on products development; only a few companies in recent years have supported biological research. Finally, other agencies — most notably the National Park Service and the Soil Conservation Service — added to the fund of forest science knowledge. The Forest Service research story tangibly begins in 1876.

In response to a recommendation from the American Association for the Advancement of Science, in 1876 Congress added $2,000 to the Agriculture budget. The amount was to support the gathering of forestry facts. The commissioner of agriculture asked Franklin B. Hough, who had prompted the AAAS petition, to take on the task. Funding would be continued, and during the next eight years, Hough and others assembled four impressive volumes that reported the state of the art for forest science and forest conditions for the United States, plus glimpses of the rest of the world.

These *Reports Upon Forestry* show us that a clear understanding existed about the need for forestry research and that the complex and diverse American ecosystem meant that the investigations would need to be regionally organized. We can also see that the level of science itself was modest; John Evelyn's *1664 Sylva: A Discourse on Trees* that drew heavily upon classical authorities had yet to be replaced. The best current science was limited to botanical descriptions contained in impressive works such as the *Silva of North America* compiled by Harvard's Charles S. Sargent.

As useful as Hough's compilations were, the process added little new knowledge. That deficiency would end when Bernhard E. Fernow was appointed in 1886 as chief of the Division of Forestry. The division itself had evolved from Hough's earlier, annually renewed efforts and was now a permanent agency. While Hough had been trained in medical science and had been a statistician for the Bureau of Census, Fernow had been trained in forestry in his native Germany. When Fernow left the agency after twelve years to found the nation's first four-year forestry school at Cornell University, he could report that the Division of Forestry had produced twenty-three major publications, twenty informational circulars, and a variety of reports.

The publications reported upon ways to conserve wood by using alternative materials, such as metal railroad ties. There were also studies of southern pine, sheep grazing, dendrology, wood decay, wood structure, and forest influences. This latter topic was especially germane to the needs of the time, as Congress was in the process of developing the federal role in American forestry. If forests contributed a positive influence on flood control and water supply, for example, then public forest reserves could be rationalized. Fernow's repeated reports and testimony to Congress on forest influences played a material role in the creation of the National Forest System.

THE PINCHOT ERA

When the politically astute Gifford Pinchot succeeded Fernow in 1898 as chief of the Division of Forestry, he would be operating in a context that was favorable to the expansion of forestry activities. By then Congress had authorized the president to reserve by proclamation forested lands in public ownership to assure future timber supplies and to maintain favorable conditions of water flows. Just how to do this required site-specific information that was generally lacking. At the same time to reassure nervous westerners that his agency was practically oriented and was not about to impede commercial development through application of unproved theories, Pinchot renamed "research" as "investigations." This minor deception would remain in place until 1915.

By 1901 the Section of Special Investigations listed "studies of commercial trees, forest fires, grazing, log scales, forests and water supply, compilation of forest histories, and the investigation of forest products." In that report, Pinchot listed forty-two individual investigations; the investigators themselves included many who would become eminent—Herman von Schrenk, George B. Sudworth, W. L. Jepson, H. D. Tiemann, Samuel J. Record, Raphael Zon, H. S. Betts, and Charles Herty.

Three decades earlier, Hough had called for a network of research facilities across the nation to capture and articulate the diverse American forest ecosystem. In 1908, Pinchot set out to create just such a system, beginning with an experiment station near Flagstaff, Arizona. Within five years, there would be field stations in Idaho, Minnesota, Washington, Colorado, and Utah, adding to the substantial effort in Washington, D.C. In addition the Forest Products Laboratory had opened in Madison, Wisconsin, in 1910, greatly bolstering the effort to understand both the fundamental properties of wood and to develop a wider range of more useful and reliable products. The ability to manufacture more products from fewer trees was and remains a key element of conservation.

For its findings to be credible, research must have a reasonable degree of independence, both in fact and in appearance. Forest Service research and its contributions to forest science are no different, and throughout its history the issue of independence from the agency's administrative goals would remain visible. A famous case in point is the forest/flood study at the Wagon Wheel Gap Experiment Station in southern Colorado. In this instance, administrators wanted research to support policy and found the money to do so. Independent scholars have questioned whether studies that might have challenged policy would be more difficult to fund. A significant contributing factor was the pragmatic requirement that early research be descriptive, as opposed to analytical, to come up with as quickly as possible the much-needed information to support rapidly expanding forest management objectives. An important step toward independence was taken in 1915 with the creation of the Branch of Research; within a decade at least most research activities were adequately buffered from day-to-day administrative needs.

THE 1920S

Since 1897 scientists had contended that forest ranges were being depleted by overgrazing. Hard information was needed on range conditions and options for management. In 1915, the Forest Service acquired two experimental ranges from the Bureau of Plant Industry; the research goal was two-fold.

The restoration, improvement, and maintenance of the basic range resource and how to obtain the greatest returns on livestock proved of central importance. An impressive compilation by James T. Jardine and Mark Anderson, *Range Management on the National Forests* quickly became the range manager's bible. As range management on national forests was highly politicized, however, research results were not always implemented or at the very least were challenged.

World War I, as would subsequent conflicts, had prompted substantial shifts in the assignments that forest scientists received. These shifts were most evident at the Forest Products Laboratory, where packaging technology to aid shipment of matériel was given high priority. Waterproof adhesives and wood-drying technology were also emphasized as the Lab's budget and personnel increased five-fold to meet the demand for war-related science and technology. But postwar reductions were even more dramatic, and it would be years before Forest Service research regained its prewar level.

By the mid 1920s it became apparent that long term studies were needed to measure the effects of human disturbance, as opposed to the ongoing and often substantial changes brought about through ecological processes and natural occurrences, such as flood and hurricane. Toward that end, in 1927 the Forest Service established the Santa Catalina Natural Area in Arizona's Coronado National Forest. The 250th such area was established in 1992 to conduct nonmanipulative research and monitoring. At the same time, the agency, through its research division, was conducting an overall and ongoing survey of the nation's forest base as means to measure quality, quantity, and conditions.

THE GREAT DEPRESSION

It is difficult to determine just how much is directly related to the financial situation, but during the Great Depression of the 1930s, the agency's research arm and its scientists had more to do and more money to do it. Three major studies stand out; one on forestland taxation, another on rangeland, and a third on Douglas-fir logging methods.

Many had long felt that property taxes on forestland provided a disincentive to long range management in that early liquidation of timber would reduce value and thus taxes. Since 1908 Professor Fred R. Fairchild at Yale had studied the issue, and in 1924 the Forest Service asked him to conduct a major investigation. The final, 681-page report, *Forest Taxation in the United States*, appeared in 1935. Fairchild stated that although taxes might be an impediment to reforestation following logging, the decision to log or not was influenced by many factors. Property taxes, though, were not a significant contributor to the decision. For whatever reason, the tax study failed to convince and quickly dropped from view. It remains an interesting historical anecdote.

Axel Brandstrom and Burt T. Kirkland investigated alternatives to clearcutting Douglas-fir. Brandstrom was assigned to the experiment station in Portland, Oregon, and Kirkland was on the forestry faculty at the University of Washington. One of their goals was to make logging profitable during the period of extraordinarily low prices of lumber during the Depression. In 1936 they coauthored a report in which they proposed that selective logging of old-growth Douglas-fir would both increase profits and also be silviculturally acceptable. The report was controversial and as a compromise was published through a private grant. Although repeatedly challenged by silvicultural experts, the report has remained visible because it offers an alternative to the clearcutting of what now are called "ancient forests."

Range use had been controversial since the 1890s, and it remained controversial. When Congress resisted Forest Service attempts to reduce overgrazing during the 1920s, range quality continued to decline. As a result, in 1932 the Forest Service began a massive study of range conditions, publishing *The Western Range* in 1936. The six-hundred page report was blunt; rangelands were seriously deteriorated for two basic reasons. The Department of the Interior had failed to live up to its management responsibilities, and also because the 1934 Taylor Grazing Act allowed the livestock industry too much autonomy. That the industry had to deal with two federal agencies— Agriculture's Forest Service and Interior's Grazing Service— compounded the situation. Those accused responded that too little rainfall was the root of the range problem and not overgrazing. Once again the issue became politicized and there was no clear resolution.

Indoors, as it were, at the Forest Products Laboratory an effort took place to fill the research void caused by private sector slowdown during the Depression. The first laminated beam appeared in Madison in 1935 and the first prefabricated house in 1937. Lab scientists also studied lignin, one of wood's primary constituents, and products that could be chemically derived from wood. The Lab published the *Wood Handbook* in 1935, an invaluable reference work.

The forest industry supported the Lab and its budget, giving it full credit for the "remarkable expansion" of the industry in the South, for bringing plywood beyond being "crude and expensive," and for its "almost revolu-

The Forest Products Laboratory located in Madison, Wisconsin, began researching wood and wood products in 1910. This building was constructed in 1932 and is now on the National Register of Historic Places.
(FOREST HISTORY SOCIETY PHOTO)

tionary" advances in wood seasoning and preservation. Industrialists believed that getting facts, proving their effectiveness, and making them available was the Lab's "basic job."

In 1936 the Oxford Paper Company offered thousands of poplar hybrids that it had been studying for over a decade to the Northeastern Experiment Station. A year later Seattle lumberman James G. Eddy donated the Eddy Tree Breeding Station in Placerville, California, to the Forest Service; both gifts assured that forest genetics would become a significant research activity. More and more often, logged-off forests would be replaced by genetically "superior" plantations.

During this period, Forest Service researchers assisted farmers who wanted to develop cooperatives. They studied silvicultural systems, devised a log-grading system, and refined an inventory method for standing timber. Other federal programs to aid farmers funded various Forest Service economics research efforts and became a significant part of the investigative budget.

The work of entomologist Paul Keen well illustrated the goal to develop "useful" processes. He studied ponderosa pine bark beetles to determine which host trees were most susceptible to attack. Soon the field foresters could compare ponderosa pine crown conditions with simple sketches that Keen had prepared, and remove those stems that would soon be lost to insect infestation. In just these sorts of way, scientists added bits and pieces to the forester's toolkit.

It could be argued that the single most valuable research program was the setting up of growth plots, species by species. It would be the next generation of scientists who would reap the benefits of having access to continuous data collected over decades. Much of this information is critical to current understanding of forests. A related effort was the establishment of additional experimental forests and ranges; eventually there would be more than eighty.

Data analysis itself came under scrutiny during the 1930s; less and less often would it be acceptable for an investigator to make observations and prepare an explanatory narrative. More and more often, the analysis would need to be "statistically valid." As early as 1931 Forest Service scientists would begin to apply mathematical procedures to both experimental design and data analysis. At the same time, forestry schools began to teach statistical methods, and throughout the 1930s there would be increased emphasis on these more rigorous processes. In 1938 when the National Research Council published its study of Forest Service research, it reported that "statistically valid results" were the order of the day. It would not be until the next generation of Forest Service scientists, however, that statistical methods were routinely applied.

WORLD WAR II AND POSTWAR EXPANSION

World War II ended the Great Depression and abruptly shifted Forest Service research efforts once again to military purposes. For example, Japanese control of southeastern Asian supplies of natural rubber required that new sources be found. Two domestic efforts were the cultivation of a Russian dandelion called kok-saghyz and guayule, a southwestern shrub. By war's end when rubber imports resumed, guayule showed promise but kok-saghyz production had yielded little more than a set of experimental tires. Of much greater benefit was research at the Forest Products Laboratory, where scientists developed low-temperature glues for laminating beams and preservatives for wood used in the tropics. But as in the Great War a generation earlier, the Lab excelled in packaging technology. The effort included the training of some fourteen thousand personnel in the efficient use of packaging materials, economy of space, and protection of contents. When peace returned, the Lab quickly shifted priorities to civilian needs, such as kiln technology to produce quality lumber for the millions of new homes demanded by returning military.

By the 1950s Forest Service research expanded rapidly into areas of basic research. Statistical design and analysis became common, and new hires more and more often held advanced degrees. Flexible schedules and the Government Employees Training Act allowed scientists already employed to acquire additional academic training. Although the majority of Forest Service research would continue to focus on solving practical problems faced by field personnel, more and more attention would be given to answering questions of a basic nature.

To the standard research fields of forest management, fire, range, forest influences/watershed, economics, products, and tropics were added pathology, entomology, engineering, marketing, habitat, and recreation—a doubling of fields accompanied by new scientists who had not received a basic education in forestry. Not only was the list of fields changing but the scientists were changing as well, with an ever-higher proportion holding Ph.D.'s and representing a broad array of scientific disciplines.

Most new fields grew out of established areas of study. For example, game management had an element of recreation about it, and with shifts in emphasis, recreation research was spun off from game. Recreation was especially difficult to establish as a recognized field, however, because many felt that it was not a critical topic. Initially recreation researchers looked at campsite spacing, site protection, facilities design, and wilderness use. By the 1960s the field was well established.

Afforestation studies in the Prairie States joined water quality studies in Alaska as new ventures. At all experiment stations across the United States, investigators looked for ways to improve both natural and artificial regeneration following logging. Treatment of competing vegetation, selection of genetically superior seedlings, and classification of soils were just a few such studies that were aimed at improving regeneration.

More traditionally a long-accepted line of investigations produced a series of monographs that treated major forest species. G. A. Pearson published on ponderosa pine, Thomas Croker on longleaf pine, Leo Issac on Douglas-fir, and so on down the list of species, adding to a basic fund of knowledge that would be used universally. Through these means, the Forest Service set the standards for nomenclature, dendrology, and silvicultural methods.

The highly productive southern forest was matched by a high level of

Bernard Eger, forest ranger on the Pedlar District of the George Washington National Forest in Virginia, examines the hardwood tree sample he obtained using an increment borer.
(USDA FOREST SERVICE PHOTO BY B. W. MUIR, 1939)

Forest Service research activity. The Southern Institute of Forest Genetics brought research on that topic to par with older genetics institutes in the Lake States and California. Hardwoods research received special attention in the South, with a wide range of silvicultural studies. Also there was a fire laboratory in Georgia, as was in Montana and California. At the fire laboratories, scientists measured the effectiveness of chemical retardants, and they studied ways to prevent lightning-caused fires. In experimental forests, water supplies received attention. In some regions, increasing water yield was the goal, while in others it was to affect distribution throughout the year. Thus, there were water studies in arid Utah and humid North Carolina that developed baseline data that remains useful.

BROADENING THE RESEARCH BASE

The timing was no doubt coincidental, but the base of forest science was broadened appreciably during the same year that all forestry institutions would be placed under heavy pressure to look beyond traditional horizons. In 1962 the McIntire-Stennis Law began pumping substantial sums into land grant universities in support of forestry research. The goal was to expand university-based forest science to the same level as that of the Forest Service. Also in 1962 Rachel Carson's *Silent Spring* was published, ushering in the environmental movement with its insistence that research go beyond finding means to the prudent husbanding of resources to include quality of life and the so-called noncommercial resources of the forest. Very high on the research agenda was the seeking of nonpersistent pesticides and the development of biological controls. The roundheaded pine beetle, the tussock moth, the larch case bearer, and dwarf mistletoe would very shortly meet their biological match.

Related to the environmental movement was a rapidly broadening interest in wilderness. In 1924 the Forest Service had established the world's first wilderness area, in part to retain examples of untrammeled tracts to serve as benchmarks. By the 1960s, wilderness had become highly politicized; Congress would debate wilderness legislation for eight years. In this context, Forest Service scientists studied wilderness to determine who used it, how it might be managed without reducing its wildness, and even to better define the term. Wilderness had become a human construct, where "man himself is a visitor who does not remain."

Research in other fields progressed. Under the general heading of forest management, genetics studies aimed at developing "improved" trees with longer fiber length, uniform wood density, and disease resistance. Silvicultural research yielded a new guide for managing hardwoods in the Central States, sought optimum stocking levels for the South's loblolly pine, and measured nutrient loss brought about by logging. Economists noted that lumber was being displaced by plywood in home construction. They also predicted that within two decades technological shifts would displace a quarter of Pacific Northwest loggers and nearly half of the region's mill workers. Too, log exports to Japan were creating higher domestic prices.

By 1967 the annual research report contained a new heading, Environmental Forestry Research, showing a responsive shift in both nomenclature and in topics. Included here was work on air pollution and its effect on tree mortality. In only a few more years, there would be much concern about cleaner air and related issues. This early research would enable the Forest Service to achieve a much broader base for its later investigations.

THE ENVIRONMENTAL DECADE

Stepping up the tempo, the 1970s began as the "Environmental Decade," with the National Environmental Policy Act that established the Council on Environmental Quality and the federal reorganization that created the Environmental Protection Agency. Federal agencies, some more quickly than others, learned that the new laws were aimed at their operations. Changes would include a combination of new emphases and shifts in existing emphases. To bring these changes to the ground level required refocused research.

Clearcutting had been debated for generations, but the debate quickly intensified under the broader aegis of environmental concerns. Forest Service silvicultural scientists prepared a report—*Silvicultural Systems for the Major Forest Types of the United States*—that showed the state of the art for all logging systems, including clearcutting. The point was, properly applied clearcutting was acceptable for some timber types under some conditions. The National Forest Management Act of 1976 permitted clearcutting under those defined conditions, a leniency that might not have occurred without scientific support. The agency's research arm was to receive more and more specific requests for operational guidelines, as the margin of error had been reduced by an order of magnitude.

In response to *Silent Spring* and related events, use of DDT and other persistent pesticides became more and more restricted. Targeted pests remained, however, and often increased. Scientists, including those in the Forest Service, tested alternative agents. During a six-year period, 130 chemical compounds were tested before determining that Zectran was the best DDT substitute. The Council on Environmental Quality set the standards, which were enforced by the Environmental Protection Agency. The Forest Service, long the leader, was now regulated by other agencies.

RECENT RESEARCH

Activities at Idaho's Priest River Experimental Forest offer a good example of the varied nature of specific research projects at this time. There were permanent growth plots to be remeasured, as were the treated and untreated pairs of plots for a fertilization study. Then there was the collaborative effort as part of the International Biological Program, where the forest provided a coniferous biome coordinating site. Investigators were looking at water from various angles—calibrating a watershed for a new study, analyzing data continuously collected in another watershed for more than thirty years, and calculating snow interception and losses through evaporation in different types of forest stands. Then there were the genetics studies and likewise seven investigations of tree diseases. Finally, scientists from the Northern Forest Fire Laboratory in Missoula had run test fires in fuel beds constructed on the forest.

Of course there were many other examples that suggest the range of activities. Pathologist George Hepting published *his Diseases of Forest and Shade Trees of the United States* in 1971, and the following year Peter Koch would publish in three volumes *Utilization of Hardwoods Growing on Southern Pine Sites*. At the Rocky Mountain Station, Jerome S. Horton produced a 192-page bibliography on the relation of water to vegetation management, while at the Northeastern Experiment Station, Normand R. Dubois was deeply immersed in his long term investigations of biological control of insects, especially the gypsy moth. In the Pacific Northwest, entomologist Robert Furniss, who many felt was the successor to Paul Keen, collaborated with Val Carolin on a project that would lead to the publication of *Insect Enemies of Western Forests*.

Improved utilization had been an important goal for the Forest Products Laboratory since its inception. This emphasis continued, by now looking at ways to utilize computer technology for wood processing. The Lab developed Best Opening Face, a method that extracted 10 percent more lumber from the ever-smaller logs that appeared at the mill. In fact, in 1970 there were 15 billion board feet of small logs converted to lumber, and thus a 10 percent increase in utilization was significant.

At the fire laboratory at Riverside, California, work was well along in developing a computer-driven management information system via Project FIRESCOPE. Contractors with background in the nation's aerospace program conducted much of the research. FIRESCOPE developed the "methods and vocabulary" for interagency cooperation in fire suppression. The Incident Command System is used worldwide to manage situations where groups of people work in complex emergency situations, such as fire, floods, and earthquakes. Observers rate FIRESCOPE as one of the agency's most productive research and application programs. Later on, for budget efficiency, the Riverside lab and its counterpart in Macon, Georgia, would be merged with that in Missoula, Montana; the one laboratory doing the work of three.

Many insect pests had plagued American forests over the years, but three species in particular were not only troublesome but persistent. Needed was a large, coordinated, and accelerated program that would not only capture the interest of Forest Service investigators, but that of its university and state

FURTHER READING

Caraway, Cleo, 1976. *A Forestry Sciences Laboratory . . . and how it grew*. North Central Forest Experiment Station.

Chapline, William R., 1944. "The History of Western Range Research" *Agricultural History* 18(1), 127–143.

Clapp, Earle H., 1926. *A National Program of Forest Research*. American Tree Association.

Clapp, Earle H., Unpublished Memoirs. n.d.

Cowlin, Robert, "Federal Forestry Research in the Pacific Northwest." Unpublished manuscript n.d.

Doig, Ivan, 1976. *Early Forestry Research: A History of the Pacific Northwest Forest and Range Experiment Station, 1925–1975*. Pacific Northwest Forest and Range Experiment Station.

Fernow, B. E., 1899. *Report Upon the Forestry Investigations of the U.S. Department of Agriculture, 1877–1898*. H. Doc. 181, 55 Cong 3.

Geier, Max G., 1998. "Forest Science Research and Scientific Communities in Alaska: A History of the Origins and Evolution in USDA Forest Service Search in Juneau, Fairbanks, and Anchorage." Pacific Northwest Research Station, U.S. Department of Agriculture, Forest Service.

Godfrey, Anthony, 1990. *Progress through Wood Research: A History of the U.S. Forest Products Laboratory—Madison, Wisconsin, 1960–1990*. Unpublished manuscript.

Harper, V. L., G. M. Jemison, C. L. Forsling, 1968. *Early Forest Service Research Administrators*. Interview by Elwood R. Maunder, Forest History Society.

Isaac, Leo A., 1967. *Douglas Fir Research in the Pacific Northwest, 1920–1956*. Interview by Amelia R. Fry, University of California.

Josephson, H. R., et al., 1982. *History of Forest Economics*. Interview by Ann Lage, University of California.

Kaufert, Frank H., and William H. Cummings, 1955. *Forestry and Related Research in North America*. Society of American Foresters.

Keck, Wendell M., 1972. *Great Basin Station—Sixty Years of Progress in Range and Watershed Research*. Intermountain Forest and Range Experiment Station Research Paper INT-118.

Merz, Robert W., 1981. *A History of the Central States Forest Experiment Station, 1927–1965*. North Central Forest Experiment Station.

Munger, Thornton T., 1967. *Forest Research in the Northwest*. Interview by Amelia R. Fry, University of California.

Munger, Thornton, T., 1929. "Fifty Years of Forest Research in the Pacific Northwest" *Oregon Historical Quarterly* (Sept. 1955): 226–247.

National Academy of Sciences, 1929. "Forestry Research."

National Research Council, 1938. "Forest Research in the United States." Processed.

National Research Council, 1909. *Forestry Research: A Mandate for Change.* National Academy Press.

Nelson, Charles A., 1971. *History of the U.S. Forest Products Laboratory, 1910–1963.* Forest Products Laboratory.

Ostrom, Carl E., 1994. *An Interview with Carl E. Ostrom.* Interview by Harold K. Steen, Forest History Society.

Price, Raymond, 1976. *History of Forest Service Research in the Central and Southern Rocky Mountain Regions, 1908–1975.* Rocky Mountain Forest and Range Experiment Station General Technical Report RM-27.

Rudolf, Paul O., 1985. *History of the Lake States Forest Experiment Station.* North Central Forest Experiment Station.

Schrepfer, Susan R., et al., 1973. *A History of the Northeastern Forest Experiment Station.* Northeastern Forest Experiment Station Technical Report NE-7.

Steen, Harold K., ed., 1994. *View From the Top: Forest Service Research by R. Keith Arnold, M. B. Dickerman, and Robert E. Buckman.* Forest History Society.

Wadsworth, Frank H., 1993. *The Evolution of Tropical Forestry: Puerto Rico and Beyond.* Interview by Harold K. Steen, Forest History Society.

Wellner, Charles A., 1976. *Frontiers of Forestry Research—Priest River Experimental Forest, 1911–1976.* Intermountain Forest and Range Experiment Station.

USDA Forest Service, 1958. *Half A Century of Research: Fort Valley Experimental Forest, 1908–1958.* Rocky Mountain Forest and Range Experiment Station Paper No. 38.

USDA Forest Service, 1961. *A National Forestry Research Program.* USDA Forest Service.

USDA Forest Service, 1964. *A National Forestry Research Program.* Miscellaneous Publication No. 965.

USDA Forest Service. *Assessment of Forestry Research: A Mandate for Change.* Processed, n.d.

NOTES

1 Drawn from: Harold K. Steen, *Forest Service Research: Finding Answers to Conservation's Questions.* Durham, NC: Forest History Society, 1998.

navigated the globe in 1829–1830, making some corrections of existing hydrographic charts and incidentally picking up some natural history specimens, and the *USS Potomac*, which made a similar voyage between 1831 and 1834. The latter vessel also made some attempts to drag the ocean bottom for shellfish in the West Indies. In neither case, however, was collecting natural history materials a principal objective of the voyage, nor was it done in any systematic fashion. In part this was owing to congressional doubts as to the appropriateness of federal support of such enterprises. The growing commercial importance of whaling, however, did suggest that it might be wise for the government to begin charting and exploring the Atlantic and Pacific oceans. The American whaling industry had begun when four Massachusetts ships went to the Pacific in 1791. This was two years after the first English vessel had sailed into the same region on a similar mission. Later, in the 1820s, New England whalers initiated intensive whaling activities in Antarctic waters.[5]

The Great United States Exploring Expedition of 1838–1842, led by Commander Charles Wilkes, was the first planned naval exploring enterprise and the most ambitious voyage undertaken by any federal government agency to that time. It included a corps of civilian naturalists, notably Titian Peale, James Dwight Dana, and Charles Pickering. This expedition visited various island groups in the Pacific, Australia, and New Zealand, and also ventured into the Antarctic. The publication of the Wilkes Expedition reports continued on well into the 1870s, and although incomplete (Congress terminated appropriations before all of the scientific volumes had been printed), they constituted the first federally-sponsored zoological research effort. Once again, however, a good many of the specimens brought back by the expedition naturalists were lost or destroyed because government facilities for their retention and curation were still lacking.[6]

Despite this fact, a substantial number of naval expeditions undertaken for scientific and economic reasons were launched in the 1840s and 1850s. Virtually all returned with at least some natural history specimens or recorded observations. They included the Naval Astronomical Expedition to the Southern Hemisphere (1846–1852); the Lynch Expedition to the Dead Sea and the River Jordan (1847–1849); the Herndon-Gibbon Expedition to the Amazon Region (1851–1852); the Page Expedition to the Rio de la Plata and Rio Paraguay (1853–1856 and 1859–1860); Commodore Perry's expedition to the China Sea and Japan (1852–1854); several expeditions undertaken to Panama (1854–1857); voyages to the Arctic under the command of Elisha Kent Kane and others in search for Sir John Franklin (1850–1855); and naval surveys of Liberia and western Africa (1852–1853). A great many, but not all, of the scientific results of these efforts were published.[7]

SPENCER F. BAIRD AND THE SMITHSONIAN'S ROLE

Four years following the establishment of the Smithsonian Institution in 1846, the twenty-seven year old Spencer Fullerton Baird, a professor at Dickinson College in Carlisle, Pennsylvania, was appointed the junior assistant secretary "to take charge of the cabinet and to act as naturalist of the institution." With Baird's arrival in Washington, the federal government first began giving some consistent attention, not only to the sponsorship of scientific efforts in zoology, but to the competent handling of the notes, journals, reports, correspondence, and not least, the specimens that resulted from them. Soon after his arrival, Baird gradually began developing the invaluable network of scientific collectors who would bring back the information and specimens which he personally curated and often described.[8]

Some of Baird's informants were young army and navy officers, a fair number of whom remained in the service until they retired. Some of the young civilians eventually entered or returned to academia or drifted off into other occupations. A modest number of the other men became and remained in federal service as government naturalists, though in many cases, it took Baird some time to find suitable positions for them. Baird was scrupulous about not spending government money when it was not appropriate to do so, though he would send a man letters of encouragement, labels for specimens, government publications and "how to" instructions (which he himself had usually written), and careful directions for wrapping and shipping specimens to his office in Washington.[9]

When Jefferson Davis, secretary of war, sent out survey parties to ascertain the most practicable route for transcontinental railways in the early 1850s, Baird ensured that young naturalists were attached to each group, failing which the young army officers in charge, many of them members of the Corps of Topographical Engineers, were carefully instructed as to how they should make the necessary observations and collections. Ultimately, the number of specimens brought back grew to the point where a federal museum of natural history was organized in 1858. Baird was named its first curator. First located in the Smithsonian "Castle," it was not moved into its own separate museum structure (now the Arts and Industries Building) until 1881. The present National Museum of Natural History building dates from 1910 and has since been expanded several times.[10]

Baird was also able to place some young scientists with certain naval expeditions. A major pre-Civil War example was the Ringgold, or North Pacific Expedition of 1853–1855. First commanded by Commander Cadwallader Ringgold, it traveled east, making scientific collecting stops at the southern tip of Africa, and also in Australia and Hong Kong. In the latter port, Ringgold was compelled to give up his command because of illness, and his second-in-command Lieutenant John Rodgers continued on to Korea, Japan, some of the neighboring Japanese-held islands, Kamchatka, the Bering Straight, and the Arctic Ocean before returning to San Francisco late in 1855. Several scientists, notably William Stimpson, a student of Louis Agassiz's who served as expedition zoologist, brought with him to Washington many thousands of specimens, principally marine species. Virtually all of these ended up in the collections of the Smithsonian, where Baird worked with Stimpson and others to complete the vitally important scientific papers describing what had been found. Congress appropriated $15,000 for the project, which got under way in 1857.

Unfortunately, the Civil War broke out just as Stimpson, Baird, and their associates had completed about two-thirds of the plant and animal reports based upon the Ringgold Expedition materials. With the energies of the federal government focused on the war, there was no hope of completing publication of the results of four years of painstaking effort. Stimpson took most of the invertebrate specimens with him to his new post as director of the Chicago Academy of Sciences. These specimens, many drawings, and his extensive collections of books and manuscripts were destroyed in the Chicago fire of 1871. Stimpson himself did not long survive that tragedy, dying later that same year. Fortunately, his descriptions of a number of the materials which had been collected did survive and were published thirty-six years later.[11]

Just after the Civil War, it became possible to attach some of Baird's young scientists to the four geological and geographic surveys of the western territories led by Dr. F. V. Hayden, Lieut. George M. Wheeler, Clarence King, and Major J. W. Powell. Beginning in the early 1870s, and for some years thereafter, the very modest National Museum facilities and staff provided the only available institutional "home" for federal natural history work. Baird himself and some of his young associates were housed in the towers of the Smithsonian "Castle" building. If a government billet was not available for his particularly promising protegees, and Baird was very conservative in such matters, he did what he could to find them private employment. He sought desirable assignments for some of his young military col-

Major John Wesley Powell talks to a Paiute Indian about the location of the water supply, during a survey in northern Arizona.
(SMITHSONIAN INSTITUTION PHOTO BY J. K. HILLERS, 1873)

laborators, handled some of their personal business for them when their duties precluded their doing so, and on occasion invested their meager funds. In many instances, he arranged to have them become part of private expeditions and collecting enterprises. Most of the people in his network were Americans, but there were also a small number of Canadians. The vast majority were men, but Baird also encouraged the handful of women interested in natural history who got in touch with him. He could not pay these people any stipends, but he could reward them in modest fashion by giving them personal credit for their collections or their other contributions in Smithsonian publications. A number of Baird's younger colleagues subsequently became leading figures in American scientific zoology in the late nineteenth and early twentieth centuries. They included Robert Ridgway and Elliott Coues, ornithologists, and Clinton Hart Merriam, a mammalogist.[12]

Baird was the first federal scientist to synthesize the available data concerning North American vertebrates and the factors governing their geographic distribution. He also was among the first zoologists in this country to insist on precision in their descriptions and measurements of specimens by his assistants and collaborators. Unable to go regularly into the field after the mid 1860s, both because of the demands of his position and occasional bouts of ill health, Baird based his conclusions and his publications on the collections made by others which he had built up over the years. His *Mammals of North America* appeared in 1859, followed by *The Birds of North America* (with John Cassin and George N. Lawrence, 2 vols., 1860). With the assistance of Thomas Brewer and Robert Ridgway, the latter one of his protegees, Baird completed *A History of North American Birds: Land Birds* (3 vols., 1874) and *The Water Birds of North America* (2 vols., 1884).[13]

One of this country's pioneering biogeographers, Baird presented a paper before the newly organized National Academy of Sciences in 1865 on "The Distribution and Migrations of American Birds," which was published the next year in the *American Journal of Science and Arts*. This was one of the very first attempts to explicate the factors governing avian zoogeography. He also found time to write accounts of the reptiles brought back to Washington by the various Pacific Railroad expeditions and to found in 1871 the United States Fish Commission, which late, in 1903, was placed within the United States Department of Commerce. In July 1939, this agency, together with the U.S. Biological Survey (discussed below), was transferred to Interior, and in June 1940, they were combined to form the U.S. Fish and Wildlife Service.[14]

Elliot Coues (1842–1899), probably the leading American ornithologist of the late nineteenth century and a stormy petrel throughout much of his career, had been one of Baird's many proteges, and their personal and professional relationship was often a troubled one. Yet Coues fittingly characterized the mid-nineteenth century period of American natural scientific endeavor the Bairdian Era. Hard working, extremely able, and personally modest, Baird was the dominant figure in the federal natural history community during the years from 1850 until his death in 1887. One cannot understand this period of American natural science without an appreciation of this man's many lasting contributions.[15]

LATE NINETEENTH AND EARLY TWENTIETH CENTURY
DEVELOPMENTS AT THE NATIONAL MUSEUM

A small number of able men were available to continue work in natural science after the death of Baird. Most did their work at or in cooperation with the Smithsonian, along lines laid down by Baird. The majority of them had some common school education, but their high degree of enthusiasm and aptitude for natural history was what had initially brought them to Baird's attention. Most of those who came to Washington developed their skills while on the job. Few American colleges or universities offered comprehensive training in the natural sciences until early in the twentieth century. A medical school education was the commonest route taken by those who did seek graduate or professional training in anatomy and physiology. Coues was a special case. He earned undergraduate and medical school degrees at Columbian, now George Washington University, in the 1860s. He then spent nearly two decades as a U.S. Army surgeon. Baird's and the Army's long-time cooperation made it possible for Coues to get military assignments which enabled him to do natural history work. Baird doubtless found this preferable to providing Coues with an institutional home at the Smithsonian (something which the younger man eagerly sought), essentially because Coues, often a very difficult person, would probably have made a poor team player.[16]

When Congress amalgamated the four separate federal geographical and geological surveys of the west into one U.S. Geological Survey under civilian leadership in 1879, the Army's long-standing role in natural history was largely eliminated, save for the completion of some outstanding paleontological projects. There being no further purpose in satisfying Coues' desire for billets which would facilitate his natural history work, his superiors decided that his future assignments would be made in compliance with the needs of the Army. Coues therefore left the Army in 1881 and essentially freelanced for the rest of his life. He nevertheless enjoyed an unusually productive career. A co-founder of the American Ornithologists' Union in 1883, his major published works included *Key to North American Birds*, which underwent five editions between 1872 and 1903, with a sixth edition appearing in 1927, nearly thirty years after his death. Other titles included *Birds of the Northwest* (1874); *Birds of the Colorado Valley* (1878); *Fur Bearing Mammals: A Monograph of the North American Mustelidae* (1877); a four part bibliography of American and British ornithological publications; several editions of the *Coues Check List of North American Birds* (1879 and 1882);

and, with a committee of other ornithologists, what in effect was the first American Ornithologists' Union Check List of American Birds (1886). His many other publications included valuable editions of the travel accounts of Lewis and Clark and other explorers of the early west.[17]

Illinois-born Robert Ridgway (1850–1929), largely self taught in ornithology, collaborated with Baird on several of the latter's major works, published several editions of his own *Manual of North American Birds* (1887, fourth edition, 1915), and completed eight volumes of *The Birds of North and Middle America* (1901–1919), a work later extended by Alexander Wetmore and others.[18]

Baird, Coues, Ridgway, and their successors laid a firm foundation for the U.S. National Museum (now the National Museum of Natural History) collections in Washington, which have since become one of the leading world assemblages of vertebrate skins, skeletal material, and alcoholic specimens, to say nothing of the massive invertebrate collections. Among the other zoologists who made their mark at the Smithsonian in the late nineteenth and early twentieth centuries may be mentioned William Healey Dall (1845–1927), a malacologist and long-time paleontologist for the U.S. Geological Survey; Frederick William True (1858–1914), an authority on whales, dolphins, and fish, who for many years was curator of mammalogy, head curator of the Department of Biology, and assistant secretary in charge of library and exchanges; George Browne Goode (1851–1896), ichthyologist, historian of science, and an assistant secretary; and Charles Doolittle Walcott (1850–1927), paleontologist, geologist, and for many years, secretary of the Smithsonian. While none of these able men may quite have caught the broader vision of Baird, their mentor, the various subdisciplines of zoology had, within several decades of Baird's death, developed to the point where no one individual could have been expected to deal effectively with all of them.[19]

Among the many later twentieth century federal biologists instrumental in making the National Museum of Natural History a world-class institution may be mentioned Alexander Wetmore (1886–1978) and S. Dillon Ripley (1913–) ornithologists and Smithsonian secretaries; Gerrit Smith Miller, Jr. (1869–1956), and Arthur Remington Kellogg (1892–1969), mammalogists; Leonhard Stjneger (1851–1943), a herpetologist, ornithologist, and mammalogist; and Waldo LaSalle Schmidt (1887–1977), carcinologist and marine biologist. Kellogg was also director of the National Museum and an assistant secretary of the Smithsonian.[20]

THE FIRST CENTURY OF
FEDERAL GOVERNMENT ENTOMOLOGY

In 1863, Townsend Glover (1813–1883) became United States Entomologist in the Department of Agriculture, a government agency then headed by a commissioner who was not of cabinet rank. For fifteen years, until his health failed, Glover collected much useful data concerning plants, plant diseases, insects, and other subjects, but he was primarily an artist, and not a collector. Western farmers in the mid-1870s dealt with an unusual number and variety of insect pests, especially the Rocky Mountain Grasshopper, and demands for federal assistance rained down on the heads of their representatives in Congress. Several of the leading authorities on the subject, notably Charles Valentine Riley, Missouri State Entomologist, Cyrus Thomas, Illinois State Entomologist, and Alpheus Spring Packard, Jr., Director of the Peabody Academy of Sciences in Salem, Massachusetts, looked into the situation, and western governors supported Riley's suggestion that a federal agency be created to deal with the problem. In March 1877, Congress created the U.S. Entomological Commission. Riley was made chief, with Packard as secretary. During its short life (1877–1882), the commission was briefly attached to the U.S. Geological and Geographical Survey of the Territories, under Dr. Ferdinand V. Hayden, then became an entity of the Department of the Interior, and was finally transferred to the Department of Agriculture. Riley resigned his position as chief after two years owing to policy disputes with the commissioner of agriculture, but returned in 1881. During the interim, the work in entomology was led by the very capable Prof. John H. Comstock of Cornell University. In 1882, soon after Riley's reappointment, Congress elevated the Entomological Commission to division status in Agriculture, and Riley was retained as chief. He remained at his post for another thirteen years.

Riley, Packard, and Thomas led a group of investigators who had previously published detailed studies of noxious insects, or who were to do so under the commission's or division's aegis. The English-born Riley (1843–1895) came to the United States at age seventeen, worked on a farm for a time and then became a writer and editor on entomological subjects. He was named the first entomologist for the State of Missouri at age twenty-five. In 1868, he established a scientific journal, *American Entomologist*, probably the first in this country to be concerned with a specific taxon. The nine annual reports he published concerning noxious, beneficial, and other insects while state entomologist between 1869 and 1877 placed him at the forefront among the American economic entomologists of his day. During this period, he also taught entomology at the universities of Missouri and Kansas and at the Missouri State Agricultural and Mechanical College.[21]

A. S. Packard (1839–1905) had been a student of Louis Agassiz' at Harvard, a founder and editor of *American Naturalist*, and a zoologist for the Hayden Survey. In the latter capacity, he had completed several studies concerning insects. Cyrus Thomas (1825–1910) had also served Hayden as an entomologist, publishing several important studies. He later completed five volumes of information concerning the noxious and beneficial insects of Illinois.

The Commission and later the Division of Entomology published several bulletins assessing the damage done by locusts and what action had been taken to try to limit the spread of this pest. Five annual reports were published between 1878 and 1890, summarizing field research and other studies done on the Rocky Mountain locust and other pests. These included information dealing with their biology, geographic distribution, and methods whereby these insects might be controlled. The reports were substantial, ranging (with one exception) from well more than 500 to nearly 1,000 pages in length.[22]

Riley laid a firm foundation for the Division of Entomology, which grew rapidly in importance during his tenure. He was responsible for much of the early work done on insecticides and fumigants, but in addition, he began and carried out a program whereby natural controls were also used to defeat agricultural pests. The Australian ladybug or Vedalia beetle, for example, was imported into the United States and proved very efficacious in controlling the cottony-cushion, or white scale, which had threatened the citrus industry. Riley performed a signal service to the French wine industry when he made it possible for healthy California grape vine stock to be substituted for French vines affected on a widespread basis by phylloxera. In 1882, Riley gave a portion of his personal collection of insects to the U.S. National Museum. Following his appointment as assistant curator of the Museum and Curator of Insects in 1885, he presented the remainder of his collection to the museum, and this became the foundation of what today is one of the world's leading collections. Riley unsuccessfully sought to secure the post of assistant secretary of agriculture in 1894 and resigned from his post as entomologist, this time for good. The following year, he died as the result of a bicycle accident.[23]

Leland O. Howard (1857–1950) was Riley's successor and another major figure in the Division, later (after 1904) the Bureau of Entomology. A graduate of Cornell University, Howard went to work for Riley in 1878, served as chief of the division and bureau from 1894 to 1927, and continued on until 1933 as principal entomologist. Howard was an authority on mosquitoes and houseflies and was co-author of a four volume work, *The Mosquitoes of North and Central America and the West Indies*, published between 1912 and 1917. He also authored *Mosquitoes, How They Live, How They Are Classified, and How They May Be Destroyed* (1901); *The Housefly—Disease Carrier* (1911); *A History of Applied Entomology—Somewhat Anecdotal* (1930); and *Fighting the Insects: The Story of An Entomologist* (1933), among other works. The bureau underwent a number of subsequent transformations, becoming the Bureau of Entomology and Plant Quarantine (1934), the Entomology Research Branch (1953), and the Entomology Research Division in 1957.[24]

THE BEGINNINGS OF FEDERALLY-SPONSORED WILDLIFE RESEARCH

In September 1883, the American Ornithologists' Union (AOU) held its inaugural meeting at the American Museum of Natural History in New York. A group of men interested in bird migration or distribution soon formed one of its volunteer committees, with New York physician C. Hart Merriam as chair. The committee solicited information about its subject from interested individuals and soon found itself inundated with data from some 1,200 observers in the United States and Canada. Merriam discussed the situation with Spencer Baird and the AOU leadership, and it was decided to appeal to Congress for support, on the grounds that the rapidly accumulating data could be of broad economic interest to the general public, particularly the agricultural community. Merriam and his colleagues argued that farmers would benefit from a clearer understanding of the nation's bird population, its distribution, food habits, and impact upon agriculture. With the support of Senator Warner Miller, a New York Republican and family friend, Congress responded by creating an Office of Economic Ornithology within the Agriculture Department, with an initial appropriation of five thousand dollars. Merriam was appointed to head it in July 1885, and his office was placed within Riley's Division of Entomology.[25]

The twenty-nine-year-old Merriam held an M.D. degree from the New York College of Physicians and Surgeons and was an established ornithologist and mammalogist, having published accounts of the birds of Connecticut and the mammals of the Adirondacks to critical acclaim. The annual appropriation for his office was doubled in 1886. Merriam was given respon-

*C. Hart Merriam, early in his career as first chief
of the United States Biological Survey.*
(U. S. FISH AND WILDLIFE SERVICE PHOTO)

sibility for mammals as well as birds and thereafter operated independently of Riley. A decade later, in 1896, Merriam's agency had become a division, and in 1905, a bureau within the Agriculture Department. Merriam and his small but able staff were soon producing a veritable blizzard of of scientific reports, circulars, popular and scientific bulletins, maps, and posters, which gave his agency a prominence in the public mind out of all proportion to its size.

In the early 1890s, Merriam developed a set of criteria for what he termed "life zones," wherein various combinations of plants and animals were to be found in particular biogeographic categories delineated primarily by climatic factors. Most of his early research was done in the region around San Francisco Mountain in Arizona during the summer of 1889. Merriam, however, was to depend heavily on his findings there when making extrapolations for the entire continent. His life zone criteria were largely based on terrain (elevation), moisture, and warm temperatures. The Biological Survey issued a series of maps delineating these zones over the next forty years which were very influential until Merriam's critics, using more sophisticated models and methodologies, developed more precise and accurate explanations for plant and animal distribution. The life zone concept was the construct, however, on which most Biological Survey publications would be based for nearly half a century, and some zoologists believe that it is still has some applicability in the western United States.[26]

With a few exceptions, Merriam's first employees were young men who had not gone to college, but who nevertheless became proficient field biologists and biogeographers under his tutelage. His collectors carefully followed a set of procedures which he had developed for specimen collection and observation. He and they also took advantage of the newly-invented cyclone, ancestor of the common snap traps of today, with which they collected many hundreds of smaller mammals hitherto unknown to science. Within less than two decades, the number of known varieties of North American mammals had jumped from 363 to well more than 1,450, the majority of which had been identified and described by Merriam and his associates. In part this was due to better collecting methods, but several other things had been taking place. In the first place, the concept of subspecific differences between organisms were first being recognized in the 1880s and later. In the second place, many zoologists had become "splitters," meaning that organisms were separated into distinct groupings on the basis of very fine physical distinctions. Merriam himself was a confirmed splitter, but modern zoologists no longer recognize many of the criteria he employed for describing species. On the other hand, even with the use of much more conservative taxonomic criteria, the number of known forms (species and subspecies) of mammals, birds, and other vertebrates has continued to climb over the past century, and Merriam's careful field methods have in part been responsible for this development.[27]

Merriam was not impressed with the laboratory approach to college instruction in biology at the turn of the century, principally because most students were given few opportunities to work with animals and plants in the field. For many years, therefore, he resisted employing college-trained applicants for positions within the Survey. This became more difficult when new Civil Service regulations obliged him to select the best-qualified people when openings became available. In 1889, he began publishing the *North American Fauna*, a continuing series of monographs on the systematics and distribution of North American vertebrates, with the western United States and Canada receiving most attention. Yet congressional appropriations for the Survey remained niggardly, never rising above $62,000 per annum (of which a mere $17,500 was allotted for salaries) during Merriam's quarter-century tenure. This was owing in part to Merriam's distaste for the politics of the appropriation process and in part to the priorities he set for the Biological Survey. Early in his tenure, Merriam decided that it was more essential to carry out a biological reconnaissance of the North American continent than it was to cater to the concerns of ranchers, farmers, and stock raisers. For twenty years, he persuaded Congress to underwrite essential basic research into the principles of biogeographical distribution and to establish the extent and variety of the vertebrate fauna. Merriam was perhaps most at home in the West, and the various *North American Fauna* issuing from his pen and those of his colleagues were pioneering efforts of considerable value.[28]

One important legacy of Merriam's tenure as leader of the Biological Survey was the several generations of biologists whose talents were developed under his aegis. Some authorities have claimed that Merriam made no point of creating a trained cadre of biologists, but the facts do not support this view. A good many of these men became major figures in early twentieth century natural science.

Vernon Bailey (1864–1942) was a Minnesota farm boy who became an authority on small mammals and ultimately served as Chief Field Naturalist under Merriam. He later married Merriam's sister Florence, herself an able ornithologist. A[lbert] K[enrick] Fisher (1856–1948), a classmate of Merriam's at the College of Physicians and Surgeons in New York, was responsible for important studies concerning birds, particularly their food habits. He served for a time as assistant chief of the Survey. Theodore Sherman Palmer (1868–1955) was primarily an ornithologist who became senior biologist with the Survey and an authority on game protection and international treaties having to do with the protection of migratory birds. After four years with Merriam (1894–1898), Gerrit Smith Miller, Jr. spent the bulk of his six-decade career as curator of mammals with the U.S. National Museum. Edward Alexander Preble (1871–1957) was responsible for a number of notable biological reconnaissances of parts of Canada and Alaska, the results of which were published in the *North American Fauna* series. Wilfred Osgood (1875–1947), a specialist on small mammals, spent over a decade with the Survey, authoring several valuable studies in the *North American Fauna* series before leaving in 1909 to join the staff of the Field Museum in Chicago. Edward William Nelson (1855–1934) was perhaps best known for his extended biological reconnaissances of Mexico with Edward Alphonso Goldman (1873–1947), though Nelson also spent some time in the Arctic. Nelson later became the third chief of the Survey.[29]

The next generation of naturalists, most of whom served under Merriam during his final years with the Biological Survey, included Hartley H. T. Jackson, a student of game birds and mammals and later chief of the Survey's Department of Wildlife Research; Arthur Brazier Howell (1886–1961), who from 1922 to 1928 was a scientific assistant with the Survey and later taught anatomy at Johns Hopkins; and Stanley Paul Young (1889–1969), an authority on predatory mammals and rodents. Waldo Lee McAtee (1883–1962) was an economic ornithologist and entomologist whose entire forty-three year career was served with the Biological Survey and Fish and Wildlife Service. Many of these individuals later served as presidents of the American Society of Mammalogists, founded by Jackson and others in 1919. Merriam was the first to hold this post, later held by Nelson, Osgood, Bailey, Jackson, Howell, Goldman, and Remington Kellogg, the latter at the U.S. National Museum. Merriam had earlier been secretary and president of the AOU, and Palmer was later AOU secretary for many years. McAtee was a treasurer of the AOU.[30]

THE FEDERAL GOVERNMENT ENTERS
THE REGULATORY PHASE

By 1905, Congress had become increasingly impatient with Merriam. He had hoped to complete the biological reconnaissance of the North American continent begun in 1885, but appropriations for his agency were too small to make that possible within a reasonable period of time. Merriam's distaste for the political process and his extreme reluctance to play the annual appropriations game with the relevant House and Senate committees angered many congressmen and made it virtually impossible for him to secure a better funding base. Constituent pressures also mandated that the focus of Merriam's activities be altered. As a result, Merriam and his staff found themselves saddled with responsibilities which he had not sought and did not want. The first of these came about following passage of the Lacey Act of 1900, which addressed several long-standing issues.

In the mid-1880s, Merriam had taken note of the ease with which exotic species from other nations had been and could continue to be introduced into the United States and the danger that such species could pose to the native fauna. The English sparrow was a notable example of a bird which rapidly adapted to conditions in this country and had spread from coast to coast within a matter of decades. The European starling was another example of a fast-spreading species. No federal action was taken immediately, in part because understanding concerning avian population dynamics was in its infancy. Several of the states, however, notably California and Oregon, undertook to quarantine various animals and birds by outlawing their entry into those states and authorizing the destruction of individual animals already present. The Lacey Act of 1900 dealt with this issue on a nation-wide basis and also expanded the powers of the Department of Agriculture (and hence of the bureau, to which they were delegated) in other ways. It was thenceforth illegal "to import into the United States any foreign wild animal or bird except under special permit from" the Department of Agriculture. Game killed in violation of local and state laws could no longer be transported across state lines. The same restrictions on the interstate movement of animals were later extended to foreign game animals and birds. The secretary of agriculture was also authorized to take steps to restore native game and other wild birds "in those parts of the United States . . . where the same have become scarce or extinct, and also to regulate the introduction of American or foreign birds and animals in localities where they have not theretofore existed." Exceptions were made for domesticated cage birds and for "natural history specimens for museums or scientific collections.[31]

Subsequent enactments for which the bureau became responsible included protection of the eggs of migratory birds (1900); regulation of the importation of eggs of game birds (1902); responsibility for the enforcement of Alaskan game laws (1902 and 1908); protection of game bird eggs in game preserves (1906); responsibility for the National Bison Range and other preserves set aside for endangered game mammals (1908); amendments to the Lacey Act extending previous provisions (1909); protection of the seal fisheries of Alaska (1910); custody over migratory birds (1913, 1918); and a more comprehensive plan for the destruction of predatory mammals (1915).

Federal efforts to control predatory animals began when the Survey became a clearinghouse for various state and local efforts to place bounties on such animals in the 1890s. T. S. Palmer of the Survey became the resident authority on such issues and on various state and county game laws as they became an increasingly important element in early attempts at wildlife management. Merriam was wary of getting the Survey involved in regulatory activity, viewing it as likely to weaken his ongoing research program, and he became increasingly restive as additional such responsibilities were added to his charge during his remaining years at the Survey's helm.

In the early 1900s, the Survey helped to develop poisoning programs for the control of pocket gophers, ground squirrels, and prairie dogs in farming and livestock-raising areas, particularly in the western states. In 1905–1906, the Forest Service requested the aid of the bureau in a study of wolf depredations on farm animals living on the borders of national forests. In 1907, it issued the first instructions for the destruction of predators, employing shooting, trapping, den-hunting, and fencing. All of these programs and those which came later proved to be highly controversial.[32]

The relationship between predatory and prey species was imperfectly understood at the beginning of the twentieth century. As livestock raising and ranching in the Southwest became an increasingly important industry in the late nineteenth century, stock raisers and farmers were more and more unwilling to accept depredations by wolves, coyotes, mountain lions, and other predatory species. Professional hunters, trappers, and poisoners were hired to reduce or eliminate these animals. The artist-naturalist Ernest Thompson Seton (1860–1946), whose art work later embellished some of the *North American Fauna* series, did this sort of work for a time in the early 1890s. "Lobo, King of the Currumpaw," a Seton story which appeared in his *Wild Animals I Have Known* (1898) provides a valuable word picture of this activity at the turn of the century. Seton based his account on his own experiences and those of contemporaries engaged in similar work.[33]

With increasing amounts of money and limited staff time being devoted to the bureau's regulatory and control work and with few increases in annual budget appropriations, the scientific studies Merriam favored had for a time to be slowed or cut back. Only after his resignation in 1910, when more politically sophisticated chiefs came to the helm of the Survey, could Congress be persuaded to make more substantive outlays for the agency's work. By 1915, these had risen from a high of $62,000 at the end of Merriam's tenure to $282,000 and, by 1920, to $809,000. This latter figure was exclusive of $36,271 earmarked for the "feeding [of] starving elk in [the] Jackson Hole region," an amount which would have been unheard of a dozen years before. Much of the additional money was given over to regulatory and control work, but H. W. Henshaw, Merriam's successor as chief from 1910 to 1916 and E. W. Nelson, who was chief from 1916 until 1927, were able to continue Merriam's tradition of state and regional biological surveys.[34]

Nelson provides a good example of the caliber of federal biologist that could result from Baird's and Merriam's mentoring. New Hampshire born, Nelson developed a strong childhood interest, which continued throughout his life, in natural history. He lost his father during the Civil War, and his mother supported her young family as an army nurse and later as a dressmaker. Educated in the Chicago public schools and a graduate of the Cook County Normal School, Nelson sought Baird's aid in furthering a career in natural history. Through Baird's good offices, he secured a job as an army enlisted weatherman in Alaska (1877–1881). This resulted in the publication of several well-received papers on Alaskan natural history and ethnology and an ethnographic collection of Eskimo artifacts still regarded today as one of the best extant. After a period of ill health in the mid-1880s, Nelson began his long career as a collector and all-round field man with the Biological Survey. His many years of field work included fourteen years of biological reconnaissance in Mexico. Increasing administrative responsibilities led to his eleven year tenure as the Survey's third chief.[35]

By the late 1920s, critics of the Biological Survey's trapping, hunting, and poisoning programs were becoming increasingly vociferous. Paul Redington, brought in from the Forest Service to head the Survey on Nelson's retirement, was the first man not promoted to the top job from within. His seven year tenure (1927–1934) was not a happy one. He bore the brunt of much of the increasing public and professional distress over predator control, which continued to enjoy strong support from the livestock industry and from Congress. The leaders of the predator control effort shied away from emphasizing any effort to exterminate predator species, rather, they insisted that wolves, for example, would continue to exist in Canada and Mexico. E. A. Goldman stated that the presence of larger predators could no longer be tolerated in modern civilization and that the Survey was merely doing what ranchers had been doing for well over half a century, though in a more precise and careful manner. The number of wolves had been markedly reduced in the continental U.S., but coyote populations rose, in large part because of the extirpation of wolves, their principal enemies. White-tailed and mule deer populations rose in many parts of the west as wolves were no longer available to cull the herds, and state conservation agencies were obliged to work harder to feed them, lest trees and other browse species be destroyed over much of the deer's range. Alarmingly, however, poison baits were destroying some livestock, birds, and mammal species not among the targets of this campaign of control. The poisoning and trapping continued, though it angered many present and former members of the Survey's staff, most notably including Merriam, and various conservation groups. In 1924 and again in 1931, the American Society of Mammalogists passed resolutions strongly disapproving of the direction being taken with the Survey's control work, and some leading members of the Society condemned the government program. The professional and personal relations of colleagues on both sides of this issue were severely strained. Ira Gabrielson, a Survey staffer and future Survey chief, left the Mammal Society because of its vigorous stand against predator poisoning. The Survey's control programs were placed under a new Division of Predator and Rodent Control in 1929. A new policy of cooperative arrangements between the Survey and state and local entities requesting assistance with their control efforts was worked out whereby much of the cost was assumed by these agencies, while the Survey provided the necessary expertise and direction. This made it possible to reallocate some of the Survey's limited fiscal resources to other work. In most respects, however, research into the factors governing wildlife populations in the United States was still in its earliest stages, and it would take time for better research techniques to be developed.[36]

Neither poison proponents nor those in opposition to its use were as yet able to offer comprehensive scientific data in support of their positions. While poisoning technology is somewhat more sophisticated today, its usage has proven to be as much a political as a scientifically divisive issue down to the present. Poisons were used in national parks from 1898 until terminated in the 1950s. Trapping had previously been discontinued as a control technique in the 1930s, and shooting has since been the preferred method. The Fish and Wildlife Service, on the other hand, has with some interruption employed poison down to the present. New poisons, such as Compound 1080 (introduced in 1947), and devices such as "coyote getters," were popular with ranchers because they were highly destructive to predators, but they also killed many non-target animals and a few innocent passersby and thus precipitated much controversy.

As the 1920s ended and the 1930s brought the Depression era and the New Deal, the Biological Survey went through a very difficult period of transition. Most of the naturalists and biologists who had worked under Merriam in the 1880s and 1890s began to retire, and a new generation of biologists, many of them university trained, began taking over the various research, regulatory, and control programs operated by the federal government, universities, and other institutions.[37]

WILDLIFE MANAGEMENT IS PROFESSIONALIZED

Aldo Leopold (1886–1948), a long-time professional forester with the U.S. Forest Service, was to prove one of the most influential thinkers in the fields of American game management and environmentalism. The Iowa-born Leopold entered the Forest Service in 1909, following completion of his master of forestry degree at Yale. Working first in the Apache National Forest of Arizona as a forest assistant, he rose quickly, becoming deputy supervisor of Carson National Forest in 1911 and supervisor in 1912. A near fatal illness sidelined him during much of 1913, but he returned to his duties, and in 1917 was promoted to assistant district forester, with responsibilities for game, fish, and recreation. At first Leopold worked in concert with traditional naturalists who argued in favor of eliminating major predators to conserve the population of deer in the Southwest. When the deer population in the Kaibab Plateau region of Arizona mushroomed in the 1920s, only to die off in large numbers because of the lack of sufficient browse, Leopold gradually began to change his point of view. Clearly, predators were essential players in helping to maintain the balance between prey species and their environment. In consequence, Leopold began pressuring the Forest Service to examine and adopt the goal of wilderness protection at the beginning of the 1920s. In 1924, the Forest Service responded by establishing the 574,000 acre Gila Wilderness area in New Mexico. Over the next half-century some fourteen million acres of land would be set aside as wilderness. Leopold continued to work out his ideas, arguing that many of the vital qualities of Americans on the old frontier might be retained in a wilderness environment. Here, too, biologists could determine what factors contributed to what Leopold was later to term a "land ethic." This concept was to become one of his most enduring contributions to the field of conservation. Man, argued Leopold, needed to expand his conception of human communities to embrace wildlife, plant life, soils, and water.[38]

In 1924, Leopold was named associate director of the new Forest Products Laboratory in Madison, Wisconsin. Here he remained until 1928, when the Sporting Arms and Ammunition Manufacturers Institute engaged Leopold to study game populations and their environment in the Upper Midwest. The resulting study, *Report on A Game Survey of the North Central States* (1931), was a pioneering effort "to appraise the chance for the practice of game management as a means to game restoration in the . . . region." In 1933, Leopold published *Game Management*, based in part on his previous research and in part on teaching methods developed during the brief temporary teaching stint at the University of Wisconsin in 1929. In this book, he articulated the views of systems ecology that thereafter provided direction for wildlife management practitioners. The year 1933 also saw Leopold's appointment as professor of wildlife management at Wisconsin, the nation's first; indeed, it can be said that he created the academic discipline. For several years, Leopold worked hard for greater emphasis on game research at the state level. When the Cooperative Wildlife Research programs were instituted by the Biological Survey in 1935, the University of Wisconsin could not participate because the Wisconsin Conservation Commission refused to take part. In 1934, President Franklin D. Roosevelt appointed Leopold, Jay N. "Ding" Darling, and Thomas Beck to the Special Committee on Wildlife Restoration, in which capacity he was able to exert some influence on emerging new concepts of management worked out within the Biological Survey and the National Park Service. Leopold was offered the post of chief of the Biological Survey when Redington stepped down in the spring of 1934, but he much preferred research to administering federal management and control programs, and the post went to Darling instead.

Leopold's posthumously published book, *A Sand County Almanac* (1949), has proved to be the most effective formulation of his ideas. This slim volume, many times reprinted, has become a classic text of American environmental thought over the past half-century. During the 1930s and 1940s, Leopold became convinced that natural controls in wildlife management were more efficacious than management by human agencies. Before his death in 1948, he was able to demonstrate that interference by man had in fact presented wildlife managers with a host of new problems which sooner or later would demand their attention and the attention of policy makers at the state and federal levels.[39]

ROLE OF THE NATIONAL PARK SERVICE SINCE 1916

From 1872, when Yellowstone, the first national park, was created, the secretary of the interior had nominal oversight over all parks. He was given no authority to control hunting or fishing by the public until 1883 and, even then, no powers of enforcement. Congress placed the Army in charge of local administration of the parks in 1886, leaving the Cavalry Branch to determine appropriate management procedures. Not until 1894, however, was legislation enacted specifically providing for the protection of wildlife. Only then could fines or imprisonment finally be levied against those found in violation of the statute. The Army continued to manage the parks until 1917, though all policy decisions were being made by the interior secretary after 1901. By 1900, there were fourteen national and military parks and national monuments. In 1916, the year that the National Park Service was created, that number had risen to forty-nine, and by 1928, seventy-four parks and monuments existed.[40]

The terms of the Organic Act of 1916, establishing the National Park Service, directed the new agency:

> To conserve the scenery and the natural and historic objects and the wild life therein and to provide for the enjoyment of the same in such manner and by such means as will leave them unimpaired for the enjoyment of future generations.

From its inception, problems arose with the various interpretations placed on this document by the leadership of the new organization. Stephen T. Mather, wealthy Borax company executive, had been a prime mover behind the establishment of the National Park Service and served as its first director from 1916 until his retirement in 1929. Mather interpreted his mandate to mean that he should maximize access to the parks by the public and stress what National Park Service Historian Richard West Sellars has termed "facade management" objectives. This could be carried to extremes, as when a winter sports carnival was begun at Yosemite in 1931, and plans to host the 1932 Winter Olympics at the same park failed by a narrow margin. On the other hand, Mather and his assistant Horace Albright, who succeeded him in 1929 and served as director until 1933, managed to defeat numerous other proposals for ambitious park development projects. The parks were by no means left "unimpaired," however. Highways, hotels, water supply systems, parking areas, and waste disposal facilities were all built.[41]

During these early years, wildlife management issues remained secondary in importance, but at least one study, *Mammals and Birds of Mount Rainier National Park*, was researched and published (1927). This was for the most part, however, an outgrowth of field work done by the Biological Survey and a Washington State College professor. During the early 1920s, Horace Albright did initiate game counts in the parks, and the life histories of game species were recorded for a few years. Park officials made distinctions between presumably more desirable animals, such as deer and elk, and predators, such as wolves and cougars. The latter were at first efficiently dispatched in line with the long-standing practices of the Army and the Biological Survey, both of which were operating under mandates laid down by Congress. Predator control work was rigorously pursued, such that wolves and cougars had nearly been extirpated from the parks by the mid-1920s. By that time, however, sportsmens' groups, professional organizations such as the American Society of Mammalogists, and nascent conservation organizations began lobbying strongly against these practices, and by the time of Mather's retirement, Park Service policies had begun to change. William Rush was employed to study large mammals at Yellowstone in 1928, at which time the Park Service had six naturalists and nine wildlife rangers on the payroll to monitor conditions in seventy-four parks and monuments. Control programs were directed toward the larger predators, but with the understanding that these were not to be eliminated altogether. The Park Service's Educational Division in Berkeley, California, sponsored the first Park Service Naturalist's Conference, held there late in 1929. This agency was subsequently reorganized into the Branch of Research and Education in 1930. Under Harold Bryant, this office began coordinating the first organized research in natural history within the Park Service.[42]

George Melendez Wright (1904–1936), a well-to-do ranger-naturalist assigned to Yosemite, had been a student of the noted zoologist Joseph Grinnell at the University of California in the 1920s. He offered to underwrite and initiate the first systematic attempt to understand vertebrate fauna in the parks. With Albright's approval, Wright, Joseph S. Dixon, a mammalogist and former associate of Grinnell's, and Ben H. Thompson, a research associate, got to work in 1929. In the summer of 1931, Wright and his colleagues secured a congressional subvention of $22,500 to continue their studies, and Wright's initiative became the Wildlife Division within Bryant's division. Over the next several years, this office produced a series of reports under the series title, *Fauna of the National Parks of the United States*. The first of these, "A Preliminary Survey of Faunal Relations in National Parks," sometimes referred to as "Fauna One," was published in 1933. The authors addressed "the existing status of wild life in the parks, analyze[d] unsatisfactory conditions, and outline[d] a proposed plan for the orderly development of wild-life management."

Wright argued that parks were "artificial units," in that space, terrain, and life cycle requirements of resident wildlife species were not given sufficient consideration when their boundaries were established. Arno Cammerer, then (1933–1940) National Park Service director, endorsed this view in 1934. A recent study of the subject (1992) suggests that parks are still unable "to

George M. Wright was the first professional biologist of the National Park Service. Carl P. Russell photographed Wright, in Yosemite National Park during the mid-1920s.
(NATIONAL PARK SERVICE HISTORICAL PHOTOGRAPH COLLECTIONS, HARPERS FERRY CENTER)

function as complete biological units," and that this remains "the most important wildlife problem in parks." Other challenges confronted by park managers in the late 1920s included the introduction of exotic species, competing requirements of animals and man, depleted habitat, management of both over and under population problems among various species, and the need to reintroduce extirpated animals. All parks received some attention within the pages of Fauna One, and needed parameters for animal management policies were set forth. In "Fauna Two," "Wildlife Management in the National Parks," (1935), Wright and his colleagues focused on plans for reintroducing species which had been extirpated in the various parks, while "Fauna Three" (1938), "Birds and Mammals of Mt. McKinley National Park, Alaska," described the wildlife of that area. "Fauna Four" (1940) was a pioneering study of the "Ecology of the Coyote in the Yellowstone," based on field work conducted during the years 1937 to 1939. The series elicited high praise from professionals. In addition, wildlife rangers had finally been assigned to every park in the system.[43]

In May 1934, Parks director Cammerer told a Special House of Representatives Committee on Conservation of Wildlife Resources that predators were "considered an integral part of the wildlife protected within national parks, and no widespread campaigns of destruction are countenanced." Only "coyotes or other predators . . . making serious inroads upon herds of game or other mammals needing special protection" were subject to limited control. Rodents were subject to poison control around buildings or in the event of epidemics, but otherwise poison was "believed to be a nonselective form of control and is banned." Wolves were among half a dozen species of fur-bearing mammals "in need of special protection because nearly exterminated." He reported that buffalo herds in Yellowstone and other parks were being carefully monitored and periodically culled, so as not to overstrain available food supplies and the patience of farmers and ranchers outside of park boundaries.[44]

Early in 1936, George Wright was killed in an auto accident while in Big Bend National Park. With his death, the once promising wildlife program lost its most articulate spokesman and the gradually increasing emphasis he had won for it. Money and staff time for science projects remained static for a time and then gradually declined. Many senior National Park Service administrators simply did not perceive biological science or game management as priorities, except insofar as the latter fit into their long-standing views concerning "facade management." In 1939, Wright's successor, Victor

H. Cahalane, was transferred to the Biological Survey with his staff and the Park Service's Wildlife Division was finally eliminated in 1940.[45]

Shortly after World War II broke out, National Park Service headquarters were transferred for the duration from Washington to Chicago to make room for temporary wartime agencies despite vehement protests from Interior Secretary Harold L. Ickes. National Park Service funding and visits to the parks by the public decreased sharply during World War II, but Newton B. Drury, Cammerer's successor as Park Service director, managed to prevent the use of Park Service properties for military purposes. Little scientific or management research could be done during this period, though "Fauna Five," Adolph Murie's "The Wolves of Mount McKinley," based on work done during 1939–1941, was published in 1944. Murie noted that Stanley P. Young, long-time head of the Biological Survey's rodent and predator control work, had recently (1942) articulated a widely-held view among "naturalists, outdoors men and other interested persons throughout the land" when he stated, "Where not in conflict with human interests, wolves may well be left alone. They form one of the most interesting groups of all mammals, and should be permitted to have a place in North American fauna." Young further stated that wolves fortunately still "exist[ed] in many areas where there is no conflict with human interests," and he doubted that they would be "crowded out" of these "by further development" for some time, "if ever." Some National Parks administrators, however, regarded Young's views, and Murie's conclusions that wolves posed no great threat to other wildlife, as akin to apostasy.[46]

The 1940s and much of the 1950s proved to be the nadir of wildlife research within the Park Service. Little could be done and much valuable ground was lost. In 1947, Cahalane's Office of National Park Research within the Fish and Wildlife Service was eliminated, and its staff returned to the Park Service to work under the Division of Interpretation on minor tasks. Cahalane and his associates vigorously agitated for more science funding in light of the millions being devoted to park refurbishment and development in the 1950s. He ultimately resigned as chief biologist in 1955 when unable to achieve any resolution of this issue, and was succeeded by Gordon Fredine. Not until 1958 did Park Service administrators begin listening to their critics and allocate the very modest sum of $26,880 for research. For several years, these annual subventions remained at the $29,000 level.[47]

With the appointment of Stuart Udall as interior secretary in 1961, matters improved, and $80,000 was appropriated that year for scientific research. Howard Stagner, head of the Mission 66 Program, which had done much to upgrade park facilities, argued forcefully that scientific research must be given the highest priority. The few National Park scientists who had survived the frustrations and budget cuts of the forties and fifties argued for an ecological perspective within a comprehensive long-range research plan. On Stagner's recommendation, Udall commissioned two studies in 1963, one by the National Academy of Sciences, the other by a Special Advisory Board headed by A. Starker Leopold, a zoologist son of Aldo Leopold, who was on the faculty of the University of California at Berkeley. The National Academy report took the Park system to task for the inadequacies of its science program and recommended a number of administrative changes, most of which were endorsed within the Interior Department, though few were ever implemented. The Leopold Board, which included Ira Gabrielson, former Biological Survey and Fish and Wildlife head; Clarence Cottam, former assistant director of the Fish and Wildlife Service, Thomas Kimball, director of the National Wildlife Federation, and Stanley Cain, professor of conservation at the University of Michigan, specifically addressed the importance of mission-related research that would result in better wildlife management.[48]

Within its still somewhat limited staff and budget, the Park Service's reconstituted Wildlife Protection Branch attempted to carry out some of the Leopold Board's recommendations, but this was done on a piecemeal basis, and a much-needed system-wide focus was largely absent. Leopold, Cain, and Sigurd Olsen, a respected environmental educator, constituted an Advisory Committee on Natural History Studies, created in 1964 to improve relations between the Park Service and the scientific fraternity. This advisory group met infrequently and had little influence, but changes nonetheless took place, notably among the veteran park ranger and scientific staff, which briefly strengthened the scientific program. Additional staff were hired because of the increase in the number of parks, and many older rangers, whose perspective on Park priorities was close to that of Mather and Albright, retired.

With many new people on staff at the various parks, most of them college trained, Robert Linn, the new chief scientist of the Division of Natural Sciences, sought to establish a cohesive science policy among the nearly forty scientists and game management people in his division. Unfortunately, many traditionalists in the National Parks administration found this new arrangement threatening, arguing that the scientific program, if directed by Linn from Washington, might interfere with broader system objectives.

Responding to these objections, George B. Hartzog, then Park Service director, made some fundamental structural changes which were totally at odds with the earlier recommendations which had been made by the advisory committee of the National Academy of Sciences. Six new regional chief scientists were appointed, who were to report to their regional National Park directors. Park scientists now had to report either to their regional directors or individual park superintendents. Scientific and game management programs were badly weakened under this arrangement and much valuable ground won in the previous decade largely lost. Effective national direction of the Park Service's scientific program was no longer practicable, and Park Service scientists found it much more difficult to communicate with one another concerning common professional challenges. Since 1971, some improvement has occurred in communication between Washington and the field, but scientific priorities are still subject to the agendas set by the park administrators in the larger parks and by those in charge of the ten regional offices.[49]

One productive new program involved the assignment of some Park Service scientists to cooperative park study units at universities within each region. The idea here is very similar to the cooperative wildlife research units devised by Darling when Biological Survey chief in the mid-1930s. Park personnel in park study units have the advantage of working with and calling on the services of academic and research colleagues who are part of the university staff and faculties. Groups of scientists can thus focus their expertise on particular challenges, and the Park Service scientists also have the use of libraries and other facilities at their institutions.[50]

CHANGES IN FISH AND WILDLIFE PROGRAMS SINCE 1934

The Iowa-born Darling, a dedicated conservationist and a registered Republican, was brought in by President Franklin D. Roosevelt to head the Biological Survey early in 1934. Staff morale improved, and Darling secured large appropriations from Congress for the establishment and maintenance of wildlife refuges while initiating the duck stamp program, which raised funds for needed management efforts. Congress enacted the Fish and Wildlife Coordination Act of 1934, directing that Interior work with other state and federal agencies to protect game and fur-bearing animals, purchase lands for wildlife conservation projects, and conduct necessary investigations in support of these objectives. Darling was instrumental in fostering the Cooperative Wildlife Research Unit Program, discussed above, which enabled federal biologists to work with universities in addressing wildlife management problems. He founded the National Wildlife Federation in 1935 and was a principal instigator and organizer of the First North American Wildlife Conference held in Washington in February 1936. By his own choice, however, Darling's tenure was brief, lasting only twenty months, as he was on leave from his full-time profession as a syndicated cartoonist.[51]

Darling was succeeded in November 1935 by Gabrielson, another Iowan, who had been promoted from within the organization. Gabrielson was the Biological Survey's last chief (1935–1939) before it was transferred to the Interior Department on July 1, 1939. One of the major events of his early tenure was the passage by Congress of the Federal Aid in Wildlife Restoration Act, commonly known as the Pittman-Robertson Act, in 1937. Carl Shoemaker, former head of the Oregon Fish and Game Commission and from 1930 to 1947 special investigator for the Senate Special Committee on Conservation of Wildlife Resources, was a prime mover of this legislation. Inasmuch as the majority of the nation's wildlife was found on state lands, he persuaded Congress that the states would have to have federal assistance to ensure that the nation's wildlife resources be most effectively managed. Under the act, the Biological Survey (later the Fish and Wildlife Service) was given the authority to approve state conservation initiatives. The federal government could underwrite up to three-quarters of the cost of a game-enhancement or mitigation project, with the states assuming responsibility for the balance. When it came to funding this initiative, attention turned to an older idea of using proceeds from a hunting stamp for wetlands purchases originally proposed by George A. Lawyer, then chief U.S. Game Warden, in 1922. Despite various attempts over the years, however, Congress would not give its approval. That concept was eventually modified by Darling and became the duck stamp program. What finally made Pittman-Robertson feasible was an eleven percent tax on firearms and ammunition (pistols were taxed at ten percent). Later, a tax on hunter-utilized bows and arrows was added. During the first half-century of its operation, approximately $1.7 billion federal dollars and $600 million in state funds were used for such purposes as buying wildlife habitats, restocking wildlife, carrying out population surveys of various game species, training hunters in firearms safety, and related purposes.[52]

In the late 1930s, the Biological Survey became something of a political football when Secretary of the Interior Ickes sought to incorporate all federal government elements concerned with natural resources into a new federal Department of Conservation. While this effort did not succeed, the Biological Survey and the Fish Commission were finally combined in Interior as the U.S. Fish and Wildlife Service late in June 1940. Gabrielson became the first director of the new entity, remaining at the helm until 1946. During World War II, he doubled as deputy coordinator of Fisheries, ensuring that seafood production remained high during the crisis. Gabrielson represented the United States at the International Whaling Conference in 1946, spent twenty-four years as president of the Wildlife Management Institute, and was a founder of the International Union for the Conservation of Nature and an organizer of the World Wildlife Fund.[53]

In the years following the Biological Survey's transformation into the Fish and Wildlife Service, the agency increasingly focused its attention on management and regulatory issues. Such scientific research as took place was essentially oriented to specific mission objectives. Merriam's program of state and regional faunal surveys came to an end, although published studies in the *North American Fauna* series have continued to appear down to the recent past. All of the old Biological Survey specimens were under the jurisdiction of the "Biological Survey Section" in the Denver Wildlife Research Center of the Fish and Wildlife Service, though they have been housed in the National Museum of Natural History building in Washington since it was erected in 1910. By 1985, the Biological Survey Section had a annual budget of only $700,000 and a staff of sixteen. When the *Washington Post* took note of the Biological Survey's centennial that year, the Biological Survey Section was described as "a struggling, dispirited bureaucratic waif far from the center of power."[54]

In the 1950s, under pressure from commercial fishing organizations, on the one hand, and conservation groups on the other, Congress again reorganized the Fish and Wildlife Service in a vain effort to please both. The Fish and Wildlife Act of 1956 created two bureaus within the Service, the Bureau of Sports Fisheries and Wildlife and the Bureau of Commercial Fisheries. This enactment also strengthened the ability of the new Bureau of Sports Fisheries and Wildlife to acquire new refuges, initiated a new nationwide policy for recreational use of fish and wildlife, and gave encouragement to the nation's commercial fishing industry. A 1958 amendment provided for consultation between Fish and Wildlife, state fish and wildlife departments, and other federal agencies sponsoring dams or other water-development construction projects. In 1966, all national wildlife refuges were brought under the aegis of the Fish and Wildlife Service.[55]

In 1970, the Bureau of Commercial Fisheries was placed in the Commerce Department, renamed the National Marine Fisheries Service, and located within the newly created National Oceanic and Atmospheric Administration. Occasionally other agencies assumed some management jurisdiction, as in 1971, when the Wild Free-Roaming Horse and Burro Act gave responsibility for these animals to the Bureau of Land Management. That agency created a Division of Wild Horses and Burros in 1979, which had charge of an adoption plan, while the National Academy of Sciences assisted with scientific studies. In 1974, the Bureau of Sports Fisheries and Wildlife once again became the U.S. Fish and Wildlife Service. It immediately assumed major responsibilities for two recently passed conservation measures, and a third was added six years later. Under the first of these, the Marine Mammal Protection Act of 1972, Fish and Wildlife was designated to protect marine mammals which were considered more land-oriented (manatees, dugongs, polar bears, sea otters, and walruses), while the Marine Fisheries Service had jurisdiction over whales, dolphins, seals, and sea lions. No hunting or killing of any of these species was to be permitted within two hundred miles of the U.S. coastline or wherever else they were protected under American law, nor were any marine mammals or products utilizing their body parts to be sold or brought into the United States. The Marine Mammal Commission, an advisory group created by the 1972 act, was to be kept regularly advised of all actions taken by both regulatory agencies.[56]

One of the more controversial conservation measures was the Endangered Species Act of 1973. Based in part on earlier measures enacted in 1966 and 1969, but this time with enforcement powers, it was designed to protect endangered and threatened species of animals, both foreign and domestic. The 1973 legislation also constituted implementation by the U.S. of the Convention on International Trade in Endangered Species of Wild Fauna and Flora (CITES), signed earlier that same year. Again, responsibility was divided between the Fish and Wildlife Service and the National Marine Fisheries Service, and the importation, export, sale or possession of animals covered by the act was made illegal. In addition, however, both agencies are obliged to design and implement recovery plans, to include cooperative arrangements with individual states. Disputes over this act's efficacy and the extent to which it has forced cancellation or postponement of major construction and other projects have continued down to the present. In addition, many private citizens and advocacy groups expressed concern about the implications for private property rights posed by this legislation.[57]

The Endangered Species Act brought with it tension in several related areas. One concerned the use of poisons in animal control. Farmers and ranchers insisted on continued federal control of prairie dogs, for example, but these animals were the principal prey to the black footed ferret, an endangered species. President Nixon discontinued use of Compound 1080 in 1972, but constituent pressure brought it back in modified "coyote getters" in 1975, and the Reagan Administration again authorized 1080 in 1981.

A second issue had to do with the reintroduction of extirpated predators. By 1978, all American wolves, for example, were classified as endangered species. Much debate followed, with many western ranchers adamantly opposed. A few red wolves were unsuccessfully introduced in the southeastern United States in the 1970s, but later efforts in the 1980s and since have proven successful. In the western United States, this has been a contentious issue. Wolves naturally reintroduced themselves into Glacier National Park in 1982, and the Fish and Wildlife Service finally made a successful reintroduction in Yellowstone Park in the winter of 1994–1995. The first pups appeared in the spring of 1995. Opposition remains extremely strong among some livestock raisers in the region, and some wolves have been shot despite the fact that they are federally protected. Fallout from this controversy doubtless helped eliminate funding for the National Biological Service late in 1996.

The Federal Aid in Fish Restoration Act (1950), often referred to as the Dingell-Johnson Act, was designed to accomplish for fish what Pittman-Robertson had been designed to do for game species. Here, the funding, configured similarly to Pittman-Robertson, came from a tax on rods and reels. Under Dingell-Johnson, some $873 million was expended between 1952 and 1988. The Fish and Wildlife Conservation, or Forsythe-Chafee, Act of 1980 covered non-game species of land and water animals in a manner similar to Pittman-Robertson. Application procedures for federal support and the federal-state split of costs was substantially identical. Far less use has been made of this legislation, however, than Pittman-Robertson or Dingell-Johnson.[58]

As of 1989, the Fish and Wildlife Service's 6,400-person staff was divided between the Washington offices, seven regional offices, and fifty-seven field offices around the nation. Much of the marine zoological research done by the Fish and Wildlife Service was carried on at Patuxent, Maryland, regarded as an eighth regional facility. There were 441 national wildlife refuges, 152 waterfowl production areas and 58 coordination sites, all under the jurisdiction of other federal agencies, wherein wildlife is protected by Fish and Wildlife, the other federal agency, and the state in which the site is located. Highly productive cooperative fish and wildlife research unit programs were in place at twenty-nine state universities in various parts of the country. Some wildlife refuges have been the subject of controversy, as when oil and gas are found underground, and the Interior Department must determine whether drilling may be permitted under certain conditions. Government-imposed regulations governing grazing by privately-owned livestock on wildlife refuges and national forest land has long been another perennially contentious issue.[59]

THE NATIONAL BIOLOGICAL SERVICE, 1993–1996

In October 1993, Secretary of the Interior Bruce Babbitt announced the formation of the National Biological Survey, to be created within his department as the result of internal reorganization. This new agency was to have responsibilities:

> for inventorying, mapping, and monitoring biotic resources, performing basic and applied research on species, groups of species, populations, and ecosystems; and providing the scientific support and technical assistance needed for management and policy decisions in DOI [Department of the Interior].

Some 1,360 scientific personnel from five divisions within Interior were initially drawn together to pursue the new agency's objectives. The bulk of this group came from Fish and Wildlife, with much smaller numbers taken from the Park Service, the Bureau of Land Management, and the Bureau of Reclamation. In all, this group represented less than seven percent of all scientists working within Interior. The vast majority were biologists and biological technicians, with a much smaller numbers of physical scientists, mathematicians, veterinarians, social scientists, and museum curators and specialists. It was estimated that forty-six percent of the budget would be allocated to biological and ecological research, with seventeen percent allocated to inventorying and monitoring of biological resources, ten percent to the Cooperative Unit program and administration, and eight percent to information transfer activities. Initial editorial comment about the new agency's objectives was generally favorable, and the program got under way at the beginning of October 1993.[60]

The idea of recreating such an agency was not new, of course; it had "been in the air" since the dissolution of the old agency more than half a century before. A 1986 study published by the Association of Systematics Collections had stressed that,

The living species of plants, animals, and microorganisms must be documented and the processes fundamental to the maintenance of the biota must be understood before we can fully utilize these organisms and knowledgeably monitor the health of our ecosystems.[61]

Babbitt had requested advice from the National Academy of Sciences in February 1993, and their report, *A Biological Survey for the Nation*, called for creation of a National Biotic Resources Information System "to make reliable biological information more accessible to diverse users." It also proposed creation of a National Partnership for Biological Survey, involving "federal, state, and local agencies, museums, academic institutions, and private organizations," which would "collect, house, assess, and provide access to the scientific information needed to understand the current state of the nation's biological resources (status), how that status is changing (trends), and the causes of the changes." Babbitt strongly endorsed this cooperative approach to research objectives, put in place within the new agency.

Babbitt himself "intend[ed that] the NBS . . . be a world class science agency," with the task of "map[ing] the Nation's biological resources." He emphasized that it must help the federal government "avoid environmental trainwrecks." Turning to the controversial issue of government "takings," Babbitt stressed that the National Biological Survey would not be a regulatory agency, and that it had no authority to seize any citizen's private property. Further, he assured Congress and the press that "no employee of the NBS will enter private property without permission from the owner." He added that he would be "grateful if the Congress would debate that issue when it considers provisions of law that do authorize regulation of private property," and indicated that "the takings issue is not relevant to the NBS [because] the NBS is not a threat to private property."[62]

Initial ecological studies by the new agency stressed that the nation had sustained critical ecosystem losses, particularly in the "Northeast, South, Midwest, and California," due perhaps to "more extensive land uses in these regions, earlier settlement by Europeans, and more intensive scientific study." They suggested that looking at ecosystem conservation was probably more "cost efficient" than the "species by species approach" of the Endangered Species Act, which did "not solve all conservation problems" and was "inefficient." As one possible approach, they cited a Nature Conservancy study which estimated that "85–90% of species can be protected by conserving samples of natural communities without separate inventory and management of each species." Other studies suggested alternative proposals, one of which would conserve a "full array of physical habitats, environmental gradients, and landscape patterns in reserves."[63]

Rumblings of discontent were heard from the outset. Both Democrats and Republicans in the most recent Democratically-controlled Congress (1993–1994) expressed concerns about the possible effects that these, and other National Biological Survey activities, might have on the free usage of private property. Some considered the new agency vulnerable because it had been created through an internal reorganization, rather than by securing formal congressional approval, a risky maneuver. Despite Babbitt's assurances on the takings issue, some critics were unconvinced, pointing to other federal activities, such as Interior's enforcement of the Endangered Species Act. Some viewed the National Biological Survey (which was renamed National Biological Service early in 1996 in an effort to mollify critics) as a potential haven for scientists interested in identifying and designating more endangered species. Many conservative congressmen argued that free use of private and commercial property should, virtually without exception, take precedence over conservation concerns. They believed that the more environmentally oriented scientists from the Biological Service, in making their reconnaissance of national wildlife resources, might employ information thus obtained to support further endangered species initiatives. While he understood their concerns, Babbitt envisioned that in virtually every case where endangered species considerations required it, private property owners and the government would be able to come up with property swaps and other solutions acceptable to both sides, and that most small property holders could often be exempted from the effects of the legislation.[64]

With the election of a Republican-controlled Congress in 1994, the opposition mounted. Continuing appropriations for the National Biological Service soon fell prey to congressmen who generally disapproved of federal funding for science, particularly for biological research and conservation. Some opponents of the Endangered Species Act could not be persuaded of its importance to wildlife. Still others saw the elimination of most science as one preferred method of reducing the budget deficit. The Biological Service program was abolished by Congress in late August 1996. Some 1,600 Biological Service personnel, however, assigned to forty Biological Service field stations across thirty-eight states, were transferred to the new Biological Resources Division of the U.S. Geological Survey at the beginning of the new fiscal year in October 1996. Geological Survey administrators assured their new associates that they would have considerable administrative autonomy, and there was some hope that some ecological and biodiversity objectives might still be achieved. Gordon Eaton, director of the Geological Survey, opined that his agency might eventually become a National Natural Resources Agency, reflecting its new, broader mandate, but Babbitt and others resisted such a change, arguing that it might invite more unwanted attention from anti-science Congressmen.[65]

As the end of the twentieth century approached, many biologists and environmentalists in and out of government were increasingly concerned about the apparent anti-science and anti-environmental stance recently taken by Congress. There were strong feelings on both sides of these issues, but congressmen probably reflected the continuing ambivalence felt by many Americans concerning the proper role of the federal government in natural science.[66]

NOTES

1 Stephen E. Ambrose, *Undoubted Courage: Meriwether Lewis, Thomas Jefferson, and the Opening of the American West* (New York: Simon and Schuster, 1996); Paul Cutright, *Lewis and Clark: Pioneering Naturalists* (Urbana: University of Illinois Press, 1969); Paul Cutright, *A History of the Lewis and Clark Journals* (Norman: University of Oklahoma Press, 1976); Alexander Mackenzie, *Voyages from Montreal, on the River St. Lawrence, through the Continent of North America, to the Frozen and Pacific Oceans; in the years 1789 and 1793. With a Preliminary Account of the Rise, Progress, and Present State of the Fur Trade of that Country.* (London: Cadell and Davies, 1801); Walter Sheppe, ed., *First Man West: Journal of His Voyage to the Pacific Coast of Canada in 1793, by Alexander Mackenzie* (Berkeley, University of California Press, 1962).

2 [Nicholas Biddle, ed.,] *The History of the Expedition Under the Command of Captains Lewis and Clark, to the sources of the Missouri, thence across the Rocky Mountains and down the River Columbia to the Pacific Ocean. Performed during the years 1804–5–6,* 2 vols., Prepared for the Press by Paul Allen, Esquire, (Philadelphia: Bradford and Inskeep, and New York: Abm. H. Inskeep, 1814); Elliott Coues, ed., *History of the Expedition Under the Command of Lewis and Clark,* 4 vols., (New York: Francis P. Harper, 1893); Raymond D. Burroughs, *The Natural History of the Lewis and Clark Expedition* (East Lansing: Michigan State University Press, 1961). The original journals appeared as Reuben Gold Thwaites, ed., *Original Journals of the Lewis and Clark Expedition,* 8 vols., (New York: Dodd Mead & Co., 1904–1905). Until the mid-nineteenth century, Philadelphia was the leading scientific city in the United States. Most of the nation's leading scientists and several of its most respected scholarly organizations were located there. An excellent guide to the manuscript collections from the American scientific expeditions sent out between 1803 to 1860 is William Stanton, ed., *American Scientific Exploration: Manuscripts in Four Phila-delphia Libraries* (Philadelphia, American Philosophical Society, 1991). In addition to the Philosophical Society, this guide covers the pertinent holdings of the Academy of Natural Sciences of Philadelphia, the Library Company of Philadelphia, and the Historical Society of Pennsylvania.

3 For Peale's Museum, see Charles C. Sellers, *Mr. Peale's Museum: Charles Willson Peale and the First Popular Museum of Natural Science and Art* (New York: Norton, 1980), the standard study. See also C. C. Sellers, *Charles Willson Peale* (New York: Scribner, 1969), the standard biography, a revision and updating of Sellers' earlier two-volume work (Philadelphia: American Philosophical Society, 1939). See also Edgar P. Richardson, Brooke Hindle, and Lillian B. Miller, *Charles Willson Peale and His World* (New York, Harry N. Abrams, 1982). Peale clearly anticipated that the nation would need a national museum of natural history early in the nineteenth century. He appealed to President Thomas Jefferson for the federal aid which might have made this possible, but Jefferson felt strongly that this was not a permissible use of federal resources. Silvio Bedini, *Thomas Jefferson, Statesman of Science* (New York: Macmillan, 1990), pp. 308–309; Sellers, *Mr. Peale's Museum*, p. 167. After Peale's death in 1827, his sons and the museum trustees kept the Museum going until the 1840s, when financial reverses forced them to sell the contents and close it. Most specimens were sold to the showman Phineas T. Barnum. Virtually all were destroyed in two conflagrations, Barnum's Philadelphia Museum fire in 1851 and his American Museum blaze in New York City in 1865. Very few specimens have survived. Virtually all are now at Harvard's Museum of Comparative Zoology. Sellers, *Mr. Peale's Museum*, chapter 9.

4 For other representative Army explorers prior to the Civil War, see, for Stephen H. Long, William H. Goetzmann, *Exploration and Empire: The Explorer and the Scientist in the Winning of the American West* (New York: Knopf, 1966), pp. 58–64; Maxine Benson, ed., *From Pittsburgh to the Rocky Mountains: Major Stephen Long's Expedition, 1819–1820* (Golden, Colo.: Fulcrum, 1988), and Richard G. Wood, *Stephen Harriman Long, Army Engineer, Explorer, Inventor* (Glendale, California: Arthur Clark, 1966); *Dictionary of American Biography* (hereinafter *DAB*) XI: p. 580; for Zebulon Pike, Donald Jackson, ed., *Journals of Zebulon Montgomery Pike, with Letters and Related Documents*, 2 vols., (Norman: University of Oklahoma Press), vol. I, pp. 344–375, *DAB* XIV: pp. 599–600; and for John C. Fremont, Allen Nevins, *Fremont: Pathmarker of the West* (New York: Longmans, 1939 and 1955), Allen Nevins, ed., *Narratives of Exploration and Adventure*, by John Charles Fremont, (New York: Longmans, 1956), Donald Jackson and Mary Lee Spence, eds., *The Expeditions of John Charles Fremont* (Urbana: University of Illinois Press, 1970–), Ferol Egan, *Fremont: Explorer for a Restless Nation* (Garden City: Doubleday, 1977), *DAB*, VII: pp. 19–23.

5 Vincent Ponko Jr., *Ships, Seas and Scientists: U.S. Naval Exploration and Discovery in the Nineteenth Century* (Annapolis: U.S. Naval Institute Press, 1974), pp. 4, 7–8; John D. Kazar, "The United States Navy and Scientific Exploration, 1837–1860," unpublished Ph.D. dissertation, University of Massachusetts, 1973; John P. Harrison, "Science and Politics: Origins and Objectives of Mid-Nineteenth Century Government Expeditions to Latin America," *Hispanic-American Historical Review* 35 (1955): 175–202; Wayne D. Rasmussen, "The United States Astronomical Expedition to Chile, 1849–1852," *Hispanic-American Historical Review,* 34 (1954): 103–113.

6 William Stanton, *The Great United States Exploring Expedition of 1838–1842* (Berkeley: University of California Press, 1975), is the standard text on this effort, the first federally-sponsored exploring expedition to take place outside the continental United States. See also Ponko, ch. 3; David B. Tyler, *The Wilkes Expedition: The First United States Exploring Expedition, (1838–1842)* (Philadelphia: American Philosophical Society, 1968); Herman J. Viola and Carolyn Margolis, eds., *Magnificent Voyagers: The U.S. Exploring Expedition, 1838–1842,* (Washington: Smithsonian Institution Press, 1985); Jessie Poesch, *Titian Ramsay Peale; 1799–1885, and his Journals of the Wilkes Expedition* (Philadelphia: American Philosophical Society, 1961). The standard bibliographical work on Expedition publications is Daniel C. Haskell, *The United States Exploring Expedition, 1838–1842, and its Publications, 1844–1878* (New York: New York Public Library, 1942), reprinted by Greenwood Press, Westport, Connecticut, 1968. Haskell makes it clear that some politicians of the day railed against the cost of printing the expedition reports. In 1861, by which time publication had been under way for seventeen years (it was to continue for another seventeen), Senator John P. Hale complained "We [have] published eleven or twelve volumes of the exploring expedition, illustrated with pictures of bugs, snakes, and reptiles. It has cost us millions of dollars to print these pictures, and now we are going to spend $10,000 to distribute them . . . The thing is all wrong, Sir." It did not help that three fires, one in the Library of Congress in 1851, and two at the plants of the Philadelphia printers and binders in 1853 and 1856, destroyed much printed material, some of which had to be done again. Haskell, p. 23.

7 Ponko, pp. 33–60, 62–79, 89–159, 179–180, 183–205, and Kazar, passim.

8 E. F. Rivinus and E. M. Youssef, *Spencer Baird of the Smithsonian* (Washington: Smithsonian Institution Press, 1993), chapters 2–3; William A. Deiss, "The Making of a Naturalist: Spencer F. Baird, The Early Years," in *From Linnaeus to Darwin: Commentaries on the History of Biology and Geology* (London: Society for the History of Natural History, 1985), pp. 141–148. Robert Ridgway, "Spencer Fullerton Baird," *The Auk* V (1888): 1–14.

9 Rivinus and Youssef, ch. 8.

10 Rivinus and Youssef, pp. 62–64, 82–85. William A. Deiss, "Spencer Fullerton Baird and His Collectors," *Journal of the Society for the Bibliography of Natural*

History 9 (1980): 635–645; Debra Lindsay, *Science in the Subarctic: Trappers, Traders, and the Smithsonian Institution* (Washington: Smithsonian Institution Press, 1993), chapters 2 and 6.

11 Rivinus and Youssef, pp. 94, 163. A. G. Mayer, "William Stimpson," in *Biographical Memoirs of the National Academy of Sciences*, vol. 8, 1919, pp. 417–433. William A. Deiss and Raymond B. Manning, "The Fate of the Invertebrate Collections of the North Pacific Exploring Expedition, 1853–1856," in *History in the Service of Systematics* (London: Society for the Bibliography of Natural History, 1981), pp. 79–85. Ron Vasile, "William Stimpson," in K. B. Sterling, R. Harmond, G. Cevasco, and L. Hammond, *Biographical Dictionary of American and Canadian Naturalists and Environmentalists* (Westport, Connecticut: Greenwood Press, 1997); *DAB*, XVIII, pp. 31–32.

12 Deiss, "Baird and His Collectors;" Harry Harris, "Robert Ridgway," *The Condor*, vol. XXX (1928): 1–118; Harry C. Oberholser, "Robert Ridgway–A Memorial Appreciation," *The Auk* L (1933): 159–169; Paul Cutright and Michael J. Brodhead, *Elliott Coues, Naturalist and Frontier Historian* (Urbana: University of Illinois Press, 1981); K. B. Sterling, *Last of the Naturalists: The Career of C. Hart Merriam*, rev.ed., (New York: Arno Press, 1977).

13 Baird's volumes on mammals and birds constituted volumes VIII and IX, respectively, of the eleven volume set entitled *Reports of Expeditions and Surveys to ascertain the most practicable and economical route for a railroad from the Mississippi River to the Pacific Ocean* (Washington: A. D. P. Nicholson, 1853–1854). There was also a second edition in twelve volumes, published between 1855 and 1861, the twelfth volume being an account of a railway survey from St. Paul, Minnesota, to Puget Sound in Washington. The numbering of Baird's two volumes was identical in both sets. Both government-issued (by A. D. P. Nicholson) and commercially published editions were printed. The bird and mammal volumes were reissued privately in slightly modified form in 1859 and 1860, respectively (by J. B. Lippincott, Philadelphia). Both the mammal and bird volumes were also reprinted in 1974 by Arno Press. Baird also contributed a section concerning reptiles to a volume dealing with fishes by Charles Girard (volume X, 1858–1859). The five volumes in the *History of North American Birds* were published by Little, Brown of Boston. The *Water Birds* portion was published in conjunction with the Museum of Comparative Zoology at Harvard and the Geological Survey of California. Baird had earlier (1870) added to, and completed, a work on the land birds of California, based on a manuscript and notes by J. G. Cooper, which had been given to Baird for revision and publication. This was the first volume of what had been projected as a two-volume set. The second volume on the water birds, however, was much expanded by Baird and his colleagues and was replaced by the two volume set completed by Baird, Brewer, and Ridgway in 1884. Ron Tyler, in an unpublished paper presented before "Surveying the Record: North American Scientific Exploration to 1900," an American Philosophical Society Conference in Philadelphia in March 1997, estimated that the cost of printing the Pacific Railway Reports (approximately $1.2 million) was roughly two and a half times the cost of the expeditions themselves ($455,000). Ron Tyler, "Illustrated Government Publications Related to the American West, 1843–1863."

14 S. F. Baird, "The Distribution and Migrations of American Birds," *American Journal of Science and Arts*, 2nd series, I (1866): 77–90, 184–192, and 337–347. Dean C. Allard, Jr., *Spencer Fullerton Baird and the U.S. Fish Commission* (New York: Arno Press, 1978). Nathaniel P. Reed and Dennis Drabelle, *The United States Fish and Wildlife Service* (Boulder: Westview Press, 1984), p. 10; Eric J. Dolin, *The U.S. Fish and Wildlife Service* (New York: Chelsea House, 1989), p. 41.

15 Elliott Coues, *Key to North American Birds*, 4th ed., (Boston: Estes and Lauriat, 1894), pp. xxv–xxvi.

16 Cutright and Brodhead, *Elliott Coues*, chapters 1–3; Edgar E. Hume, *Ornithologists of the United States Army Medical Corps* (Baltimore: Johns Hopkins University Press, 1942), ch. IV.

17 Cutright and Brodhead, *Elliott Coues*, chs. XVII–XXVI.

18 Harris, "Robert Ridgway," *DAB*, vol. XV, pp. 598–599.

19 For True, see *DAB*, XIX, p. 5; for Goode, see David Starr Jordan, "Biographical Sketch of George Brown Goode, in George Brown Goode, ed., *The Smithsonian Institution, 1846–1896* (Washington: The De Vinne Press, 1897), pp. 501–515; Constance McLaughlin Green, *Washington: Capital City, 1879–1950* (Princeton: Princeton University Press, 1962), pp. 378–379. For Walcott, see Ellis L. Yochelson, "Charles Doolittle Walcott," *National Academy of Sciences Biographical Memoirs*, vol. 39, (New York: Columbia University Press, 1967), and Paul H. Oehser, *The Smithsonian Institution* (New York: Praeger, 1970, pp. 53–58). For Dall, see *DAB*, V, pp. 35–36, and Lindsay, *Science in the Subarctic*, pp. 108–110, 112–114, 118–119.

20 For Wetmore, see Paul Oehser, "In Memoriam: Alexander Wetmore," *The Auk* 97 (1980): 608–613; for Gerrit Smith Miller, Jr., see H. H. Shamel, et al., "Gerrit Smith Miller, Jr.," *Journal of Mammalogy* 35 (1954): 317–344; for A. Remington Kellogg, see Henry W. Setzer, "A. Remington Kellogg, 1892–1969," *Journal of Mammalogy* 58 (1977), pp. 251–253. for Stejneger, see Alexander Wetmore, "Biographical Memoir of Leonhard Hess Stejneger, 1851–1943," *National Academy of Sciences Biographical Memoirs*, vol. XXIV, (Washington: National Academy of Sciences, 1945), pp. 145–170. for Schmitt, see Richard E. Blackwelder, *The Zest for Life: or Waldo Had a Pretty Good Run: The Life of Waldo LaSalle Schmitt* (Lawrence, Kansas: Allen Press, 1979).

21 Arnold Mallis, *American Entomologists* (New Brunswick: Rutgers University Press, 1971), pp. 50–52, 61–69, 69–79, 126–138, 296–302.

22 Mallis, pp. 50–52, 296–302.

23 Mallis, pp. 69–79.

24 Mallis, pp. 79–86.

25 Jenks Cameron, *The Bureau of Biological Survey: Its History, Activities, and*

Organization (Baltimore: Johns Hopkins University Press, 1929), pp. 18–21; K. B. Sterling, *Last of the Naturalists: The Career of C. Hart Merriam*, rev. ed., (New York: Arno Press, 1977), pp. 56–65; K. B. Sterling, "Builders of the Biological Survey, 1885–1930," *Journal of Forest History* 33 (1989): 180–187. Elmer R. Birney and Jerry R. Choate, eds., *Seventy-Five Years of Mammalogy (1919–1994)*, [n.p.], American Society of Mammalogists, Special Publication no. 11, 1994, pp. 28,30.

26 Cameron, pp. 26–28; Sterling, *Last of the Naturalists*, Chapters 1–2 and pp. 204–235; Sterling, "Builders," pp. 183–184. Merriam's principal publications concerning his biogeographic theories included: "Results of a Biological Survey of the San Francisco Mountain Region and Desert of the Little Colorado, Arizona," *North American Fauna*, no. 3, (Washington: U.S. Department of Agriculture, 1890); "The Geographic Distribution of Life in North America, with special reference to the Mammalia," *Annual Report of the Board of Regents of the Smithsonian Institution for 1891* (Washington: Smithsonian Institution, 1893), pp. 365–415; "Laws of Temperature Control of the Geographic Distribution of Terrestrial Animals and Plants," *National Geographic Magazine* VI (December 1894): 229–238; and "Geographic Distribution of Animals and Plants in North America," *Yearbook of the Department of Agriculture, 1894* (Washington: Government Printing Office, 1895), pp. 207–214. Later authorities using more sophisticated modeling techniques substantially revised some of Merriam's conclusions beginning in the 1930s.

27 Sterling, *Last of the Naturalists*, chapter 4, and pp. 167–169. C. Hart Merriam, "Suggestions for a new method of discriminating between species and subspecies," *Science* 5, new series, (1897), pp. 754–756.

28 On Merriam's standards for his staff, Sterling, *Last of the Naturalists*, pp. 88, 199–201, 321, 374n; Sterling "Builders," p. 186; C. Hart Merriam, "Biology in Our Colleges: A Plea for a Broader and More Liberal Biology," *Science* 21 (1893): 352–355. On the North American Fauna series, see Sterling, *Last of the Naturalists*, pp. 99–100, 102, 212, 285.

29 On Bailey, see Howard Zahniser, obituary of Bailey in *Science* 96 (1942): 6–7, brief obituary of Bailey in *Journal of Mammalogy* 23 (1942): 244, also notes on Bailey in the Albert Kenrick Fisher papers, Library of Congress Manuscript Division; on Fisher, Francis M. Uhler, "In Memoriam: Albert Kenrick Fisher," *The Auk* 68 (1951): 210–213; on Palmer, W. L. McAtee, "In Memoriam: Theodore Sherman Palmer," *The Auk* 73 (1956): 367–377, also notes on Palmer in A. K. Fisher Papers, Manuscript Division, Library of Congress; on Preble, W. L. McAtee, "Edward A. Preble," *The Auk* 79: 730–742; on Goldman, S. P. Young, "Edward Alphonso Goldman," *Journal of Mammalogy* 28 (1): 91–109, 1947; on Nelson, E. A. Goldman, "Edward William Nelson—Naturalist, 1855–1934," *The Auk* 52 (1962): 135–148, April, 1935; Notes on Bailey, Fisher, Palmer, Nelson, and Goldman in Waldo Lee McAtee Papers, Manuscript Division, Library of Congress.

30 John W. Aldrich, "In Memoriam: Hartley Harrad Thompson Jackson (1881–1976)," *Journal of Mammalogy* 58 (1977): 691–694; L. Little, "Alfred Brazier Howell,

1866–1961," *Journal of Mammalogy* 49 (1968): 732–742; H. W. Setzer, "A. Remington Kellogg, 1892–1969," *Journal of Mammalogy* 58 (1977): 251–253; R. H. Manville, "Stanley Paul Young, 1889–1969," *Journal of Mammalogy* 51 (1970): 131–141. On the founding of the American Society of Mammalogists, see Donald F. Hoffmeister, "A History of the American Society of Mammalogists" in *50th Anniversary Celebration* [Conference Program], (New York: American Museum of Natural History, 1969), pp. 8–14; Jackson, Howell, and Kellogg are also covered in Birney and Choate, *Seventy-Five Years*, pp. 38–40 and 42–43; see also D. F. Hoffmeister and K. B. Sterling, "Origin," in Birney and Choate, *Seventy-Five Years*, pp. 11–20.

31 On the English sparrow, see Eugene Kinkead, "In Numbers Too Numerous to Count," *The New Yorker*, May 22, 1978, pp. 40–88. The Lacey Act has been augmented and amended a number of times, most recently in the 1980s. See [Alan Pilisoul, et.al.], *Digest of Federal Resource Laws of Interest to the United States Fish and Wildlife Service*, (Washington: U.S. Department of Interior, 1992), pp. 42–43. Sterling, *Last of the Naturalists*, pp. 80–81; Dolin, pp. 31–32.

32 [Michael J. Bean], *The Evolution of National Wildlife Law* (Washington: Council on Environmental Quality, 1977), chapters 2,3; Dolin, pp. 32–37; Cameron, pp. 45–64,70–124; Reed and Drabelle, pp. 8–9; Thomas A. Lund, *American Wildlife Law* (Berkeley: University of California Press, 1980), pp. 84–88. F. Wayne King, "The Wildlife Trade," in H.P Brokaw, ed., *Wildlife in America: Contributions to an Understanding of American Wildlife and its Conservation* (Washington: Council on Environmental Quality, 1978), p. 254. Justice Oliver Wendell Holmes spoke out strongly in favor of federal jurisdiction over migratory birds in *Missouri v. Holland* (1920). "Wild birds are not in the possession of anyone; and possession is the beginning of ownership. The whole foundation of the State's rights is the presence within their jurisdiction of birds that yesterday had not arrived, tomorrow may be in another State, and in a week a thousand miles away. . . . Here a national question of very nearly the first magnitude is involved . . . But for the [Migratory Bird] treaty and statute there soon might be no birds for any powers to deal with. We see nothing in the Constitution that compels the government to sit by while . . . the protectors of our forests and our crops are destroyed. It is not sufficient to rely upon the States. The reliance is in vain. . . ." Max Lerner ed., *The Mind and Faith of Justice Holmes* (Boston: Little Brown, 1943), pp. 273–278.

33 Ernest Thompson Seton, "Lobo, King of the Currumpaw," in his *Wild Animals I Have Known* (New York: Scribner, 1898), pp. 17–54.

34 Cameron, pp. 312, Edward William Nelson, "Henry Wetherbee Henshaw—Naturalist, 1850–1930," *The Auk* XLIX (1932): 399–427.

35 E. A. Goldman, "Nelson," E. A. Goldman, "Edward William Nelson—The Formative Years," and Paul H. Oehser, "Nelson and Goldman: Naturalists of the Old School, With Some Reminiscences of the Biological Survey Four Decades Ago," in Paul H. Oehser, ed., *Lower California and Its Natural Resources*, by E. W. Nelson, (Riverside, California: Manessier, 1966), pp. iii–vii and viii–xvi. K. B. Sterling, "Two Pioneering American Mammalogists in Mexico: The Field Investigations of Edward William Nelson and Edward Alphonso Goldman, 1892–1906," in Michael A. Mares and David J. Schmidly, eds., *Latin American Mammalogy: History, Biodiversity, and Conservation* (Norman: University of Oklahoma Press, 1991), pp. 33–47.

36 Thomas R. Dunlap, *Saving America's Wildlife* (Princeton: Princeton University Press, 1988), chapters 4–5. A typical statement of the Biological Survey's view on predator control in the late 1920s is: Albert Day and Almer Nelson, "Wildlife Conservation and Control in Wyoming under the Leadership of the United States Biological Survey, 1928," in Rick McIntyre, ed., *War Against the Wolf* (Stillwater, Minnesota: Voyageur Press, 1995), pp. 192–199. Predator extirpation in Yellowstone Park and other parks is discussed in McIntyre, pp. 201–208 and 210–213. This volume has a number of other excerpts from reports on the subject during the late nineteenth and twentieth centuries. Stanley P. Young estimated the number of wolves killed by the Biological Survey between 1915 and 1941 at 24,132. Stanley P. Young, *The Wolf in North American History* (Caldwell, Idaho: Caxton Press, 1946), p. 146. In addition to the American Society of Mammalogists' debate on predator control in 1924, there was another symposium on the subject at the 1930 annual meeting in New York in May 1930. See *Journal of Mammalogy* 11 (1930): 325–389. See also, Horace M. Albright, "The National Park Service's Policy on Predatory Mammals," *Journal of Mammalogy* 12 (1931): 185–186; "Report of the Committee on Problems of Predatory Mammal Control," *Journal of Mammalogy* 12 (1931): 340–344. See also Sterling, *Last of the Naturalists*, pp. 309–313, for a discussion of C. H. Merriam's involvement in predator control issues in the 1920s and 1930s.

37 Dunlap, *Saving America's Wildlife*, pp. 39–40, 48–61.

38 Susan Flader, *Thinking Like a Mountain: Aldo Leopold and the Evolution of an Ecological Attitude toward Deer, Wolves, and Foxes* (Columbia: University of Missouri Press, 1974), pp. 7–18. For some of Leopold's earlier ideas concerning predator control, see McIntyre, pp. 187–192.

39 Flader, pp. 18–35; Curt Meine, *Aldo Leopold: His Life and Work* (Madison: University of Wisconsin Press, 1988), chapters 12–21. Studies of the national forests include Harold K. Steen, *The United States Forest Service: A History* (Seattle: University of Washington Press, 1976), and Michael Frome, *The Forest Service* (New York: Praeger, 1974).

40 The standard history of the national parks is Alfred Runte, *National Parks: The American Experience* (Lincoln: University of Nebraska Press, 3rd ed., 1997). Wildlife is discussed on pp. 109–140, *passim*. Runte's bibliography is most helpful. R. Gerald Wright, *Wildlife Research and Management in the National Parks* (Urbana: University of Illinois Press, 1992), pp. 3–12. Richard W. Sellars, "Manipulating Nature's Paradise: National Park Management under Stephen T. Mather, 1916–1929," *Montana* 43 (1993): 3–13. See also Richard W. Sellars, *Preserving Nature in National Parks: A History* (New Haven: Yale University Press, 1997).

41 Wright, pp. 12–14. Richard W. Sellars, "The Rise and Decline of Ecological Attitudes in National Park Management, 1929–1940," Part I, *The George Wright Forum* 10 (1993): 74; Sellars, "Manipulating Nature's Paradise," p. 6.

42 Sellars, "Manipulating Nature's Paradise," pp. 3–8. Walter P. Taylor and W. T. Shaw, *Mammals and Birds of Mount Rainier National Park* (Washington: United States Department of the Interior, 1927).

43 Wright, pp. 13–19 and chapter 3; Sellars, "Rise and Decline of Environmental Attitudes," part 1, pp. 59–67, 70. George M. Wright, Joseph S. Dixon, Ben H. Thompson, "A Preliminary Survey of Faunal Relations in National Parks," [Fauna One], *Fauna of the National Parks of the United States* (Washington: U.S. Department of the Interior, 1933). Other early volumes in this series published by the Interior Department were George M. Wright and Ben H. Thompson, "Wildlife Management in the United States," [Fauna Two], 1935; Joseph S. Dixon, "Birds and Mammals of Mt. McKinley National Park, Alaska," [Fauna Three], 1938; Adolph Murie, "Ecology of the Coyote in the Yellowstone," [Fauna Four], 1940.

44 U.S. Congress, *House Special Committee on Conservation of Wildlife*, "Conservation of Wildlife," Hearing, 73rd Congress, 2nd Session, 4 May 1934, (Washington: Government Printing Office, 1934), pp. 125, 129.

45 Sellars, "The Rise and Decline of Ecological Attitudes," Part I, p. 69. R. W. Sellars, "The Rise and Fall of Ecological Attitudes in National Park Management, 1929–1940, Part III, *The George Wright Forum* 10 (1993): 51–52; Wright, pp. 17,18.

46 Thomas R. Dunlap, "Wildlife and Endangered Species," in Michael J. Lacey, *Government and Environmental Politics: Essays on Historical Developments Since World War II* (Washington: The Woodrow Wilson Center Press and Johns Hopkins Press, 1991), p. 229, note 7. Wright, p. 21. Adolph Murie, "The Wolves of Mount McKinley," [Fauna Five], *Fauna of the National Parks of the United States* (Washington: U.S. Department of Interior, 1944), pp. xiv, 221, 230–232. Timothy Rawson has argued that the controversy over wolves and Dall sheep in the Mt. McKinley area was a struggle between opposing pressure groups between the 1930s and 1950s. One side consisted of many Alaskans and hunter-sportsmen from the lower 48 states, the other conservationists and some biologists who thought justification for wolf control unpersuasive. Rawson, "Science, Public Opinion, and the Wolves of McKinley Park," paper presented before the American Society for Environmental History, Baltimore, March 1997.

47 Wright, pp. 22, 23.

48 Wright, pp. 24–28. A. Starker Leopold, Stanley A. Cain, Clarence M. Cottam, Ira N. Gabrielson, Thomas L. Kimball, "Wildlife Management in the National Parks," in *Reports of the Special Advisory Board on Wildlife Management for the Secretary of the Interior, 1963–1968* (Washington: Wildlife Management Institute, 1969), [no pagination]. *The Fauna of the National Parks of the United States* continued with Ralph E. and Florence B. Welles, "The Bighorn of Death Valley," [Fauna Six], (Washington: U.S. Department of the Interior, 1961).

49 Wright, pp. 28–30; Leopold, et al., "Predator and Rodent Control in the United States," in *Reports of the Special Advisory Board*, 9 March 1964; A. S. Leopold, Clarence Cottam, Ian Mc T. Cowan, Ira N. Gabrielson, Thomas L. Kimball, "The National Wildlife Refuge System [1968]," in same. Wildlife research personnel in the national park system were transferred to the new National Biological Survey in 1993. This agency became the Biological Services Division of the U.S. Geological Survey in the fall of 1996. See endnotes 64 and 65, below. Frederick H. Wagner, et al., *Wildlife Policies in the U.S. National Parks* (Washington: Island Press, 1995), p. 95, Richard W. Sellars, "Science or Scenery," *Wilderness* 52 (1989): 29–37.

50 Wright, p. 31 and chapter 11. Current structural features and wildlife management problems in the national park system as they existed prior to the creation of the National Biological Survey in 1993 are discussed in Wagner, et al., *Wildlife Policies in the U.S. National Parks*, and in Wright, chapter 11. Richard W. Sellars, "A House Divided: The National Park Service and Environmental Leadership," paper delivered before American Society for Environmental History, Baltimore, March 1997.

51 David L. Lendt, *Ding: The Life of Jay Norwood Darling*, (Ames: Iowa State University Press, 1979), chapters 11–13.

52 Harmon Kallman, et al., *Restoring America's Wildlife, 1937–1987: The First 50 Years of the Federal Aid in Wildlife Restoration (Pittman-Robertson) Act* (Washington: U.S. Department of Interior, 1987), is a standard source. For the origins of Pittman-Robertson, see pp. 3–16. An obituary of Ira Gabrielson appeared in *The Auk* 102 (1985): 865–868.

53 Gabrielson obituary in *The Auk*; biographical sketch of Gabrielson in Richard H. Stroud, ed., *National Leaders of American Conservation*, (Washington: Smithsonian Institution Press, 1985), pp. 162–163. Gabrielson was the author of three important texts, including *Wildlife Conservation* (New York: Macmillan, 1941); *Wildlife Refuges* (New York: Macmillan, 1943); and *Wildlife Management* (New York: Macmillan, 1951). For a discussion of Harold L. Ickes' plans for a federal Department of Conservation, see: T. H. Watkins, *Righteous Pilgrim: The Life and Times of Harold L. Ickes, 1874–1952* (New York: Henry Holt, 1990), Part VII.

54 Boyce Rensberger, "Biological Survey an Orphan at 100," *Washington Post*, 1 July 1985.

55 Dolin, pp. 46–49.

56 W. L. Hobart, ed., *Baird's Legacy: The History and Accomplishments of NOAA's National Marine Fisheries Service, 1871–1996*, ([n.p.]: U.S. Department of Commerce, 1995), p. 36; *Third Report to Congress: Administration of the Wild Free- Roaming Horse and Burro Act*, (Washington: U.S. Department of the Interior, 1980), pp. v, 3 ,30; Dolin, pp. 52–54.

57 Reed and Drabelle, pp. 87–103; Dolin, pp. 55–63, 89–91. Nearly a thousand endangered species were listed at the end of the 1980s, of which slightly less than half (492) were located in the United States. Wright, chapter 8; Dunlap, *Saving*

Wildlife, p. 170; Washington Post, 7 May 1995; Peter Steinhart, *The Company of Wolves* (New York: Knopf, 1995), pp. 4–9. F. Wayne King, "The Wildlife Trade," p. 254.

58 Dolin, pp. 44–45, 65, 96.

59 Dolin, pp. 15, 80.

60 "The National Biological Service Fact Sheet," (Washington: U.S. Department of Interior, 8 June 1993). "The National Biological Service: Excerpted Editorial Quotes," (Washington: U.S. Department of the Interior, 8 June 1993).

61 *U.S. Congress, House, Subcommittee on Technology, Environment, and Aviation and the Subcommittee on Investigations and Oversight of the Committee on Science, Space, and Technology, The National Biological Survey Act of 1993, Hearing*, 103rd Congress, 1st Session, 14 September 1993, (Washington: Government Printing Office, pp. 8, 39–40); Ke Chung Kim and Lloyd Knutsen, eds., *Foundations for a National Biological Survey* (Lawrence, Kansas: Association of Systematics Collections, Holcomb Research Institute, and Illinois Natural History Survey, 1986), pp. 3–4.

62 National Research Council, *A Biological Survey for the Nation* (Washington: National Academy of Sciences Press, 1993), pp. 5–8 and 12–22.

63 Hearing, *The National Biological Survey Act of 1993*, pp. 12–13, 16–17.

64 Hearing, *The National Biological Survey Act of 1993*, pp. 30, 41; Daniel S. Greenberg, "Scientists Must Join the Fray," *Technology Review* 99 (1996): 57–58; Colin Macilwain, "U.S. Geological Survey Picks Up the NBS Pieces," *Nature* 302 (22 August 1996), p. 658.

65 Macilwain, p. 658; Eliot Marshall, "Shaking Up the Lab Rats," *Discover* 17 (1996): 64; Bob Benenson, "Conferee's Interior Initiatives May Get Clinton's Veto," *Congressional Quarterly Weekly Report* 53 (23 September 1995), pp. 2883–2884; "Biodivisiveness," *Economist* 334 (25 February 1995): 85; Elizabeth A. Palmer, "Work on Interior Bill Drags on with Hot Issues Still to Go," *Congressional Quarterly Weekly Report* 53 (15 July 1995): 2055–2056. "Biological Services Division, U.S. Biological Survey," Reston, Virginia, U.S. Geological Survey, 1 October, 1996, 2 pp.; "Biological Services Division, Strategic Science Plan," Reston, Virginia, U.S. Geological Survey, 3 September 1996, 13 pp.; Bruce Babbitt, "Transfer of the National Biological Service to the U.S. Geological Survey as a new Biological Services Division," Order no. 3202, News Release, U.S. Geological Survey, U.S. Department of the Interior, 30 September 1996, 2 pp.

66 Dunlap, "The Federal Government, Wildlife, and Endangered Species," p. 228.

DOUGLAS HELMS

CONTRIBUTIONS OF THE SOIL CONSERVATION SERVICE TO FOREST SCIENCE

Among the most important legacies of Hugh Hammond Bennett's shepherding of the early soil conservation movement was his insistence on the multidisciplinarity of the conservation job. Bennett and the early Soil Erosion Service staff recruited engineers, foresters, agronomists, biologists, economists, and anthropologists for the pioneering soil conservation demonstration projects where teams worked with farmers to rearrange their fields and operations both for the benefit of productivity and for conservation. Staff working in the demonstrations projects of the 1930s accumulated knowledge through trial and error. Then, as now, performances sometimes fell short of the ideal of giving equal attention to the utilization of all disciplines; but the objective of considering all possible alternatives remains a good operating guideline.[1]

The work being carried out by the Soil Conservation Service (SCS) presented new opportunities for foresters. The intersection of opportunity and innovative thinkers produced contributions to the science of forestry. Conversely, that for much of its history SCS has not had broad forest research authorities and the emphasis has been overwhelmingly on operations meant some restrictions in the potential for contributions. This paper will concentrate on the contributions of SCS personnel to forest science, while recognizing that separate history exists of the impact of SCS's work on privately owned forests.

The task facing SCS foresters predetermined some of the questions asked of research and empirical observation. Foresters teamed with other specialists to arrange farms to fit the needs of the farmer and nature. One task involved reforesting eroded, degraded sites such as gullies, revegetating stream banks, or planting seedlings in former fields that should be retired from cultivated land. The second major part of the work was to enlist the farmer's cooperation in managing the land still in forests, usually second-growth, unmanaged forests. The farmers might want to improve timber stands for a variety of reasons: reduction of erosion, wildlife habitat, providing timber and fire wood for farm, and providing annual income as any other crop might.

NURSERY OPERATIONS AND PLANT MATERIALS CENTERS

Early demonstration projects were located in areas needing special attention to erosion problems, including reforesting of cropland. At a time when there were few commercial outlets for planting materials, a great need existed for nursery stock of conifers, hardwoods, shrubs, and grasses. SCS acquired some seed from federal, state, and commercial nurseries, but also undertook massive collection of native seed for nursery stock production and direct planting. By 1939 the nurseries in the Service had a capacity to produce between 150 to 200 million trees and shrubs annually.[2]

In addition to the production of planting stock, some nurseries also included an observational nursery unit. Nursery managers collected native seed and grew woody species while observing them for their habits of growth, soil-binding properties, drought resistance, ease of propagation and transplanting, and general adaptability for use in the soil conservation program.[3] SCS nurseries also acquired and introduced seeds and plants from the Division of Plant Exploration and Introduction of the Bureau of Plant Industry. Nurseries made interregional transfers of material both to supply the needed planting material and to make observations.

The experimental and observational aspect of the nursery work led to the domestication of many native species. That is, the nurseries developed the propagation recommendations. The primary developer of the method, Dr. Franklin J. Crider, who had begun his SCS career in the West, was called to Washington to head the Nursery Division in 1936.[4] There he gained Hugh Bennett's endorsement of a nationwide system of observational studies "of native and exotic plants—to determine their adaptations, characteristics, propagation propensities, essential cultural requirements, and possible or potential usefulness for soil and moisture conservation and related utility purposes."[5] Crider worked out cooperative agreements with the state experi-

*The Plant Materials Center in Manhattan, Kansas, helped select
woody material for reforestation and conservation.*
(USDA SOIL CONSERVATION SERVICE PHOTO BY H. L. GAMBLE, 1962)

ment stations and the Bureau of Plant Industry covering the nursery observational work, which he saw as "occupying an essential intermediary place
between research and practical field application. . . ."[6]

As the commercial nurseries grew in the United States, SCS nurseries came
under increasing pressure to stop supplying seeds and plants to farmers. In
the early 1950s, USDA's administration directed SCS to cease its nursery operations, but allowed the Service to retain the observation and selecting of
plants for conservation uses, a service which has been called the plant materials program. The Natural Resources Conservation Service, formerly
the Soil Conservation Service, now makes formal releases of cultivars which
a center has selected, tested, and documented. Releases are made with a state
agricultural experiment station through the state certifying agency. Herbaceous plants are registered with the Crop Science Society of America. The
Pullman, Washington, nursery (later plant materials center) initiated this
method in 1943 under the leadership of Atlee L. Hafenrichter and John
Schwendiman.[7]

Until very recently no system existed for certifying seeds of woody species (registration did exist for vegetatively propagated wood species). SCS requested the Association of Seed Certifying Agencies to develop definitions and standards for registering source-identified seed. Now a system is in place for certifying seed-propagated woody species. SCS makes the seed and plant selections available to commercial nurseries for propagation and increase. Since the inauguration of the formal system of plant releases, the Soil Conservation Service has released eighty-eight woody cultivars. SCS also released considerably more grasses, legumes, and forbs than woody plants.[8]

WILDLIFE AND FORESTRY CONNECTION

The Soil Erosion Service, partly due to the suggestion of Aldo Leopold, hired a biologist to work on its first demonstration project at Coon Valley, Wisconsin. The biologist sought to enhance the habitat for wildlife through special plantings, but also through the selection of plants that benefited wildlife while having other uses for erosion control or timber. Both the biologist and the forester had need of guides to plants that might be suitable to particular uses. With this in mind, Ernest Holt, head of the Section of Wildlife Management, suggested that one of his first recruits, William Van Dersal, undertake the compilation of a list of native woody plants that had uses for wildlife and erosion control.

The list relied upon research by university and herbarium authorities as well as experts in the Bureau of Biological Survey, the U.S. Forest Service, Bureau of Plant Industry, and SCS. Again the mission of the SCS prompted the need for this document. While much scientific literature existed, the practitioner in the field could turn to no single source. Van Dersal first made his list available as a mimeographed SCS technical publication, *Handbook of Native Woody Plants Of The United States*. Later, USDA issued a hard-bound version, *Native Woody Plants Of The United States: Their Erosion Control and Wildlife Values*. Van Dersal had a definite preference for native species; he followed H. L. Shantz's dictum "Where one looks, nature has pointed the way to recovery." In most cases SCS needed materials susceptible to extensive planting; they would not be cultivated, fertilized, or otherwise tended. Given this set of circumstances, his advice to the technician for whom the volume was intended would be "that our first choice of species to plant on a given site should be these which have already been tested for that site, as natives to the region in question. Our second choice would be species which are not natives; that is, exotic or introduced plants. If it is certainly known that there is no native species which can be grown in a chosen site, then we are justified in turning to exotic forms."[9]

The wildlife and nursery sections, beginning in 1937, started a nationwide project to select suitable wild planting materials, including but not limited to woody plants. Regional biologists were to compile lists of plants known or thought to combine values for soil cover and wildlife food. After the lists had been paired down, the regional nursery made observational plantings and field-tested promising materials. In this manner SCS selected numerous plants that it promoted for wildlife habitation.[10]

Edward Graham, another of Holt's early employees, had graduate training in botany, a discipline which fitted well with the emphasis on selecting plants that provided food and habitat. The prolific Graham authored several books. Of *Natural Principles of Land Use* Leopold wrote, "I have seen no better sketch of forest ecology than Dr. Graham presents . . . the net effect is coherent and convincing, and the reader is literally given the meat of the author's wide erudition in this field."[11]

REFORESTATION AND SURVIVAL STUDIES

The nurseries and foresters worked closely on adjusting nursery production to the trees that survived best on the farm. Nurseries made inter-regional transfers of materials. Often the areas replanted and the trees used were the first tested for many localities. A forester, likely Ted Plair, assessed the situation in the Pacific southwest region in the late 1930s.

> The management of farm woodlands, their reforestation, and the afforestation of certain arable lands in California and Nevada present a variety of new problems. Prior to the advent of the Soil Conservation Service, little or no attention had been given these by anyone. Practically all of the work of an investigative nature on related problems was restricted to National Forest and large commercial timber areas.[12]

For example, an SCS forester at the Yerrington, Nevada, office claimed that Mason Valley and Smith Valley soil conservation districts supervised the first large-scale plantings in Lyon County, Nevada. The observation studies focused on impacts of fertilization, cultivation, or no treatment on survival and growth.[13]

The forestry division of SCS standardized the survival studies and issued national instructions on preparing the annual "Reports On Survival and

Plantation Status."[14] The reforestation experience, often on eroded areas, contained surprises. Having seen the effects of insects, bacteria, rabbits, grasshoppers, fire, and wind on the seedlings, forester John Preston admitted after six or seven years of experience that, "There is much more to the ecological and the biological side of the problem than was ever realized. . . ."[15] The need for artificial reforestation still existed, but the foresters would no longer approach it so sanguinely and would continue to seek answers for increased success. Much of the information on adaptation of species and site was not published in the professional forestry literature, but it eventually found its way into the "technical guides" for field operations. Technical guides, now maintained on a state basis, contain variations for particular portions of each state. Also, some of the early survival studies listed soil type, based on the nomenclature of the National Cooperative Soil Survey. In this, one can detect some of the genesis of the soil site interpretations.

HILLCULTURE DIVISION

The Hillculture Division dedicated itself to research and promotion of the use of trees and shrubs on hilly land. The wood, fruit, and nuts were meant to provide income on an annual basis, or at least more frequently than most timber production did. The Hillculture, Nursery, and Forestry divisions selected several cultivars of black locust that were particularly prized for fence posts. In addition to black locust they collected shipmast locust from Long Island.[16] The Hillculture Division developed a cooperative agreement with the Forest Pathology Laboratory for indexing the durability of the various strains. Field testing for use as fence posts supplemented the laboratory testing.[17]

The forestry division of the southeastern region held a "black locust clinic" in the fall of 1939. The next year the division distributed Dr. T. J. Grant's mimeographed report "A Method for Judging Quality of Planting Sites for Black Locust, Based on Field Clinic Data," to the other regions. In the southeast the guide would be used to help farm planners in recommending the use of black locust on specific sites. But they also requested that the other regional foresters "take personal interest in this problem of developing a practicable method for judging the quality of planting sites for black locust. . . ."[18]

MANAGEMENT OF SECOND-GROWTH STANDS

In much of the United States, little work had been done with management of the second-growth forests that predominated on farms. Although there had been some concern about assistance to private woodland owners, too often the impulse had been to apply commercial forestry methods to farms, when, in fact, farmers had diverse needs for on-farm products and annual income. The work should be planned to fit the seasonal distribution of the farmer's labor. Despite the reforestation of eroded crop land or degraded sites, the real mission of the SCS, according to John F. Preston, head of SCS's forestry work, would be the management of second-growth forests on farms. He wrote, "I think that all workers will agree that the crux of the farm forestry problem is the management of existing timber in farmwoods by farmers under a policy of sustained yield."[19] The forester played a role in the rehabilitation work of controlling gullies and reforesting eroded field, but for many foresters the preventive side of conservation, rather than rehabilitation and restoration, lay in demonstrating to farmers the value of their wooded land. Commenting on the absence of planning in a document about forested areas of California, regional forester Ted Plair wrote, "I believe that public servants can render more valuable service by planning so as to protect and preserve the national resources than can ever be done in rebuilding and reclaiming areas laid waste by those who *mine*."[20]

The nationwide character of the work of the Soil Conservation Service gave some of the foresters an opportunity to do pioneering work on management of second-growth stands on private lands. Despite the research work of the Forest Service and the state and private forestry network there remained numerous questions about the silvicultural methods and the economic possibilities for production of wood products for on-farm use or income.

In the Pacific Coast region, SCS foresters saw potential for turning the bark of the cascara tree into a commercial crop for farmers. Native Americans used the extract of the bark as a stimulant and cathartic and passed along the knowledge to the Spanish priests of California. Latter-day gatherers of the bark often obtained it by trespass, giving little thought to conservation for future production. SCS foresters worked out a gathering method that would allow harvest, conserve the tree, and produce nursery stock for farmers. SCS supplied the farmers with necessary planting, management, and harvesting information.[21]

When William J. Lloyd went to work in Snohomish County in western

A soil conservationist evaluates the effectiveness of woody vegetation used to heal a gully in California.

(USDA SOIL CONSERVATION SERVICE PHOTO, FROM THE NATIONAL ARCHIVES, PACIFIC REGION, SAN BRUNO, CA)

Washington, he found stands of second-growth red alder. A minor species nationally, it was a major species in this particular county. Lloyd and other SCS personnel in the area worked out thinning methods for red alder. The Forest Service's experiment stations typically produced yield tables for commercial species, but had not produced one for red alder. The experiment station appointed Lloyd chair of the committee which eventually produced yield tables for red alder. It became a major source of baby furniture. As the grain takes stain well, the wood is often used as a substitute for other woods in furniture.[22] Red alder had a more profitable future than cascara.

FOREST MANAGEMENT TOOLS FOR
TECHNICIANS AND FARMERS

The organizational structure of the Soil Conservation Service greatly influenced the search for ways to assist land owners. During the early days of demonstration projects, it remained possible for teams of specialists to work on recommendations to farmers. In 1937, SCS began to work through conservation districts, which would eventually blanket the country. It became obvious that each district could not have the concentration of specialists that existed under the demonstration-project concept. The projects had covered only a small portion of the farms. Henceforth, the Service placed a great deal of emphasis on developing tools in agronomy, forestry, pasture management, wildlife management, and other disciplines that the generalist, the soil conservationist and the farmer, could use. John Preston summarized the problem in a 1940 memorandum to the regional conservators. SCS foresters had to recognize that "the majority of farm plans will have to be made by men not especially trained in forestry. It is therefore incumbent upon the foresters to develop further short cuts in technical methods applicable to the preparation of management plans for farm woodlands."[23] This was the motivation for developing tools for management.

In the realm of forestry, SCS developed several research-based tools that a soil conservationist with some training could use to assist land owners. SCS tried to include recommendations for forest management in the overall conservation farm plan. Some of the early concepts were relatively simple, such as attaching representative photographs on which trees were identified for weeding, thinning, or harvesting.[24] On the other end of the spectrum the methods SCS foresters had learned in forestry schools of computing basal area were much too complicated for inclusion in the conservation plan for a farmer to follow. The need existed for science-based tools that both SCS technicians and farmers could use. Homer C. Mitchell supplied one such tool. An Ohio native, Mitchell earned both bachelor and master degrees in forestry from the University of Michigan and had worked for the Great Southern Lumber Company, the Extension Service in Mississippi, and as the assistant state forester in Mississippi. He joined the Soil Conservation Service in 1936 and served as the regional forester at Fort Worth from then until 1954.

From Mitchell's earliest employment he sought a tool that would be understandable to the farmer, a particular solution to the conundrum presented in trying to persuade farmers to view the woodland as part of the farming operation and to treat it accordingly. From his office in Fort Worth, he wrote to John F. Preston, head of the SCS woodland section, "we have no cooperator in this region who has the technical ability to understand and execute a management plan similar to those samples recently sent to this office from Regions 1 and 3." A new method would have to be invented. "Regulation of cut in farm woodlands has never been applied successfully in the South so there appears to be no heresy involved in rejecting standard textbook methods and attempting to substitute something usable by the class of woodland owners represented by our cooperators."[25]

Despite the considerable number of foresters SCS had in the late 1930s, Mitchell doubted that they had the right methods for translating assistance and recommendations into a system that the farmer or the SCS employee could use. He wrote, "These early foresters were not unmindful of their responsibilities but, regrettably, concepts of farm forestry and techniques adaptable to the job were unequal to the opportunity. . . . The outstanding weakness in forestry procedures was lack of a method for growing stock control which could be applied expeditiously by foresters themselves or could readily be explained to and applied by non-forestry technicians and farmers."[26]

This prompted Mitchell to develop the D+ system, which he introduced to the profession with an article in the 1943 volume of the *Journal of Forestry*, "Regulation of Farm Woodlands by Rule of Thumb." By 1945 the methods had become the standard practice in the southern pine region. The empirical formula was actually $G=(D+x)^2$, which expressed the space required by each tree for optimum growth (one common misapplication has been to think of D+ as a linear, rather than an areal expression). Mitchell arrived at the formula by long-term observation and by studying the tabulations in *Volume, Yield and Stand Tables for Second-growth Southern Pines*.[27] Such a system was particularly applicable in southern pine which tended to grow in even-age stands. G represented the growing space needed for the tree. D represented the dbh or diameter at breast height (4½ feet high). X was a constant for the particular species, 6 in the case of southern pine. Thus $(D+x)^2$ was an expression of the growing space required for the development of a normal stand of southern pine.[28]

In July 1949, Mitchell and Edward B. Williams of the East Texas Branch of the Southern Forest Experiment Station developed a cooperative experiment between the Forest Service and the Soil Conservation Service to study and compare results from the D+ system to the conventional selection cutting method advocated by the Forest Service. Forest Service personnel selected a fifty-five-year-old even-age stand that had not been managed on the San Jacinto Experimental Forest.[29] After fifteen years, John J. Stransky, research forester of the Forest Service, and Edwin C. Wilbur, woodland conservationist for SCS, concluded, "The numerous comparisons of the two cutting methods revealed no difference between treatments with regard to yield, growth, residual stand structure, or reproduction. There appears to be little indication that future results will differ more between treatments."[30]

Mitchell later led the effort which resulted in applying the D+ system to other forest types, namely oak, northern hardwoods, spruce, and fir. Another of his efforts was developing the woodland information stick which measured diameters and heights; thereby making possible an estimate of the volume of the stand.

Mitchell's contributions became seminal events in a now well-accepted movement that recognized the need to provide science-based tools which could be taught to, and applied by, those not professionally trained as foresters. The D+ method also fit well the ideal that SCS promoted of the all-age stand that provided a variety of wood products and environmental values.

ZIG-ZAG TRANSECT

The zig-zag transect was another tool the Service developed to help its foresters work with woodland owners. William Lloyd, forester for the corn belt states, began working on it in 1961. Lloyd modified the woodland information stick to be used in conjunction with the zig-zag transect method. After selecting twenty trees using the transect method, the forester could refer to the stocking tables on the woodland information stick to develop a thinning plan.[31] The Service developed this system in the mid-1960s.

Curiously the main opposition to the simplified tools came not from

research foresters but from others who gave advice to timberland land owners. Lloyd recalled his experience with the information stick:

> . . . there are those in the forestry fraternity who feel that the more
> mysterious you make it, the better it is, and professionals sometimes
> take a dim view of simplifying things so that anybody can see and do
> them. . . . Gradually we introduced it (the woodland information
> stick) in those nine corn belt states. And every place that we did, we
> immediately had a fight with the state forester. He would appeal to the
> Forest Experiment Station; they'd go over it all and say, 'It's technically
> correct. It uses our material, uses our data, it's right.' And that would
> end the fight. It wouldn't stop all the feeling, . . . Whenever we had
> problems, it was with the operational people, not with the research
> people, and the reason was that in developing both the zig-zag transect
> system and woodland information stick, I went to the research people
> for their data. As I developed the procedures, I went back to the re-
> search people and said, 'Do you see anything wrong with it? Am I
> doing it right? How would you correct it?' The research people gave
> me lots of help. Yes, this is kind of an interesting thing, that we were
> always supported by the research branch, always. Never any other way.
> But the operations people gave us fits at times.[32]

SOIL SITE CORRELATION FOR FORESTRY

SCS had leeway to make some contributions in areas where its authorities
permitted developments in forest science. The merger of the National Cooperative Soil Survey into the Soil Conservation Service in 1952 set the stage
for a national system for development of woodland site information in soil
surveys. Such information was only one of several types of interpretation
included in soil surveys. The soil survey, usually presented on a county basis, was an American invention. Milton Whitney and his innovative staff of
the Bureau of Soils in the U. S. Department of Agriculture, through trial
and error, developed the essential elements of the survey. Originally, justification of public expenditures on soil surveys rested largely on their benefit to assist farmers and agricultural production, but other "interpretations"
were added. The soil scientists developed the basic descriptive unit in the soil
survey, the soil type, which would eventually prove useful in correlating soils
characteristics to forest cover.

The founder of the Soil Conservation Service, Bennett, had joined the

Bureau of Soils in 1903. His passion for soil conservation came from his days as a soil surveyor when he began comparing the soil profiles on eroded fields to those on virgin, forested sites. The necessity of relating tree growth to soil type received emphasis at the October 1936 meeting of the SCS regional foresters.[33] Some of the early tree survival studies did in fact give soil type. It seems, however, that a reaction to the land capability classification actually spurred the increased attention to soil-forest relationships.

The Service developed a land capability classification designed to help plan conservation for the farm. The classes of land I through VIII progressed from the land having no limitations for use as cropland through classes that have some limitations requiring conservation practices. The upper classes should not be used for cropland, but might provide pasture or woodland. The limitations that the land capability classification addressed were not necessarily significant limitations to tree growth. Nor did the absence of limitations for crop production necessarily correlate to superior timber land. According to Carrow T. Prout, Jr., one of scs's chief foresters, it was this discussion which prompted the first soil-site index correlations in the United States.[34]

Plair, the regional forester for the Pacific Northwest office at Portland, Oregon, launched a new effort at correlating soils to forest site classes in 1945 when scs established plots near Grays Harbor, Washington, on lands of the Weyerhaeuser Timber Company as well as on other privately owned lands. Not the first studies to assess the relationship of root development to soil, the effort started at Grays Harbor had one important distinction. It carefully correlated the information gathered to accepted soil types established by the National Cooperative Soil Survey.[35] The field investigators, W. W. Hill, Albert Arnst, and R. M. Bond, summarized their findings about the forests on residual soils in Lewis County in "Methods of Correlating Soils with Douglas-fir Site Quality." The land capability classification had value in determining the land, in many second-growth forests, that could safely be converted to cropland. But the land capability classification did not identify the soils best-suited for forests, since sites that were not suited to cropland were often as good for Douglas-fir as the best cropland. There was little or no correlation.[36] Stanley P. Gessel of the University of Washington and William J. Lloyd of the Soil Conservation Service office at Sedro Woolley, Washington, studied glaciated soils between Puget Sound and the Cascades. They grouped soil types of similar texture, depth, and presence or absence of impermeable layers to assess the effects on Douglas-fir site quality.[37] These studies made in western Washington in the 1940s sparked interest throughout the nation in correlating tree growth to soils. Shortly after publication of the articles several universities and other federal agencies added soils experts to forestry staffs.[38]

In 1953 the Soil Conservation Service employed Dr. T. S. Coile of Duke University, who had already done research on forest soil site relationships, to initiate similar site quality studies in conservation districts in the southeast.[39] Dr. Earl L. Stone, Jr., of Cornell University gave similar help in the northeast.[40] Following extensive observations and measurements, the SCS began making the information available to users in a series of progress reports beginning with *Soil Survey Interpretations For Woodland Conservation: Forested Coastal Plain, Arkansas, 1959.*[41] By 1968, measurements had been made on 22,000 plots involving approximately 2,900 soil types and 24 species of trees or forest types. The data base now contains about 33,200 records.[42] The interpretive material has become a standard part of the soil survey.

WINDBREAKS

Numerous individuals and agencies contributed to the technology of windbreak establishment. The Bureau of Plant Industry's experiment station at Mandan, North Dakota, produced much of the experimental data on windbreaks in the Great Plains. The Prairie States Forestry Project, better known as the shelterbelt project, essentially produced small forests within the Great Plains, which changed the visual environment while lessening the impacts of wind and drought. Many of the windbreaks were as much as twenty rows wide. Since the conclusion of the shelterbelt project, research has continued on the use of windbreaks, both for cropland and for farmsteads. The Forest Service, the Agricultural Research Service, as well as the Soil Conservation Service, have been monitoring windbreaks on farms. The research results and empirical observations are used in SCS technical guides.[43]

SCS studies have demonstrated the feasibility of drip irrigation systems for windbreaks. Israeli scientists made many of the early developments in drip irrigation systems for woody plants, and, in the United States, the Agricultural Research Service and the Extension Service have worked on drip irrigation. SCS developed designs that were practicable for the field, developed standards and promoted drip irrigation. Acceptance of drip irrigation rapidly accelerated in the early 1980's. SCS contributed to this growth by designing systems in the field. In 1981 SCS foresters surveyed the survival rates of windbreaks using drip irrigation systems as compared to other windbreaks

Rows of trees between fields act as windbreaks, slowing the force of the wind and preventing soil erosion near Bismarck, North Dakota.
(USDA SOIL CONSERVATION SERVICE PHOTO, 1988)

in Kansas and Nebraska. Ninety-five percent of the drip-irrigated windbreaks had a survival rate of 90 to 100 percent. No detectable difference in survival existed due to soil texture, site preparation, species, or rates. While drip irrigation alone kept the trees alive, weeding was also important to growth rate. At the time of the study no recommendations had been developed on rates. The study established the rates used today, as well as ending most skepticism about the value of drip irrigation in establishing windbreaks.[44]

During the 1970s the removal of windbreaks planted by the Prairies States Forestry Project began to receive attention. In response to a General Accounting Office report, SCS undertook a comprehensive field windbreak survey in five plains states in 1977. The study enumerated the reasons for removal and the extent of new plantings. Coincident with the effort, SCS also developed its concept of renovation of older windbreaks. The guidelines available in SCS technical guides were then amended to include information that would help farmers renovate and restore existing windbreaks rather than completely remove them.[45]

Farmers disliked the loss of productivity along field borders due to competition from the roots of windbreak trees. Root pruning has a long history of research in the United States. Although scs did not do basic research on this, through its field trials and guidelines in technical guides, scs promoted farmer acceptance of root pruning in the Great Plains.[46] Also there has been considerable research into the overall productivity of windbreak-protected fields compared to unprotected ones. scs crop yield studies, especially in Fulton County, Indiana, and Comstock, Minnesota, supported the proposition that any loss of production to windbreak's competition for water was more than compensated by protection of crops from wind and drying.[47]

In the late 1970s the Soil Conservation Service developed the twin-row, high-density windbreak, tested cooperatively with South Dakota State University. The main advantages derived from faster growth in height and faster fill in of rows and reduction in the time that weed control was needed.[48] As fabric materials that allowed rainfall infiltration, while suppressing weeds, became available, scs developed specifications using such material in establishing windbreaks.[49]

WINDBREAK-SOIL SITES

Probably scs's greatest contribution to windbreak technology has been to extend the forest-soil site concept to windbreaks. From 1957 until 1962 scs informally studied the soil-windbreaks relationships by measuring the growth of various species and relating those to soil type. Beginning in 1962 Bill D. Seay headed a team to study 1,100 plots in 11 states. The sites selected were 15- to 39-year-old windbreaks. When the study concluded in 1966 the Service was ready to incorporate results in technical guides for conservation districts and in soil surveys. In the early 1970 scs began computerization of field data with the objective of following the development of windbreaks from planting to maturity.[50] The windbreak-soil sites data base now contains more than 3,000 records and observations.[51]

CONCLUSION

During the late nineteenth and early twentieth centuries the U.S. Department of Agriculture built a vast research organization, a portion of which was devoted to forestry. The New Deal, with its objective of direct assistance to farmers, created a number of "action" agencies, among them the Soil Conservation Service. Despite the earlier contributions of usda to science, scs found unsatisfied needs for scientific knowledge. The contributions of

SCS to forest science have been directly related to its mission to bring practical concepts and practices for conserving the environment to the nation's farmlands.

NOTES

1 The late Terry Johnson, national forester of SCS, was his usual gracious self in giving the author suggestions and citations for developments that should be included in this article. The author also wishes to thank current Natural Resources Conservation Service employees, Gary Nordstrom, Marc Safley, Keith Ticknor; and former Soil Conservation Service employees Theodore B. Plair and Curtis Sharp. Faye Helms Griffin of Simpsonville, South Carolina, provided editorial suggestions.

2 Hugh Hammond Bennett, *Soil Conservation* (New York: McGraw-Hill Book Company, Inc., 1939), p. 427.

3 *Report Of The Chief Of The Soil Conservation Service*, 1936, p. 28.

4 Maurice E. Heath, "Contribution Of Soil Conservation Nurseries To Soil And Moisture Conservation," Presented at the American Society of Agronomy meetings, Milwaukee, Wisconsin, October 26, 1949, p. 2. Typescript in History Office, Natural Resources Conservation Service, Washington, D.C.

5 Franklin J. Crider, Memorandum, Projecting Nursery Observation Work, March 2, 1939, File 225.1 Conservation Nurseries, Santa Paula Regional Office, Record Group 114, Records of the Soil Conservation Service, National Archives, San Bruno, California. Hereinafter the abbreviations RG 114 will be used to indicate the record group.

6 Franklin J. Crider to all regional conservators, "Projecting Nursery Observational Work," March 2, 1939, File 225.1 Conservation Nurseries, Santa Paula Regional Office, RG 114, National Archives, San Bruno, California.

7 *Oral History Interview with John L. Schwendiman* (Washington, D. C.: Soil Conservation Service, 1981), p. 3.

8 Telephone interview with Curtis Sharp, September 12, 1995; J. Scott Peterson and W. Curtis Sharp, *Improved Conservation Plant Material Releases by SCS and Cooperators Through December 1993* (Washington, D. C.: Soil Conservation Service, National Plant Material Center, 1994), pp. 19–22.

9 William R. Van Dersal, *Handbook of Native Woody Plants Of The United States.* Soil Conservation Service Technical Publication No. 11 (Washington, D. C.: Soil Conservation Service, 1936), pp. 9–11.

10 William R. Van Dersal to Regional Biologists, August 24, 1937, File 227.1 Woodland Management, Santa Paula Regional Office, RG 114, National Archives, San Bruno, California.

11 Aldo Leopold, review of Edward H. Graham, *Natural Principles of Land Use, Soil Conservation* 10 (August 1944): 38–39.

12 "Annual Report, 1938–39," Pacific Southwest Region, RG 114, National Archives, San Bruno, California.

13 Jack L. Reveal, "Observation Report on the Effect of Cultivation on the Survival and Height Growth of Tree Seedlings in Field Planting in Western Nevada," File 735, Woodland Management, Yerrington, Nevada Area Office, RG 114, National Archives, San Bruno, California.

14 John F. Preston, "Results Of The Soil Conservation Service Program Of Planting Trees and Shrubs," *Journal of Forestry* 37(1939): 19–22; John F. Preston to Regional Foresters, March 11, 1937, File R-227.1 Survival Data, Fort Worth Regional Office, RG 114, National Archives, Fort Worth; H. M. Sebring to John F. Preston, May 31, 1940, File 735 Woodland Management, Spartanburg Regional Office, RG 114, National Archives, Atlanta, Georgia; John F. Preston to Regional Conservator, May 13, 1940, File 735 Woodland Management, Berkeley Regional Office, RG 114, San Bruno, California.

15 John F. Preston, "The Woodland Management Program of the Soil Conservation Service," *Journal of Forestry* 41 (1943): 403.

16 Henry Hopp, *Methods of Distinguishing Between Shipmast and Common Forms of Black Locust on Long Island, N.Y.* USDA Technical Bulletin No. 742 (Washington D. C.: U. S. Department of Agriculture, January 1941), pp. 1–23.

17 Henry Hopp to H. M. Sebring, May 11, 1940, File 735 Woodland Management, Spartanburg Regional Office, RG 114, National Archives, Atlanta, Georgia.

18 A. E. Fivaz, Memorandum for Regional Conservators, August 29, 1940, File 735 Woodland Management, Spartanburg Regional Office, RG 114, National Archives, Atlanta, Georgia.

19 John F. Preston, "The Field for Farm Forestry in the Farm Conservation Program of the Soil Conservation Service," *Journal of Forestry* 40 (1942): 291.

20 T. B. Plair, "Comments on Report On Reforestation and Afforestation Needs, Los Angeles County, July 8, 1939" File 227.1 Woodland Management, Santa Paula Regional Office, RG 114, National Archives, San Bruno, California.

21 Albert Arnst, "Cascara—A Crop from West Coast Tree Farms," *Journal of Forestry* 43 (November 1945): 805–811.

22 *Oral History Interview with William J. Lloyd, September 23, 1981.* (Washington, D. C.: Soil Conservation Service, 1981), pp. 21–24.

23 John F. Preston to regional foresters, June 19, 1940, File 735 Woodland Management, Spartanburg Regional Office, RG 114, National Archives, Atlanta, Georgia.

24 Albert Arnst, "Visual Guides for Farm Forest Management," *Journal of Forestry* 44 (October 1946): 731–734.

25 Homer C. Mitchell to John F. Preston, August 31, 1937, File 227.1, Fort Worth Regional Office, RG 114, National Archives, Fort Worth, Texas.

26 "Working Plan For A Study Of A Method Of Cutting Based on D+ Spacing," Forest Ecologist Office, Natural Resources Conservation Service, Washington, D.C.

27 *Volume, Yield and Stand Tables for Second-growth Southern Pines*, U. S. Department of Agriculture Miscellaneous Publication Number 50 (1929); Homer C. Mitchell, *A Guide To Stocking Southern Pine Stands* (Washington, D. C.: Soil Conservation Service, September 1962), pp. 12–13.

28 Homer C. Mitchell, *Tree Diameter As A Guide To Thinning Southern Pine* (Washington, D. C.: Soil Conservation Service, August 1958), pp. 1–2.

29 Homer C. Mitchell to State Conservationists and Woodland Conservationists, Southeastern States, November 8, 1955. A compilation of mimeographed documents relating to the experiment. Copies in the Forest Ecologist's office, Natural Resources Conservation Service, Washington, D.C.

30 John J. Stransky and Edwin C. Wilbur, *15–Year Progress Report: A Method of Cutting Based Upon D+ Spacing* (Nacogdoches, Texas: U. S. Forest Service, 1964), p. 47.

31 Lloyd Interview, pp. 50–52; *The Woodland Information Stick and Woodland Inventory Procedures* (Soil Conservation Service, 1977), pp. 1–21.

32 Lloyd Interview, pp. 53–54.

33 Notes On Regional Foresters' Meeting, October 19 to 25, 1939, File, Regional Foresters Meeting, Amarillo Regional Office, RG 114, National Archives, Fort Worth, Texas.

34 Carrow T. Prout, Jr., "Woodland Site Index Ratings of Soils," *Soil Conservation* 34 (August 1968): 7–8.

35 Telephone interview with Theodore B. Plair, August 18, 1995.

36 W. W. Hill, Albert Arnst, R. M. Bond, "Method of Correlating Soils with Douglas-Fir Site Quality," *Journal of Forestry* 46 (November 1948): 835–841.

37 Stanley P. Gessel and William J. Lloyd, "Effect of Some Physical Soil Properties on Douglas-Fir Site Quality," *Journal of Forestry* 48 (June 1950): 405–410; Lloyd Interview, pp. 17–18.

38 Telephone interview with Theodore B. Plair, October 6, 1995.

39 Telephone interview with Plair, August 18, 1995; *Oral History Interview with Theodore B. Plair, September 8, 1981* (Washington, D. C.: Soil Conservation Service, 1981), p. 12.

40 Prout, "Wood site index ratings of soils," pp. 7–8.

41 Hartzell C. Dean and James M. Case, *Soil Survey Interpretations For Woodland Conservation: Forested Coastal Plain, Arkansas* (Washington, D. C.: Soil Conservation Service, 1959), pp. 1–54.

42 Information from Rick Nesser and Keith Ticknor, Natural Resources Conservation Service, Lincoln, Nebraska.

43 David L. Hint, "Farmstead Windbreaks," *Shelterbelts On The Great Plains: Proceedings of the Symposium*, Great Plains Agricultural Council Publication No. 78.

(Great Plains Agricultural Council, 1976), pp. 95–97; "The Spacing of Trees in Farm Forestry Plantings," *Farm Forestry For The Northern Great Plains* (Soil Conservation Service, 1946), pp. 111–120.

44 Keith Ticknor, "Effect of Drip Watering Systems on Establishment and Growth of Trees and Shrubs in Nebraska," in *National Forestry Workshop Proceedings, Atlanta, Georgia, September 13–17, 1982* (Washington, D. C.: Soil Conservation Service, 1982), pp. 1–13.

45 Soil Conservation Service Technical Guide, Huron, South Dakota. See also *Action Needed to Discourage Removal of Trees That Shelter Cropland In The Great Plains*. RED-75-375 (Washington, D. C. : General Accounting Office, 1975); and *Field Windbreak Removals in Five Great Plains States, 1970–1975* (Washington, D. C.: Soil Conservation Service, 1975).

46 Elmer R. Umland, "Root Pruning As A Management Technique," in *Windbreak Management,* Great Plains Agricultural Council Publication No. 92 (Norfolk, Nebraska: Great Plains Agricultural Council, 1979, pp. 107–109.

47 Manuscript crop yield studies supplied by Keith Ticknor, scs, Lincoln, Nebraska; John Kort, "Benefits of Windbreaks to Field and Forage Crops," *Agriculture, Ecosystems and Environment* 22/23 (1988): 165–190.

48 "Changes in Windbreak Design Increase Flexibility, Bring Other Benefits," *Soil and Water Conservation News* (December 1981): 10–12; Also, for other forestry recommendations, see the scs Technical Guide for each state.

49 Technical Guide for Kansas, 380–13 through 14.

50 C. T. Prout, "Windbreaks in SCS," July 2, 1974, 4 page typescript in office of Forest Ecologist, NRCS, Washington D.C.; Soils Memorandum 64, Soil Survey Interpretations—Rating Soils for Shelterbelts and Windbreaks, June 26, 1967.

51 Information supplied by Rick Nesser and Keith Ticknor, Natural Resources Conservation Service, Lincoln, Nebraska.

RICHARD WEST SELLARS[1]

SCIENCE AND NATURAL RESOURCE MANAGEMENT IN THE NATIONAL PARK SERVICE, 1929–1940

When National Park Service biologists under the leadership of George M. Wright began their survey of park wildlife in 1929, the Service had been in existence nearly thirteen years, yet it had never systematically studied the parks' flora and fauna, nor had it articulated a comprehensive set of policies for the management of nature in the parks. By chance, publication of the biologists' survey report, *Fauna of the National Parks of the United States*, known as Fauna No. 1, came in 1933, shortly after Congress created the Civilian Conservation Corps (CCC). Particularly through the CCC program, funds soon became available for national park biologists to implement the policies recommended in Fauna No. 1. Thus, Fauna No. 1 provided policies and the CCC funds for the Park Service to conduct its own natural resource management.

During the era of its first director, Stephen T. Mather (1916–1929), the Service had relied heavily on scientific expertise from other federal bureaus, such as the Biological Survey, Bureau of Fisheries, and Bureau of Plant Industry. Now it began to develop its own cadre of scientists, who were "park-oriented," as Park Service biologist Lowell Sumner later expressed it. Reflecting on the emergence of biological research and management in the Service, Sumner also recalled that Fauna No. 1 had quickly become the "working 'bible' for all park biologists."[2] This report truly represented the state of the knowledge for national park biological management in the 1930s. Although the report did infuse more ecologically sensitive thinking into national park activities and was soon declared official policy, implementation of its recommendations was frequently disputed and never fully realized.

As Park Service biological programs built up in the 1930s, a tension developed between traditional management which focused on scenery and public enjoyment of the parks versus that which was based on the newly formulated concerns of the wildlife biologists. This tension had no real precedent, since the scientific, ecological perspective had not previously been expounded to any degree within the Park Service.[3] Indeed, more than that of any other professional group in National Park Service history, the wildlife biologists' vision of the national parks challenged traditional management practices of manipulating natural resources to ensure public enjoyment—practices which had been accepted as standard procedure during the Mather era. The biologists stressed ecological preservation and would let nature take its course, except when manipulation of the resources was deemed necessary for ecological purposes. Yet, because of already powerful traditions within the Park Service, the wildlife biologists frequently encountered conflict and compromise (and often total rejection) in their efforts to change management. The conflicts over natural resource management that arose within the Park Service during the 1930s provided a prelude to similar conflicts that would arise in the 1960s, involving many issues which remain meaningful even at the end of the twentieth century.

Among Fauna No. 1's recommendations, two proved most fundamental: The Park Service should base its natural resource management on scientific research, including conducting "complete faunal investigations . . . in each park at the earliest possible date." And, secondly, each species should be left to "carry on its struggle for existence unaided" unless threatened with extinction in a park. In effect, the remaining recommendations qualified or elaborated upon these two basic tenets, with specific statements on such concerns as protection of predators, artificial feeding of ungulates, protection of ungulate range, removal of exotic species, and restoration of extirpated native species.[4]

Regarding scientific research, the national park naturalists (forerunners of modern-day park "interpreters") had noted at their 1929 conference that scientific data on the parks' natural history was "almost infinitesimal." This disheartening situation began to change that very year, as preparation of Fauna No. 1 got under way. Following completion of Fauna No. 1, scientific

*The first chief naturalist conference was held at the University of
California, Berkeley, campus, in November 1929. Pictured in the back
row, left to right are: Brockman, Coffman, Bean, and McKee.
Yeager, Russell, Hall, and Harwell stand in the front row.*
(NATIONAL PARK SERVICE HISTORICAL PHOTOGRAPH COLLECTIONS,
HARPERS FERRY CENTER)

research continued under the guidance of George Wright, head of the Park
Service's newly created Wildlife Division. Sumner later estimated that during the 1930s about half of the biologists' work involved research and wildlife
management, while the other half was devoted to review and comment on
proposed development projects (many of them being CCC projects). He calculated that prior to World War II the biologists had produced perhaps one
thousand reports. Having joined the Service in 1935, Sumner estimated that
he himself prepared about one hundred seventy-five reports before the war
began.[5]

The wildlife biologists conducted research on subjects such as bison, elk,
and bird life at Wind Cave; white-tailed deer and winter birds in
Shenandoah; grazing mammals in Rocky Mountain; and deer and bighorn
in Glacier National Park. Park naturalists contributed further to the gath-

ering of information, such as at Great Smoky Mountains, where plant specimens of about two thousand species were collected by the mid-1930s.[6] Given the large number of documents prepared and the limited number of biologists in the Park Service (about twenty-seven at the most), only a few of the reports and studies could have been truly in-depth works. Among the most thorough were Joseph Dixon's *Birds and Mammals of Mount McKinley National Park* (1938), published as number three in the Fauna Series, and Adolph Murie's *Ecology of the Coyote in the Yellowstone* (1940, Fauna No. 4). Murie's next major study, *The Wolves of Mount McKinley* (Fauna No. 5), was begun in 1939 and published in 1944.[7]

RESEARCH RESERVES

An important element of the biologists' programs during the 1930s was the establishment of "research reserves"—areas within national parks designated to be used for scientific research only. Likely at the urging of the Ecological Society of America and leading biologists such as the Carnegie Institution's John C. Merriam, who feared the disappearance of all unmodified natural areas in the United States, the Park Service in the mid-1920s gradually began to develop a research reserve program. In 1927, Yosemite National Park designated approximately seven square miles of high mountain country north of Tuolumne Meadows as a "wilderness reserve," later termed a research reserve, the first of its kind in the national park system.[8]

At their November 1929 conference, the park naturalists discussed the reserves, and concluded that they should be permanently set aside and primarily for scientific study. These areas were to be, as the naturalists phrased it, "as little influenced by human use and occupation as conditions permit." Park Service director Horace M. Albright followed up in the spring of 1931 by issuing a research reserve policy to "preserve permanently" selected natural areas "in as nearly as possible unmodified condition free from external influences." In effect, the areas would help meet Fauna No 1's recommendation for each species (whether flora or fauna) to "carry on its struggle for existence unaided." The reserves were to be entered only in case of emergency or by special permit; and, as a further means of protection, their location was not to be publicized.[9]

The research reserves emerged in the 1930s as the most preservation-oriented land use category the Park Service had yet devised—an important philosophical and policy descendent of Congress' mandate to leave the national parks "unimpaired" and much more restrictive than the traditional policy of allowing park backcountry to be developed with horse and foot trails. The reserves served also as precursors to national park wilderness areas established under the Wilderness Act of 1964. Designations such as primitive, primeval, wilderness, virgin, and roadless were used at times in association with the reserves.[10] In biologist George Wright's view, the reserves' greatest value lay in providing scientists the opportunity to learn what certain portions of the parks were like in their original, unmodified condition—a "primitive picture" that would provide a basis of knowledge to benefit all future research. He also believed that the reserves would not become "an actuality" until their flora and fauna had been surveyed. To Wright, setting aside the reserves was a "most immediate urgency" that should be accomplished before further biological modifications took place.[11]

The research reserves became an integral part of park management in March 1932, when Director Albright asked that they be formally designated through the cooperation of the park superintendents and naturalists and the Washington office. He requested that the superintendents indicate the location of the reserves in the five-year park development plans (master plans), and he assigned the wildlife biologists responsibility for gathering information and tracking the progress of this program. By 1933, research reserves had been designated in Yellowstone, Sequoia, Grand Canyon, and Lassen Volcanic national parks. Others followed, in Great Smoky Mountains, Glacier, Mount Rainier, Rocky Mountain, Zion, as well as Yosemite, for a total of twenty-eight designations in ten parks.[12]

The research reserve idea, however, worked better in theory than in practice. The wildlife biologists apparently did not participate in the actual selection of many of the reserves, likely because a number of the areas were designated while the biologists were busy completing Fauna No. 1 and because the biologists were unable to gain a meaningful role in the master planning process. As late as February 1934, the Wildlife Division seemed poorly informed on the exact location and character of many of the reserves; moreover, on those they knew something about, George Wright noted that some of the areas did not lend themselves to becoming worthwhile research areas—indications that the biologists had little input in designating the reserves. A reserve in Lassen Volcanic National Park proved no more than a strip of land three-quarters of a mile wide and about five miles long; while two of Grand Canyon's reserves were so close to the park boundary that activities outside the park were certain to affect their biotic makeup. Noting the potentially serious effects of external influences on the reserves,

Wright advocated the establishment of "buffer areas" around the parks (including additional winter range for wildlife), rather than "withdrawing further and further within the park" to create reserves.[13] Like the parks themselves, the reserves were not truly satisfactory biological units.

Expressing deep concern about the reserve program, Victor H. Cahalane, Wright's assistant division chief, wrote in September 1935 of the problem of selecting research reserves in parks so "artificialized and mechanized." To Cahalane, the difficulty of finding even relatively small unaltered research areas to be specially protected indicated the extent to which the Service had already failed to meet its basic mandate to protect the parks' wilderness character. Cahalane asserted that Glacier National Park had no pristine area worthy of becoming a research reserve. This had occurred "not by reason of a network of roads" in Glacier, but because:

all streams now contain exotic species of fish, because the wolverine and fisher have been exterminated from the entire park and the bison and antelope from the east side, and because exotic plants . . . have been carried to practically every corner of the park.

Recognizing the existing problems with "pristine" areas in the parks, Cahalane called for a "show-down on this matter of preservation of the greatest resource of the National Park Service—the wilderness."[14]

But beyond the difficulty of identifying largely unaltered natural areas to be designated research reserves, decisions made wholly within the Park Service produced the reserves, and thus were subject to administrative discretion and vulnerable to shifting philosophies of management. The reserves had no specific mandate from Congress. They could be supported, ignored, or, as happened to Andrews Bald research reserve in Great Smoky Mountains National Park, created and then summarily abolished. Indeed, the "show-down" that occurred over Andrews Bald went directly against the scientists' recommendations and reflected the Park Service's traditional disregard for scientific research. The outcome proved an ominous portent for the science programs overall.

Designated a research reserve in the mid-1930s, Andrews Bald was one of several reserves in Great Smoky Mountains intended to be strictly preserved so that "ecological and other scientific studies" could be conducted on a long-range basis, especially to determine natural plant succession. (The "grassy balds"—open, mountain-top areas of grasses and low-growing shrubs, and without tall trees—provided one of the primary scenic features in the Smokies and were then and remain of special scientific interest). In

The meadow conditions on Andrews Bald, Great Smoky Mountains National Park, prior to the windstorm of 1936.
(NATIONAL PARK SERVICE HISTORICAL PHOTOGRAPH COLLECTIONS, HARPERS FERRY CENTER. GEORGE A. GRANT, PHOTOGRAPHER, 1931)

early April 1936, a terrific wind storm knocked down trees in the vicinity of Andrews Bald and within the established reserve, precipitating a sharp debate in the Service as to how to manage the area.

Blown over by the storm, dead and dying trees cluttered the landscape and, in the minds of the superintendent and most of his staff, constituted a fire hazard.[15] Superintendent J. R. Eakin wanted a cleanup, as did the park's rangers and foresters. In a letter to Park Service director Arno B. Cammerer, Eakin stressed the potential fire problems. Reflecting an ongoing disagreement over what to do with naturally downed trees, the superintendent noted pointedly that "again," the Wildlife Division and the naturalists were "not concerned with fire protection" and the danger that might arise if the dead trees remained in place.[16] Particularly concerned about scenery, Frank E. Mattson, the park's resident landscape architect, argued for cleanup of the windfall, stating that because the bald attracted so many sightseers it should be treated "much as a trailside or roadside" area.[17]

By contrast, the wildlife biologists (supported by park naturalist Arthur Stupka) advocated special consideration for the reserves, so that "ecological and other scientific studies . . . may be started and continued thru the years to come." They urged that the downed trees be left untouched. Although recognizing the fire prevention concerns, the biologists argued that the wind storm was a natural phenomenon and that cleanup of the area would "thwart the objectives" of Andrews Bald research reserve. Still, Eakin believed the area constituted a serious fire hazard and, in an exchange of correspondence with the Washington office, insisted that the damaged trees should be cleared.[18]

In a stinging reply to Eakin, Acting Director Arthur E. Demaray finally granted permission to clear the downed trees, but added that the Andrews Bald Biotic Research Area was thereby abolished. He further stated that "I wish to call your attention to several factors which you seem to have overlooked"—the reserve had been approved by Eakin himself, it was included in the park's master plan, and preservation of such areas was "an established policy of the Service." In the Acting Director's view, the superintendent's insistence was forcing a change in the official use of the area from research and strict preservation to recreation: "The reason the research area is now abolished is that you have convinced us you made an error in approving its establishment. Its apparent proper use is primarily recreational."[19]

Andrews Bald illustrated the vulnerability of the reserves to administrative discretion and, as well, the vulnerability of research itself. An area committed to serve research purposes over a long period of time was subject to sudden modification as a result of internal decision-making. Indeed, the urge to clear the damaged trees was not truly based on whim. It reflected the deep-seated, traditional allegiance of the superintendents, foresters, and landscape architects to preserving national park scenery and accommodating public use, while generally evidencing not much interest in science.

Even though the director's policy pronouncement of 1931 supported the research reserves and they represented the bureau's strongest commitment to preservation of natural conditions, the Park Service eventually disregarded the entire program. Certainly most reserves did not vanish in as confrontational way as did Andrews Bald, yet Sumner later recalled that the research reserve program came to be largely ignored, beginning about the time of World War II. The Park Service itself acknowledged in 1963 that the reserves were "dormant," and that many of the areas had "remained 'on the shelf,' awaiting a more favorable period for their utilization."[20] (This statement came at the very time Park Service leadership was withholding genuine support for the proposed Wilderness Act because it did not want to lose administrative discretion over national park backcountry.)

While it may seem that ignoring the research reserve program meant that these areas would be left alone and thus remain in an unaltered condition, this very likely did not prove the case. With the program untended, and the reserves in effect forgotten, these areas of special research value were likely to be altered through such practices as fire protection (for example, the removal of dead trees from Andrews Bald), forest disease control, grazing, and fish stocking and harvesting. The neglected research reserves stood subject to the kinds of modifications which concerned Wright in the early 1930s when he stressed the "most immediate urgency" of establishing the reserves.[21]

RANGE MANAGEMENT AND CONCERN FOR THE UNGULATES

In contrast to the research reserve program, the biologists believed that in other instances it was necessary to interfere with nature and (as stated in Fauna No. 1) assist certain species to combat the "harmful effects of human influence" in order to restore the parks' "primitive state." Fauna No. 1 also specifically called for preservation of ungulate range, and advocated that a park's "deteriorated range" should be "brought back to [its] original productiveness."[22] During the 1930s, of all the Park Service's attempts to interfere with nature, the manipulation of Yellowstone's "northern elk herd" received the greatest attention and ultimately became the most controversial.

To many familiar with Yellowstone, the park's northern elk herd seemed to have become so large that it was overgrazing its range. The resulting deterioration appeared adversely to affect use of the range by competing ungulates, such as deer and pronghorn. Concurring with this assessment, the wildlife biologists determined that the population of Yellowstone's northern elk herd needed to be reduced, in line with Fauna No. 1's recommendations. Reducing animal populations was not new to the Park Service, given the long-running predator control activities, and (beginning in the mid-1920s) the slaughtering of limited numbers of Yellowstone's Lamar Valley bison herd for population control. In addition, although concerns about over-population of elk had evolved by the early twentieth century and the park had practiced limited elk removal for more than a decade, no concerted reduction program prior to that encouraged by the wildlife biologists seems to have existed.

Reduction involved shooting large numbers of the park's northern herd, which mostly inhabited the Yellowstone and Lamar river basins. For humane reasons, shooting the animals seemed far preferable to allowing them to die of winter kill when heavy snows restricted their range; furthermore, reduction could bring the population to a specified level. As believed at the time, this plan would prevent overgrazing and deterioration of the winter range and benefit all grazing species. The elk reduction program thus sufficed as the principal management strategy for the park's grazing animals, with the exception of bison.

The wildlife biologists concluded that "human influence" had caused the winter range problems in Yellowstone. This state of the knowledge in the 1930s (which decades later would become intensely disputed) was based on several fundamental assumptions: National Park Service scientists and managers believed that, prior to Anglo-American settlement of the valleys to the north of the park, the herd had wintered in those valleys, and after the park was established its protected elk population had expanded enormously. They also believed that the elk population had crashed in the period 1917-1920 and that this dramatic decline had been caused by range deterioration through overgrazing. With drought conditions affecting the range in the late 1920s and early 30s, and with elk populations believed to have increased due to protection in the park, a second population crash was seen as imminent—one which the Wildlife Division expected to bring on "hideous starvation and wastage."[23]

In 1931 Park Service biologists Joseph Dixon and Ben Thompson (working with George Wright on Fauna No. 1) had participated in a reconnaissance of the deer population irruption in the Kaibab National Forest, north of Grand Canyon. Their report asserted that an over-population of deer threatened the national forest and recommended reducing the deer herds. Likely influenced by what seemed to have happened in the Kaibab, the biologists made their recommendation that Yellowstone's elk population also be reduced. And in a February 1934 report documented with numerous photographs (and reprinted in Fauna No. 2), the Wildlife Division announced that, as a result of an overpopulation of elk, Yellowstone's northern range had been overused to the point that it was in "deplorable" condition. The biologists believed that the situation had worsened since they first saw the area in 1929 and that it now threatened the survival of other animals dependent upon the range. The report argued that the overpopulated elk herd was on the "brink of disaster" and warned that the next hard winter would cause starvation and death for thousands of elk.[24]

Indeed, the elk reduction program had strong, apparently unanimous, support among the Park Service's wildlife biologists. Their statements and reports did not equivocate on the wisdom of artificially lowering Yellowstone's elk population. Commenting in the late winter of 1935 that, without reductions the elk problems would continue—the "old winter range ghost will be walking again"—Wright himself saw the program as critical to the success of the park's wildlife and range management.[25] Also, Olaus Murie, who had overseen the Bureau of Biological Survey's elk management in Jackson Hole, south of Yellowstone, provided supporting insights on the northern herd. He urged reducing the herd, as did his brother, Adolph, a respected National Park Service scientist. In late December 1934, just before the first big reduction began, Olaus Murie wrote to Ben Thompson approving elk reduction, noting that "if carefully handled it will be successful," and adding that he looked forward "with great interest to the outcome of the experiment."[26]

Beyond their own observations, the biologists based their elk policy on research conducted in the region in the 1920s and early 1930s by U.S. Forest Service biologist W. M. Rush, whose work was privately funded with money obtained by then Park Service director Albright. Rush's conclusions supported the biologists' views.[27] Also, since they believed that longer hunting seasons and increased bag limits in Montana and on adjacent Forest Service lands would provide only limited help, the biologists recommended that the park itself be involved in the reduction to ensure that the proper number of elk would be taken each winter. As the biologists noted, until the desired population level was reached, Yellowstone must be prepared "to slaughter elk as it does buffalo."[28]

Much more cautious, however, was the opinion of Joseph Grinnell, head of the University of California's Museum of Vertebrate Zoology and mentor to numerous Park Service biologists. Asked by Director Cammerer to comment on the proposed reduction, Grinnell observed that the elk situation in Yellowstone was "truly disturbing from any point of view." He remarked on the "damage" which he believed elk grazing had done to the winter range, and agreed that human influences had been an important factor in bringing on the situation. Although he carefully avoided criticizing the decisions of his former students and close friends, Grinnell withheld support for the reduction program. Rather, he expressed hope that the killing of any park animals, including predators as well as elk, would become a thing of the past. In his summation, Grinnell advocated "adjustments through natural processes" to restore the "primeval biotic set-up."[29] More than the Park Service biologists of the 1930s, Grinnell expressed faith in allowing "natural processes" to control elk populations, with aggressive measures taken to reduce adverse human influences on the animals. He thus voiced elk management policies that the Service would eventually put into effect, after the reduction program had been underway for more than three decades.

Reduction began in January 1935, with Yellowstone's rangers shooting the elk and preparing their carcasses for shipment to tribes on nearby reservations. With the intention of reducing elk population to the range's "carrying capacity," the Park Service's goal of killing three thousand elk the first winter included animals to be taken outside of the park under Forest Service and Montana State Fish and Game Department regulations liberalized to increase the number killed by hunters.[30] During the first reduction effort, hunters on lands adjacent to Yellowstone took 2,598 elk (up from only one hundred thirty-six the previous year), and park rangers killed six hundred sixty-seven (up from only eleven in 1934), for a total of nearly thirty-three hundred.[31]

Responding to an inquiry from the American Museum of Natural History in March 1935, Wright expressed relief that the Park Service itself had not had to kill large numbers of elk during the initial reduction; yet he wrote that "we are glad to have established a satisfactory precedent" regarding the "propriety of direct control" in the national parks. Yet, even after further reduction in 1936, Adolph Murie studied Yellowstone's range and found it "undoubtedly worse" than it had been in six or seven years. Murie recommended that the kill be increased to four thousand the following winter. A lengthy 1938 report by Yellowstone ranger Rudolph L. Grimm again confirmed the belief that the range was overgrazed and advocated continued reduction.[32]

With a "satisfactory precedent" established in the mid-1930s, Yellowstone's elk reduction program began its long history, with the policy eventually applied in other areas, particularly Rocky Mountain National Park. At the end of the decade, the wildlife biologists reported that the "basic and most important problem" at Yellowstone continued to be the condition of the park's range. "As in the past," they asserted, the abundance of elk "depletes the forage of other ungulates using the same range."[33] Although he did not speak out aggressively against the reduction program, Grinnell continued to oppose it, writing to Cammerer in January 1939 that he did not approve of regulating "the numbers of certain animals in certain Parks."[34] Grinnell urged that the Service submit the problem to a group of specially trained ecologists. (This approach, when implemented in the early 1960s, resulted in the "Leopold Report," which clearly recommended that the reduction policy be continued, not terminated. Only later, in 1967–68, did the Park Service change its elk policy to the "natural processes" concept, in line with Grinnell's ideas. This strategy itself became very hotly debated.)[35]

BISON MANAGEMENT

As with elk management, bison management in the 1930s did not create discord between the wildlife biologists and other Park Service personnel. Moreover, throughout the decade, management of bison in Yellowstone's Lamar Valley (the herd of most concern to the Park Service) remained more intensive and varied than that given the park's elk. Using domestic livestock ranching methods first developed by the U.S. Army then expanded during Director Mather's time, bison management changed little during the decade. With operations still headquartered at the Buffalo Ranch along the Lamar River, bison work primarily involved rounding up and corralling the herd in the winter for feeding, vaccination (for hemorrhagic septicemia), and for removal of excess animals (or those not wanted for breeding) by slaughtering or shipping them live to other areas.[36]

Principally, the lack of discord resulted from the wildlife biologists' acceptance of the need to manipulate the herd for ecological purposes. In fact, in Fauna No. 1 the biologists had little to recommend regarding bison management, stating only that winter feeding of the animals was "absolutely necessary." Yet, regarding *all* park fauna, the report's recommendations called for putting threatened species on a "self-sustaining basis" when such measures as feeding were no longer necessary. Similar counsel appeared in Fauna No. 2. Noting that bison had been saved from extinction in the park by intensive management, the latter report urged returning this species to its "wild state" to the degree that the "inherent limitations" of each park would permit. The biologists believed that such measures as winter feeding and slaughtering would have to continue until "artificial management" was no longer necessary.[37]

Based upon recommendations made during the late 1920s and early 30s, the park sought to keep Yellowstone's Lamar Valley herd limited in size, at first seeking a population level of one thousand animals, then eight hundred beginning about 1934—levels believed within the "carrying capacity" of the bison range and what the Buffalo Ranch facilities could accommodate.[38] But even by the following year, some expressed concern that the population remained much too high. Harlow B. Mills, a biologist at Montana State College who had worked in Yellowstone, wrote an extensive report on wildlife conditions in the park in 1935, recommending that the Lamar Valley herd be reduced to "100 or less animals." Mills believed there were likely too many bison in Yellowstone and that the current population was probably greater than under primitive conditions. The ranching operations seemed to be a loss of "energy, time, and money." And while Yellowstone had helped save America's bison from extinction, Mills added that the bison "has been saved and there is now no necessity of fearing that the species will disappear." But, despite Mills' recommendations, the Service maintained the population level at close to eight hundred through the remainder of the 1930s.[39]

The methods used to maintain the desired population appeared in Fauna No. 2, which also provided statistics on bison losses in recent decades: Since the Army began its bison management in 1902, six hundred eighty-two of the animals had been slaughtered, two hundreed seventy-nine shipped live, and forty-eight "outlaws and cripples" destroyed. In addition, one hundred twenty-four bison had died from disease during this period.[40] In 1935, the year Fauna No. 2 was published, George Wright expressed his considerable displeasure with live shipping, whether of bison or elk and whether to other national parks or to state or local parks. He believed that such activity involved the "inadvised mixing of related forms and the liberation of certain species in areas unsuited to their requirements," which brought "great and irreparable damage in many instances."[41]

Regardless of the wildlife biologists' disapproval, live shipping remained a regular activity in the parks, as did slaughtering and occasional destruction of "outlaws." Yellowstone superintendent Edmund Rogers reported in late 1937 that fifty-nine bison, including "some old animals that we wish to take from the herd," were being held for live shipment. The park planned shipments to the Springfield, Massachusetts, zoo; to an individual in Wolf Creek, Montana; and to Prince Ri Gin in Korea. In addition, bison carcasses were intended to be sent to the Wind River Agency in Wyoming, for distribution to local Indians. In Wind Cave National Park, where until the mid-1930s the Bureau of Biological Survey had been in charge of wildlife management, efforts began to reduce bison and elk to satisfactory numbers. The Service reported the following year that both Wind Cave and Platt national parks were reducing their bison populations, mainly by shipping carcasses to nearby Indian tribes.

These live shipments or distributions of carcasses may not have provided much political advantage, but the shipment of buffalo robes was at times partly intended to reap political gain. Recognizing this possibility, Director Cammerer wrote Secretary of the Interior Harold L. Ickes in 1936 that disposition of the hides "to friends of the Service and the Department, upon their special request, has been and will be helpful in maintaining a special interest in matters relating to this Department and the Service." In this regard, Yellowstone superintendent Rogers noted that requests for hides had been received from a number of persons, some of them highly placed, such as Senator Robert F. Wagner of New York and Clyde A. Tolson of the Federal Bureau of Investigation.[42]

ANIMAL ENCLOSURES

Wind Cave and Platt shared another management practice with Yellowstone, in that these parks set up fenced-in areas for wildlife (particularly bison) to be viewed by the public. Only a few hundred acres in size, Platt had no choice but to build a display area for viewing bison, originally shipped in from a nearby wildlife preserve. The Park Service took over wildlife management in Wind Cave with fences already in place, and despite expressed intentions to remove the fences, continued to maintain an animal enclosure for the public's benefit.[43] As to Yellowstone's bison, Director Albright had stated in 1929 his determination to make the animals "more accessible to the visiting public." The problem, as he saw it, was how to manage the bison population "under nearly natural conditions and at the same time get it near the main highways where it can be easily and safely observed."[44]

Predictably, the biologists opposed enclosing park wildlife behind fences. In 1931, Wright made his opposition clear to Albright, pointedly reminding the director that the purpose of park wildlife "does not end with their being seen by every tourist" and that people see many of these animals "when the circus comes to town." To Wright and his fellow biologists, an animal enclosure had the appearance of a "game farm," an inappropriate display of park wildlife to the public.[45]

Grinnell's remarks to Director Cammerer in 1933 reflected Wright's position, after Yosemite's fenced-in Tule elk herd (*not* native to the park) had been returned to their native habitat in California's Owens Valley. Keeping a close watch on Yosemite's wildlife management, Grinnell wrote Cammerer applauding Superintendent Charles Thomson's decision to remove Tule elk from the park. And, in reference to overall national park policy, Grinnell added that parks were not places "in which to maintain any sorts of animals in captivity," adding that it was the "free-living native wild animal life that . . . gives such rich opportunity for seeing and studying." Moreover, he took it for granted that maintaining free roaming wild animals was the Service's "general policy."[46]

Grinnell was mistaken, however, as to the Service's true policy on wild-life enclosures. Yellowstone's most ambitious effort to display bison came in 1935, only two years after Grinnell's letter to Cammerer, when the park established "Antelope Creek Buffalo Pasture," an approximately five hundred thirty-acre tract south of Tower Falls in the northeast part of the park. Located along the park's main tourist road, the pasture accommodated about thirty bison and included a five-acre "show corral," to assure visitors a chance to see the animals.[47] Remaining an important part of the park's wildlife display for several years, the Antelope Creek enclosure would be discontinued in the 1940s by Director Newton B. Drury—causing a heated controversy over the very policy issues that Grinnell and the other wildlife biologists had raised.

PREDATOR CONTROL

The Park Service in the 1930s faced the problem of what to do with native predators—a matter of great concern to the wildlife biologists, who urged that the remaining predators be protected. Again, the Service's actions in this regard exposed internal disagreements over policy, and revealed difficulties which the biologists encountered in seeking to change traditional practices. Already by 1931, when Director Albright announced the policy of limiting predator control to that which was absolutely necessary, wolves and mountain lions (major predators believed to have kept populations of the more favored species reduced) were virtually eradicated from all national parks in the contiguous forty-eight states.

Accordingly, the new policy had only limited effectiveness. Of the triumvirate of carnivores most targeted for reduction by the Park Service in past decades (wolves, mountain lions, and coyotes), only the coyote remained in substantial numbers, other than in the Alaska parks which had populations of wolves. And, despite the new predator policies, during most of the decade coyotes continued to be hunted, mainly on an occasional basis, and limited control of wolves was undertaken in the Alaska parks.[48]

Indeed, the 1931 predator policy itself reflected traditional biases against the coyote. Rather than a flat prohibition, the policy stated that there would be "no widespread campaign" against predators, and that "coyotes and other predators" would be shot only when they endangered other species. Thus, the policy did not totally eliminate predator control; rather it only restricted

101

Caught in a steel trap in Yellowstone National Park, this coyote was one of many exterminated by the National Park Service's predator control policies.
(YELLOWSTONE NATIONAL PARK PHOTO, 1929)

control (no "widespread" campaigns)—and it specifically identified the coyote as a potential target, the only species so designated. Moreover, at the 1932 superintendents' conference, a lengthy discussion of predator policy focused mainly on how to deal with coyotes. The consensus was that coyotes were to be subject to "local control"—i.e., reducing this species would be a matter of each superintendent's discretion. In fact, two biologists attending the meeting, Joseph Dixon and Harold Bryant, conceded that coyote reduction might at times be necessary.

By far, the strongest support for control of the coyotes came from the ranks of park management. Albright wanted to control coyotes when they do damage to "more useful species." He particularly feared that without the current "intensive" control of coyotes, there would soon be no antelope in Yellowstone. Roger Toll, Yellowstone's superintendent, concurred. To Toll, a herd of antelope and deer was "more valuable than a herd of coyotes;" and he stated that rather than predators, the elk, deer, and antelope "were the type of animal the park was for."[49]

With support from leaders such as Albright and Toll, "wholesale coyote killing" (in the words of a Park Service report) continued in Yellowstone until the fall of 1933.[50] Earlier that same year, in Fauna No. 1, Wright's team of wildlife biologists had declared a more rigid predator policy than before —perhaps a factor in easing Yellowstone's aggressive coyote control. As stated in Fauna No. 1, predators were to be "special charges" of the National Park Service, killed only when the prey species was "in immediate danger of extermination"—and then only if the predator species itself was not endangered.[51]

In truth, the 1930s witnessed a decline in the killing of coyotes. Under the guidance of Sequoia superintendent John R. White, biologist Harold Bryant, and especially Wright, the Service began to rely on "increased scientific data rather than ancestral prejudice" to address the predator issue.[52] In November 1934, Director Cammerer issued a prohibition of all predator control without written authority obtained from his office. Yet the following year, in Fauna No. 2, Wright and Thompson acknowledged that coyote management remained controversial. They defined Park Service policy as allowing "judicious control of coyotes" to be undertaken in any park with the necessary authorization from Washington.[53]

Ongoing coyote control clearly demonstrated that these predators were not altogether "special charges" of the Park Service. Particularly in Yellowstone, efforts to reduce coyote populations continued, although apparently with less zeal after 1933. A matter-of-fact report in March 1935 revealed a cavalier attitude toward eliminating coyotes, as one ranger described how he discovered a pair of coyotes copulating "just at daylight," near lower Slough Creek; then (although aware that he had never seen coyotes do this before) he shot one of the animals dead.[54] By contrast, some Yellowstone staff doubted the wisdom of continued coyote control. In April 1935, Assistant Chief Ranger Frank W. Childs recommended that the park suspend the killing of coyotes for at least two years, with the intention of carefully studying the resulting effect on prey populations. Childs and others recognized the conflicts between, on the one hand, efforts to reduce elk populations, and on the other, killing predators that themselves were presumed to help reduce the numbers of elk. He suggested that scientific research might prove that discontinuing coyote control permanently would be best for the "general wildlife balance" in the park.[55] Evidence indicates that the park eased up on coyote control in 1935, but by 1937 considerable interest in further coyote reduction had developed.[56]

Pressure on the National Park Service to reduce its predator populations stemmed from several factors, including demands for protection of the spectacular game species so that they could be enjoyed in the parks and hunted on lands adjacent to the parks, and demands for protection of livestock on adjacent lands. Concern for the game species and domestic livestock kept the Park Service under constant pressure from sportsmen's clubs and livestock growers associations to reduce or entirely remove major carnivores from the parks. In November 1935, Crater Lake superintendent David H. Canfield responded to the Southern Oregon Livestock Association's "sweeping condemnation" of predatory animals in national park areas. The association was particularly anxious about coyotes in the vicinity of Lava Beds National Monument (a park under Canfield's supervision); and Canfield stated that the wildlife problems of the area would be addressed through scientific research. Subsequent research on coyotes in Lava Beds supported protection of these predators rather than control.[57]

The Service's policy for protection of predators, although flawed in its implementation, nevertheless contributed to sportsmen's associations and other groups opposing new national park initiatives for the Kings Canyon area in California and Olympic Mountains in Washington.[58] As elsewhere, such groups wanted the predators in these areas eliminated to protect game species. Resentment over the Service's policies motivated the California state legislature to petition Congress to force strict predator reduction in the national parks, but to no avail. This proposal would have been, in the words of Grinnell, who had long opposed predator control, a "calamity" to those "who see in national park administration the last chance of saving to the future entire *species* of certain animal groups." Viewing predators in an ecological context, Grinnell wrote to Cammerer of the need to protect the "biotic mosaic" of each park, including predators. The Service should protect the whole "biotic superorganism uninjured—to the benefit of *all* its constituent species and populations" (emphasis Grinnell's).[59]

In addition to pressure from outside organizations, repeated recommendations that some predator populations be reduced came from within Park Service circles, such as from Albright. Maintaining a keen interest in national park management long after he resigned from the Service—indeed until his death in 1987—Albright seemed most alarmed about what effect suspension of coyote control would have on the spectacular grazing species, for instance antelope. Although Albright had established the Wildlife Division after Wright had funded the initial wildlife survey, the former director remained intensely interested in assuring public enjoyment of the parks' more popular animals and steadfastly loyal to the Park Service's traditional management practices.

Albright strongly and plainly worded his letters to Director Cammerer on predators and antelope. In October 1937, the former director wrote that he deplored the ongoing, as yet inconclusive studies of the coyote's impact on Yellowstone's antelope population. He advocated "open war" on coyotes for the purpose of studying stomach contents to determine how much coyotes fed on antelope. In fact, he urged reducing the coyote population under almost any pretext, stating that, in spite of Park Service policy or the results of the studies of coyote stomachs, he would:

> [C]ontinue to kill coyotes on the antelope range for the reason that the coyotes are of no possible advantage in that part of the park, can rarely be seen by tourists . . . while on the other hand there will always be danger of depleting the antelope herd. It must be remembered that one of the animals most interesting to tourists is the antelope. . . .

Albright also feared that, if protected, the coyotes would "over-run adjacent country," causing conflict with land managers and owners outside of the park.[60]

When Albright made these remarks, the Park Service was beginning its most in-depth research to date on coyotes as predators. In line with recommendations from the wildlife biologists and from the park itself (such as ranger Childs' suggestions), Adolph Murie initiated in 1937 a study of Yellowstone's coyotes, at a time when there was renewed interest in predator control in the park. Murie's findings, entitled *Ecology of the Coyote in the Yellowstone*, were published in 1940 as the fourth in the Wildlife Division's "Fauna Series" (Fauna No. 4). His research indicated that coyote predation did not appreciably affect prey populations—having, for instance, only a "negligible" impact on elk populations. Murie noted that in view of the National Park Service's "high purpose" of preserving "selected samples of primitive America," the parks' flora and fauna should be subjected to "minimal disturbance." He concluded that coyote control was "not advisable under present conditions."[61]

Coming from one of the most outspoken Park Service biologists, Murie's conclusions drew severe criticism from those within the Service who did not want to see coyotes protected. Indeed, it appears that some individuals in top management wanted Murie fired.[62] Moreover, already aware of Murie's findings and the Wildlife Division's opposition to coyote reduction, Albright wrote Cammerer in January 1939, reiterating his disagreement with the biologists. Believing nothing to be gained "either in wildlife management or in service to the public" by protecting the coyotes, Albright stated that, if not controlled very strictly, "powerful predators" such as the coyote were certain to menace the "more desirable species of wildlife." But despite the criticism, Murie's findings gained support from Cammerer, who opposed further coyote reduction. As Cammerer stated in his 1939 annual report, the coyote was a "natural and desirable component of the primitive biotic picture," not affecting the well-being of any of its prey species, and "not requiring any control at present"—words that sound as if they were written by Murie himself.[63]

Cammerer also noted in his 1939 report that Murie had begun long-range studies of the wolves in Mount McKinley National Park. Public pressure for wolf control in McKinley (which resulted from fear that this predator was reducing Dall sheep and other popular wildlife populations) prompted Murie's study, which would extend into the early 1940s. As with the coyotes in Yellowstone, the Service sought to establish a scientific basis for its treatment of Mount McKinley's wolves. Again, however, Albright's comments on this matter revealed the differences between the wildlife biologists' recommendations and traditional Park Service attitudes. In his January 1939 letter to Cammerer, the former director stated that he found it "very difficult" to accept the idea of protecting McKinley's wolf population in the "territory of the beautiful Dall sheep." Albright believed the Park Service was taking a "grave risk" in spending so much time and effort caring for predators, a responsibility which in his opinion "does not or need not fall on the National Park Service at all."[64]

Writing to Cammerer in May 1939, Park Service biologist David Madsen reflected on the state of national park predator management near the close of the decade. Noting the ambivalence that still existed, Madsen observed that:

> In one breath we say that it is a good thing to have large predators present in the park to control what would otherwise be an over supply of our large mammals; and in the next breath we state that the large predators in particular the coyotes are not a factor in reducing the antelope in Yellowstone Park.

Madsen cited Murie's belief that the Service was troubled with "confused thinking" and did not have a "philosophical point of view" on predators. In part, Madsen attributed this indecisive attitude to a lack of scientific information, affecting all Service personnel, both managers and biologists. He saw a "need for enlightenment" on the predator issue, to help the Service handle the "crossfire" between the scientists and such groups as sportsmen and livestock owners.[65]

Although influenced by the wildlife biologists (who found support from park management at different levels, such as from Director Cammerer or Yellowstone ranger Childs), the Park Service moved slowly and erratically during the 1930s toward a more scientific understanding of predator and prey populations and the discontinuance of predator control. Murie's work at Yellowstone and Mount McKinley, and the coyote studies at Lava Beds, evidenced a willingness in the Service to use scientific research to address specific predator concerns. Nevertheless, as Madsen recognized, a strong ambivalence existed. Traditional biases which favored the popular game species over important carnivores, and agitation from livestock owners' and sportsmen's organizations, countered the scientific perspective within the Park Service. Such pressure would continue to affect predator management in the national parks.

FISH MANAGEMENT

Similar to practices during the Mather era, the Park Service's fish management under Albright's and Cammerer's leadership primarily intended to enhance sport fishing as a means of providing for public enjoyment of the parks. The Service took considerable pride in maintaining high-quality fishing in the national parks, even though it involved harvesting and consumption of native park fauna and the introduction of exotic species. In its management of fish, more than any other natural resource, the Park Service grossly violated known ecological principles. Yet so deeply entrenched was the tradition of fishing national park rivers and lakes that the wildlife biologists themselves seemed ambivalent and did not categorically challenge management practices.

That these practices contradicted the idea of preserving park wildlife in its natural state was, however, clearly recognized. In Fauna No. 1, the wildlife biologists noted in a section suitably entitled "Conflicts With Fish Culture" that fishing in parks was an "important exception to general policy." Yet, granting the long-established fish management practices, they conceded that the benefits to park visitors overruled the "disadvantages which are incidentally incurred" by allowing fishing.[66]

Already, in 1928, five years before Fauna No. 1 appeared, the Park Service had detailed a biologist from the Bureau of Fisheries to become the Service's specialist in "fish culture" and coordinate with the Bureau in raising fish and planting them in park lakes and streams. The specialist was probably Madsen, who by the early 1930s, was in fact working with the Park Service, on detail from the Bureau. Like his fellow biologists, Madsen recognized that the Park Service's fish management was "entirely inconsistent" with other wildlife policy. Yet as a fish culture specialist he predictably appreciated the popularity of fishing in the parks and stated that the sport should be "maintained and in some instances developed to the highest point possible in the interest of the visiting public."[67]

In an effort, moreover, to improve fishing elsewhere in the country, the Service regularly shipped fish eggs to areas outside the parks—thus its manipulation of fish populations and distribution extended far beyond national park boundaries. The Yellowstone Lake Hatchery became particularly active, shipping millions of native and non-native fish eggs to numerous states and some foreign countries.[68] In maintaining the sport for the visiting public, and in shipping eggs to areas outside of the parks, the Service continued Director Mather's policy of extensive reliance on expertise in the Bureau of Fisheries and the state game and fish departments—offices which shared the Service's interest in promoting sport fishing.

Early in 1935, just as Madsen was being converted to permanent Park Service employment, assigned to the Wildlife Division, he reviewed the fish cultural activities in the national parks. Madsen observed that in the past "other agencies" had run national park fish programs, and, in fact, often with very little direction from the Service. He wrote that the Bureau of Fisheries had managed fish culture in Glacier, Mount Rainier, Yellowstone, and Grand Teton, while state offices had overseen the work in the national parks of California and in Crater Lake and Rocky Mountain national parks. The Park Service, however, had recently begun asserting a greater voice in fish management, by using park rangers to do the planting (and by hiring Madsen), thereby assuming greater control over what species were planted, and where. But Madsen urged that the Service take charge of "all fish cultural activities" in the parks, in the same way that it oversaw other activities which were "properly the function of the Park Service."[69] His greater concern seemed to have been to exert control over the fish programs, rather than change policy.

Nevertheless, although Park Service biologists seem to have voiced only limited opposition to fishing in the national parks, apparently not recommending banning fishing altogether, Madsen and the other biologists proved largely responsible for the slight modifications in the Service's fish policy that did occur in the 1930s. As a fish culture expert who encouraged fishing in the parks, Madsen still acknowledged that "indiscriminate introduction" of non-native fish had adversely altered the natural conditions of park lakes and streams—a concern shared by the other biologists.[70] Fauna No. 1 contained clear recommendations to reduce populations of exotic species already present in the parks and to prevent the invasion of other exotics. In addition, the report advocated setting aside one watershed in each park to assure "preservation of the aquatic biota in its undisturbed primitive state." No introduction of fish or fish food would be allowed in any of these watersheds, except as might naturally occur; and fishing would be permitted, but only if it did not "deplete the existing stock."[71]

Overall, since apparently no strong push developed to eliminate fishing and fish culture in the national parks, the concerns about exotic species and the recommendation to keep selected park watersheds in an "undisturbed primitive state" became the only factors likely to be affected by a policy change. Thus when Director Cammerer issued the National Park Service's first written policy for fisheries management (in April 1936, and almost certainly prepared by the biologists), it dealt primarily with the question of exotic fish species, and, to a lesser degree, the idea of leaving some park waters in their natural condition. That fish cultural activities would continue in parks was a given in the new policy—in fact, the document's introduction specifically stated that it was a policy for "fish planting and distribution." Still, the policy favored protection of native species, emphasizing that the intent was to "prohibit the wider distribution" of exotics within park waters. Among other points, exotic species could not be introduced in waters where only native fish existed; and in waters where exotic and native fish *both* existed, the native species were to be "definitely encouraged."[72]

The new policy contained, however, significant deviations from the protection of native species and restrictions on exotics—deviations that left substantial options open to park managers and thereby reduced the degree of true change from earlier policy. Despite the concern about "indiscriminate introduction," stocking was allowed in waters previously barren of game fish, based on the Service's judgment whether or not a lake or stream was of "greater value without the presence of fishermen." And in waters where exotic species proved "best suited to the environment and have proven of higher value for fishing purposes than native species," stocking of exotics could continue if approved by both the park superintendent and the director. Subsequently, Cammerer refined this last point in his 1936 annual report by specifying that native species would be "favored" in waters where such species "are of equal or superior value from the standpoint of fishing."[73]

The new fish management policy thus allowed continued alteration of national park aquatic conditions for utilitarian purposes—i.e., the promotion of sport fishing and the enhancement of public enjoyment. As during the Mather era, fish management remained essentially commodity based, with stocking and harvesting on a massive scale. And the Service continued to plant exotic species in large numbers in such waters as Yellowstone's Madison, Firehole, and Yellowstone rivers in the years following issuance of the 1936 policy. In some instances, as at Mammoth Beaver Ponds in the Yellowstone River drainage, previously fishless lakes were first stocked about the time the policy was declared, and such stocking continued for years afterward.[74] Not even mentioned in the new policy, the shipment of millions of fish eggs (including both native and exotic species) from national parks to non-park areas continued undiminished throughout this period. Director Cammerer reported in 1937 that twenty million rainbow and Loch Leven trout eggs (both exotic species) were collected near Yellowstone's west boundary, with only one-fifth of them returned to park waters, the rest shipped elsewhere.[75]

Indeed, the Park Service's first detailed fisheries policy—which would remain essentially unchanged for two decades—had limited effect on fish management in the parks. Park Service biologist Carl Russell's remarks to the North American Wildlife Federation in March 1937 reflected the continuity in national park fish management when he asserted that the new policies would mean continued "maintenance of good fishing," and that the Service was "definitely" committed to fishing as a "recreational activity in parks." Similar observations came from other biologists. Cahalane commented in 1939 that the Service deemed fishing to be acceptable because of the "readily replaceable nature of fish resources" and because sport fishing results in "recreational benefits far outweighing any possible impairment of natural conditions." But, evidencing the ambivalence among the biologists, Cahalane also stated that the National Park Service had a responsibility to address the contradictions "existing between use of fish resources and of other natural resources within the parks."[76] Due to the very deeply ingrained acceptance of angling in national park waters, however, the contradictions in fish policies would remain mostly unresolved. And with widespread acceptance of fish stocking and harvesting, as sanctioned by the 1936 policy, extensive manipulation of park fish populations and distribution to areas outside of the parks would continue long after issuance of the policy.

PROTECTING THE FORESTS

Similar to fish management, the treatment of national park forests stood at odds with known ecological principles. Nevertheless, traditional forest practices endured. The entire emphasis fell on maintaining green, attractive forests, despite strong challenges to this policy by the wildlife biologists, who wished to adhere to the current ecological principles which they articulated. The debates over forestry policies highlighted fundamental differences between the wildlife biologists and much of the rest of the Park Service, with the biologists' views of park management being far ahead of the times. The failure of their challenge to forest management showed the weakness of the biologists' position within a very traditional organization and, conversely, the considerable bureaucratic strength which the foresters had developed in the Park Service.

National park forestry operations expanded tremendously during the 1930s, receiving far more funds and support from the New Deal's emergency relief programs than any other natural resource management activity in the parks. So important proved forestry in the overall work of the Civilian Conservation Corps that the CCC was at times referred to as "Roosevelt's Tree Army." And, as the 1916 National Park Service Act itself had done, the 1933 act creating the CCC specifically called for protection of the forests. Among the CCC's other responsibilities, Congress mandated that it would protect the forests from fires, insects, and disease damage—goals which fit perfectly those of most national park managers.[77]

In his 1933 annual report, Albright's comments on the initial work of the CCC foreshadowed the virtual explosion of national park forestry. The director stated that the newly established CCC crews were accomplishing "work that had been needed greatly for years," but which had been "impossible" under ordinary appropriations:

Especially has the fire hazard been reduced and the appearance of forest stands greatly improved by clean-up work along many miles of park highways; many areas of unsightly burns have been cleared; miles of fire trails and truck trails have been constructed for the protection of the park forests and excellent work accomplished in insect control and blister-rust control and in other lines of forest protection; improvements have been made in the construction and development of telephone lines, fire lookouts, and guard cabins; and landscaping and erosion control has been undertaken.[78]

Park Service forestry programs of the 1930s came under the direction of John Coffman, hired from the U.S. Forest Service in 1928 and placed in the Division of Education and Forestry, supervised by Ansel Hall. That same year, with assistance from the recently established, multi-bureau Forest Protection Board, which the Park Service had joined, Coffman and Hall drafted the Park Service's first formal forestry management statement, declared official policy by Director Albright in 1931. And during the buildup of CCC-funded forestry programs in 1933, Director Cammerer designated Coffman the Service's "Chief Forester," in charge of the newly created Division of Forestry, separate from Hall's educational work.[79] The 1931 forestry management policies promulgated by Albright provided guidance for the Park Service throughout the decade, and beyond. Under the new policies the park forests were to be "as *completely protected* as possible" against fire, insects, fungi, and "grazing by domestic animals," among other threats. This comprehensive protection was to be extended to "*all* park areas," such as those associated with "brush, grass, or other cover" [italics in the original].[80] The CCC provided the Service with sufficient manpower to implement these forestry policies. Armed with new policies and staffed by thousands of CCC enrollees, Coffman's forestry programs became an increasingly important force in national park operations during the New Deal era.

The forest management practices drew frequent and sometimes barbed criticism from Wright and the other wildlife biologists. Central to the biologists' concerns were the various "pre-fire" protection activities. They objected to the Park Service building fire roads through natural areas or clearing hazardous dead trees and snags which contributed to the fuel buildup and increased the possibility of fire (for example, the insistence on clearing storm-damaged and dead trees from the Andrews Bald research reserve in Great Smoky Mountains). Pre-fire development particularly affected some national park areas. On the North Rim of Grand Canyon, fire protection preparations by the CCC included improvement of existing roads; and construction of primitive fire-access roads and trails, lookout towers, warehouses, a fire cache, maintenance shops, residences, telephone lines, and water ponds.[81]

Significantly, although the Park Service established a Wildlife Division in the 1930s and (mostly using CCC funds) hired about twenty-seven wildlife biologists, the bureau did not hire plant biologists *per se*. Also, the management practices of the U.S. Forest Service deeply influenced the Park Service foresters (not known as biologists or botanists, but as foresters), particularly regarding control of forest fires, insects, and disease. Such forest protection concerns dominated Park Service thinking regarding plant life, leaving the wildlife biologists, by default, to deal with many plant biology issues. And as evidence of their broad ecological interests, the biologists did not shrink from the task. Moreover, they advocated ecological-attuned forest management, placing them in direct conflict with Park Service foresters.

Indeed, the wildlife biologists never agreed with the forest management policies made official in 1931. Although forests were not the focus of Wright's initial wildlife survey, preserving natural habitat, including plants, was recognized as fundamental to successful park management. In direct contradiction to ongoing Park Service forestry practices, Fauna No 1 declared that park forests should be left in a natural condition: "It is necessary that the trees be left to accumulate dead limbs and rot in the trunks; [and] that the forest floor become littered. . . ."[82] Nevertheless, the CCC programs provided funds and manpower for extensive clearing of forest underbrush and dead trees, and this clearing became of increasing concern to the biologists.

Among other clearing work, roadside clearing, a widespread practice in national parks, was intended as a fire protection measure, but became equally important, in the words of a Park Service manual, as a means "to improve the appearance of the immediate landscape of the main drive" through parks. A conflicting view came from Wright, who wrote Director Cammerer early in 1934 of the need to consider "all sides of the question" regarding clearing of hazardous debris along park roadsides, including the concern for "wild life values." Wright realized that clearing dead limbs and trees affected habitat, and he urged that the Service "reconsider" and determine "exactly under what conditions and in what parks road-side clean-up is a benefit and to what extent it should be carried on." He also told Cammerer that the biologists had discussed this matter with park superintendents and rangers and that it was "amazing to discover that there was anything but unanimity of opinion on the value of this work." Some superintendents and rangers recognized the impacts on natural conditions, while others believed cleanup did not help prevent fires.[83] Nevertheless, the Park Service rank and file sufficiently accepted clearing so that it remained a common practice in the parks.

An even stronger opinion than Wright's came from Adolph Murie in the summer of 1935, during an extended debate over whether or not to clear a twelve-square-mile area on Glacier National Park's west slope, just north of McDonald Creek, an area covered with damaged trees as a result of a recent fire. With many of the trees only partially burned, the tract seemed ripe for another fire, which could spread to adjacent, unburned forests. A meeting

in the park in July provoked strong disagreement on the propriety of cutting and removing all of the dead trees, whether standing or down. The
contentious debates reflected sharp divergence between the wildlife biologists and the foresters on fire protection and on overall national park policy
and philosophy.

Following the July 1935 meeting in Glacier, Murie reported to the Wildlife Division in Washington his intense opposition to the proposed clearing.
In a lengthy letter, Murie wrote that the burned area was still a natural area,
and he questioned the desirability of "removing a natural habitat from a
national park." With roads for trucks, bulldozers, and other equipment, the
clearing operation would cause "gross destruction," which, he believed,
would interfere with the normal cycles of forest decay and growth, creating
instead a "highly artificial appearance of logged-off lands." The removal of
the trees would reduce the area's organic material and its soil fertility and
cause drying of the soil and increased erosion. Moreover, this large clearing
project would be a precedent to justify "almost any kind of landscape manipulation" in the future. "For what purposes," Murie asked, "do we deem
it proper to destroy a natural state?" His answer was that almost no purpose
justified such destruction. He concluded his argument with an opinion
surely unheard of in national park management before the wildlife biologists
began their work under Wright:

> To those interested in preserving wilderness, destroying a natural
> condition in a burn is just as sacrilegious as destroying a green forest.
> The dead forest which it is proposed to destroy is the forest we should
> set out to protect.[84]

Murie's remarks were quickly challenged. Lawrence F. Cook, head of
Coffman's forestry operations in the western parks, had also attended the
July meeting in Glacier. Cook found Murie's report "rather typical"—and
took a directly opposite position, fearing the long-term loss of green forests.
"Nature," he commented, "goes to extremes if left alone." He reported that
"gross destruction" had been done by the fire itself, despite the Service's best
protection efforts, which were carried out with trained employees working
under professional plans and with good equipment. In addition to adequate
detection, fire protection depended on "easy access" into the forests and the
"reduction of potential fuel" through clearing—both of which would result
from the proposed work in Glacier. Cook anticipated a rapid recovery of
forest growth, but only if the area were cleared of dead trees so it would not

be burned over by another, more damaging fire. Seeking to protect the beauty of the forests, he also recognized that this part of Glacier received intensive use; it was seen, he claimed, "by more travelers than any other in the park." Thus, Cook argued that the question was not whether to allow nature to take its course in the national parks, but to what extent the Service "must modify conditions to retain as nearly a natural forest condition as possible for the enjoyment of future generations."[85]

In a separate memorandum to Coffman, written the same day, Cook reflected on his concern that the Service's foresters had been accused of being "destroyers of the natural." Not only the biologists but other Park Service officials, including some superintendents, rangers, and landscape architects had criticized their promotion of physical development for fire protection, such as truck trails and fire lookouts, and their efforts to clear forests of fuel hazards. Cook insisted, however, that the foresters sought to preserve the "natural values" of the parks, while also providing for the "greatest use and enjoyment of the parks with the least destruction." He summed up his credo of national park management, and fire protection in particular:

> The parks have long since passed the time when nature can be left to itself to take care of the area. Man has already and will continue to affect the natural conditions of the areas, and it is just as much a part of the Service Policy to provide for their enjoyment as it is to preserve the natural conditions. There is no longer any such thing as a balance of nature in our parks—man has modified it. We must carry on a policy of compensatory management of the areas.

"Forest protection," he added, is a "very necessary part of this management." Without protection, the Service faced the destruction of "any semblance of biological balance, and scenic or recreational values, as well as the forests with which we are charged." Certainly Cook's views prevailed within the Service. But, before any significant clearing could get underway in the area north of McDonald Creek, the huge Heaven's Peak fire swept through Glacier in 1936, drawing attention from McDonald Creek and likely meaning that the disputed cleanup was never completed.[86]

Indeed, the Park Service's biologists and foresters all believed they were seeking to preserve "natural values," which would allow for the "greatest use and enjoyment of the parks with the least destruction." But the two groups were operating from fundamentally different perceptions as to exactly what constituted "natural values" and what constituted "destruction" in national parks. Murie opposed the extensive alterations which resulted from the Service's fire protection methods employed before, during, and after fires. And in his letter on the proposed clearing in Glacier, he concluded that:

> My feeling concerning any of this manipulation is that no national park should bear the artificial imprint of any man's action of this sort. We have been asked to keep things natural; let us try to do so.[87]

Cook, by contrast, had written Chief Forester Coffman that human modifications to national parks meant there was no longer a "balance of nature" -- thus his argument for "compensatory management," including determined efforts to protect the forests. His compensatory management would also preserve the beauty of the forests, so important to the public's enjoyment of the parks. Cook's philosophy of national park management reflected the Service's forestry policies as well as its overall management practices. And with funds and manpower coming from the CCC program, the Service continued its intensive protection and suppression activities, very much against Murie's wishes.[88]

The biologists' and foresters' different approaches to national park management manifested themselves in disagreements over other aspects of forestry. Continuing practices of the Mather era as stated in the 1931 forest policies, both Albright and Cammerer supported aggressive war against forest insects and disease, regularly calling upon the Bureau of Entomology and the Bureau of Plant Industry for expert assistance. In his last annual report (1933), Director Albright noted that "successful campaigns" had been waged against insects in park forests, ending or reducing several major epidemics. The Service, he said, had sought to eradicate infestations of the bark beetle in Yosemite and Crater Lake and the mountain-pine beetle in Sequoia National Park. Both Glacier and Yellowstone faced insect infestations of such magnitude that studies were being made to determine if control efforts were practicable. It seemed to Albright that the national parks were truly under siege from insects, as well as from disease. Among the many threats, the disease known as blister rust was "spreading rapidly," threatening the western parks. "Unless checked," Albright reported, it was "only a matter of time" before blister rust would invade the white pine forests of Glacier and the sugar and white pines of the California parks.[89] As with fire protection, the CCC provided the Park Service with funds and manpower to wage intensive campaigns against forest insects and disease.

Again, however, the wildlife biologists challenged these efforts. Wright wrote Director Cammerer in August 1935 regarding use of the New Deal work relief programs to the greatest advantage, but he cautioned against too much "zeal for accomplishment," particularly in insect and disease control. Generally, the biologists directed their criticism toward widespread control efforts, while accepting limited control in and around park development. Wright would largely confine control to "heavily utilized areas" most frequented by visitors. The pinon pine scale infection in Colorado National Monument was, he pointed out, a natural phenomenon which seemed "best to leave undisturbed" outside of developed areas. Similarly, reporting on CCC work in Grand Canyon during 1935, Cahalane commented that the Wildlife Division "disapproves of insect control, outside of developed areas," unless a native plant was threatened with extinction.[90]

Much more critical comments came from Murie, who, after a visit to Mount Rainier in 1935, strongly objected to the Park Service's disease and insect control. Murie acknowledged to Wright that "possibly some effort" was necessary to save "certain outstanding forests." But he opposed extensive control, emphasizing that in its forest management the Service should not "play nursemaid more than is essential." Since beetles were native insects and ribes native plants (currants and gooseberries which serve as an alternate host to the blister rust fungus—the reason the foresters sought to eradicate ribes), Murie advocated leaving them alone and "permitting natural events to take their course. . . . the cure is about as bad as the disease." Ribes proved, in his words, "just as desirable in the flora as is pine," and Murie concluded that "justification for destroying a species in an area should be overwhelming before any action is taken."[91]

Arguments such as Murie's did not at all sway the foresters. In his letters to Coffman on fire management, Cook rebutted the biologists' position and defended the Service's forest disease and insect control policies as an essential part of park management. Just as with fire suppression, the foresters believed that "some modification," including insect control, "is necessary to preserve for the future the living values of the parks." And indeed, forest insect and disease control continued especially strong while CCC money and manpower remained available. Late in the decade, Cammerer reported on aggressive blister rust control and beetle eradication in a number of parks, noting the support of the Bureau of Entomology and that all control was carried out through the CCC program.[92] The termination of the CCC just after World War II began drastically reduced the resources available to the Park Service for control work—but the policies remained in force.

LEADERSHIP IN NATIONAL PARK POLICY
AND OPERATIONS

During the 1930s, guidance of the Service's natural resource management had become the responsibility of two professions, forestry and wildlife biology, and they often clashed over the basic principles and the specifics of national park management. The wildlife biologists had found a voice in national park policy and operations, but so had the foresters, who were able to continue their practices despite the biologists' objections. Decades later, Sumner reflected that "even George Wright was unable to make much progress" in establishing more ecologically sound forest management.[93] Indeed, the biologists' criticism of various forest practices had little effect on the Service's management policies—a reflection that the Park Service leadership did not seriously question the foresters' practices. The policies on forest fires, insects, and disease aimed at maintaining the beauty of the parks and thereby enhancing public enjoyment, thus bringing the foresters much more into the mainstream of national park thinking than were the wildlife biologists. Moreover, CCC money and by the mandate of the act establishing the CCC, much less by the National Park Service Act itself, backed the foresters.[94]

At the end of Cammerer's directorship, the biologists' influence was in eclipse, and the foresters were truly in the ascendancy. The Park Service's official organizational chart, revised in mid-1941 (a year and a half after Interior Secretary Ickes transferred the wildlife biologists to the Bureau of Biological Survey), showed the Branch of Forestry with no less than three divisions: Tree Preservation, Protection and Personnel Training, and Administration and General Forestry.[95]

The policies of the U.S. Forest Service, furthermore, heavily influenced foresters entering the Park Service in the 1930s and subsequent decades. Individuals such as Chief Forester Coffman had worked with the Forest Service before employment by the Park Service. Also, many national park rangers who did not have the specific title of forester nevertheless had been trained in forestry at such schools as Colorado A&M College. The so-called "ranger factory," which was just coming into being at Colorado A&M by the late 1930s and would flourish during the ensuing decades, trained young men to become national park rangers under a program administered by the forestry school.[96]

Altogether, an alliance was building between the Park Service's foresters and rangers (they would be combined organizationally in the mid-1950s).

That these two groups fed directly into top leadership positions, in charge of national park policy and operations, bolstered the strength of this alliance. With an increasing number of forestry graduates attracted into the ranks of the National Park Service, the profession was evolving into one of the most influential in the organization. By the end of the decade (with the few remaining wildlife biologists transferred to the Biological Survey and Fauna No. 1's influence on national park management swiftly declining) the foresters' bureaucratic power had begun to rival that of the landscape architects and engineers under Thomas C. Vint and Conrad L. Wirth, whose authority had also been greatly enhanced by the New Deal programs.[97] Although not always in full accord, the foresters, rangers, landscape architects, and engineers formed the core of National Park Service leadership and would dominate national park philosophy and operations for decades.

NOTES

KEY TO ENDNOTE ABBREVIATIONS

BL	Bancroft Library, University of California at Berkeley
GRSM	Great Smoky Mountains National Park Archives
Hartzog Papers	George B. Hartzog Papers, Clemson University
HFLA	Harpers Ferry Library and Archives, National Park Service
Leopold Papers	A. Starker Leopold Papers, Department of Forestry and Resource Management, University of California, Berkeley
MVZ-UC	Museum of Vertebrate Zoology, University of California
RG79	Record Group 79, Records of the National Park Service, National Archives
YELL	Yellowstone National Park Archives
YOSE	Yosemite National Park Archives

1 Richard West Sellars is a historian with the National Park Service in Santa Fe, NM. This article appeared in *The George Wright Forum*, 10 (1993), as Part II of a three-part history of natural resource management in the national parks during the 1930s, and in Richard West Sellars, *Preserving Nature in the National Parks: A History* (New Haven: Yale University Press, 1997).

2 Lowell Sumner, "Biological Research and Management in the National Park Service: A History," *The George Wright Forum* 1 (1983): 6, 10.

3 For a discussion of the Mather era, see Richard West Sellars, "Manipulating Nature's Paradise: National Park Management Under Stephen T. Mather, 1916–1929," *Montana The Magazine of Western History* 43 (1993): 2–13.

4 George M. Wright, Joseph S. Dixon, and Ben H. Thompson, *Fauna of the National Parks of the United States; A Preliminary Survey of Faunal Relations in National Parks*, Contributions of Wildlife Survey, Fauna Series No. 1 (Washington, D.C.: GPO, 1933), pp. 147–148.

5 National Park Service, "Proceedings," First Park Naturalists' Training Conference, Berkeley, California, 1–30 November 1929, typescript, 152, HFLA; Sumner "Biological Research and Management," 11.

6 Victor H. Cahalane, "Activities of the National Park Service in Wildlife Conservation," (ca. 1935), typescript, Central Classified File, RG79; *Annual Report of the Secretary of the Interior for the Fiscal Year Ending June 30, 1936* (Washington, D.C.: GPO, 1936), p. 123.

7 Sumner, "Biological Research and Management," 11; Joseph S. Dixon, *Birds and Mammals of Mount McKinley National Park*, Fauna Series No. 3 (Washington, D.C.: National Park Service, 1938); Adolph Murie, *Ecology of the Coyote in the Yellowstone*, Fauna Series No. 4 (Washington, D.C.: National Park Service, 1940); Adolph Murie, *The Wolves of Mount McKinley*, Fauna Series No. 5 (Washington, D.C.: GPO, 1944).

8 Harold C. Bryant, "A Nature Preserve for Yosemite," *Yosemite Nature Notes* VI (1927): 46–48. John Merriam's interest in research reserves is found in Merriam to Members of the Committee on Educational Problems in National Parks, 12 February 1930, with attachments, Entry 17, RG79.

9 National Park Service, "Proceedings," First Park Naturalists' Training Conference, 169, 171–174. Albright's policy on research reserves is stated in Arno B. Cammerer to All Superintendents and Custodians, 27 May 1931, with attachment, Research Reserves file, YOSE. The Fauna No. 1 quote is in Wright, Dixon, and Thompson, *Fauna of the National Parks* (1933), 147.

10 See for instance The Director to Wild Life Survey, 4 March 1932, Entry 35, RG79; and Arno B. Cammerer, "Maintenance of the Primeval in National Parks," ca. 1934, typescript, HFLA. As conceived, the research reserves were analogous to the "primitive areas" being designated in the national forests, although there is no indication that the idea was borrowed directly from the U.S. Forest Service.

11 George M. Wright to The Director, 14 March 1932, Entry 35, RG79.

12 The Director to Wild Life Survey, 4 March 1932; George M. Wright, "Research Areas," 1933, typescript, Entry 34, RG79; Kendeigh, "Research Areas in the National Parks," pp. 236–238.

13 Wright to The Director, 14 March 1932; Wright, "Research Areas"; Thompson to Cammerer, 23 February 1934; and U.S. National Park Service, Wild Life Division, "Report for February, 1934," Classified File, RG79. Comments on buffer zones for the national parks are also found in Wright and Thompson, *Fauna of the National Parks* (1935), p. 109.

14 Victor H. Cahalane to George M. Wright, 7 September 1935, Entry 34, RG79.

15 H. W. Jennison, Memorandum for Superintendent J. R. Eakin, 21 July 1936, Balds file, GRSM.

16 J. R. Eakin to The Director, 27 July 1936, Balds file, GRSM.

17 Frank E. Mattson, Memo for Mr. Eakin, 27 July 1936, Balds file, GRSM.

18 H. W. Jennison, Memorandum for Supt. J. R. Eakin, 21 July 1936, Balds file, GRSM; Eakin to The Director, 27 July 1936.

19 A. E. Demaray to J. R. Eakin, 4 September 1936, Balds file, GRSM.

20 Sumner, "Biological Research and Management," 10–11. In his history of wildlife management, Gerald Wright states that there is "no evidence" that the reserves were ever used as intended. R. Gerald Wright, *Wildlife Research and Management in the National Parks* (Urbana: University of Illinois Press, 1992), pp. 19–20. The 1960s perception is found in Conrad L. Wirth, Memorandum to All Field Offices, 15 April 1963, HFLA.

21 Wright to The Director, 14 March 1932. Keith R. Langdon, natural resource management specialist in Great Smoky Mountains National Park, recently commented on the considerable value Andrews Bald and other research reserves could have had for today's efforts to understand and manage the park's natural resources: If the park had maintained the reserves as originally intended, he stated, we would be "in the cat bird's seat." Moreover, Langdon stated that the park had no record of Andrews Bald Research Reserve. Probably it had been forgotten long before. Personal communication with Keith R. Langdon, 18 July 1991.

22 Wright, Dixon, and Thompson, *Fauna of the National Parks*, pp. 4, 147–148.

23 Wildlife Division to the Director of the National Park Service, "Report Upon Winter Range of the Northern Yellowstone Elk Herd and a Suggested Program For Its Restoration," 28 February 1934, reprinted in George Wright and Ben Thompson, *Fauna of the National Parks of the United States*, Fauna Series No. 2 (Washington, D.C.: GPO, 1935), p. 85; Douglas B. Houston, *The Northern Yellowstone Elk: Ecology and Management* (New York: Macmillan Publishing Co., 1982), pp. 24–25; and Don Despain, Douglas Houston, Mary Meagher, and Paul Schullery, *Wildlife in Transition: Man and Nature on Yellowstone's Northern Range* (Boulder, Colorado: Roberts Rinehart, 1986), pp. 22–24. See also Arno B. Cammerer to Joseph Grinnell, 10 December 1934, with attachment, Arno B. Cammerer files, MVZ-UC; and Victor H. Cahalane, "Wildlife Surpluses in the National Parks," in *Transactions of the Sixth North American Wildlife Conference*, (Washington, D.C.: American Wildlife Institute, 1941), pp. 357–358. Houston's detailed analysis of the management of the park's northern elk herd, *The Northern Yellowstone Elk*, pp. 12–15, refutes the belief that a population crash occurred in 1917–1920.

24 Thomas R. Dunlap, *Saving America's Wildlife* (Princeton: Princeton University Press, 1988), p. 69; Wright and Thompson, *Fauna of the National Parks*, pp. 85–86.

25 George M. Wright to H. E. Anthony, 15 March 1935, George M. Wright files,

MVZ-UC. Victor Cahalane later indicated that outside support for the reduction program existed, but that there was "constant protest by a few local organizations." He was not specific, however, as to which organizations or individuals supported or opposed reduction. Victor H. Cahalane, "Elk Management and Herd Reduction — Yellowstone National Park," *Transactions of the Eighth North American Wildlife Conference* (Washington, D.C.: American Wildlife Institute, 1943) pp. 95–97.

26 Olaus J. Murie to Ben H. Thompson, 27 December 1934, Entry 7, RG79 (copy from files of William E. Brown); Adolph Murie to Victor H. Cahalane, 26 July 1936, YELL.

27 Wright, Dixon, and Thompson, *Fauna of the National Parks*, p. 118. Albright mentions securing private funds for Rush's research in Horace M. Albright to the Director, 18 October 1937, Central Classified File, RG79.

28 Wildlife Division to the Director, "Report Upon Winter Range of the Northern Yellowstone Elk Herd," 85–86; Arno B. Cammerer, Memorandum for Assistant Secretary Walters, 21 November 1933, Central Classified File, RG79. The Park Service also saw overgrazing as a "landscape problem," and Fauna No. 2 advocated close cooperation between the wildlife biologists and landscape architects to address this concern. Wright, Dixon, and Thompson, *Fauna of the National Parks*, pp. 109–120. It does not appear, however, that the landscape architects became much involved.

29 Joseph Grinnell to Arno B. Cammerer 26 December 1934, Arno B. Cammerer files, MVZ-UC.

30 Cammerer to Grinnell 10 December 1934.

31 A list of annual elk "removals" from 1923 to 1979, including those taken by hunters near the park, is found in Houston, *Northern Yellowstone Elk*, pp. 16–17.

32 Wright to Anthony, 15 March 1935. Murie to Cahalane, 26 July 1936. Rudolph L. Grimm, "Northern Yellowstone Winter Range Studies," 1938, typescript, pp. 28–29, YELL. Although convinced that the range was still overgrazed, Grimm perceived that some "range recovery" had occurred, particularly in the two years just before he wrote his report. He credited, however, "favorable climatic conditions," i.e. the end of the drought (rather than the elk reduction program), as the "agency most responsible for the improvement of the range plant cover." p.27.

33 National Park Service, *Wildlife Conditions in National Parks, 1939*, Conservation Bulletin No. 3, (Washington, D.C.: GPO, 1939), p. 8. Other parks which eventually initiated limited control programs included Yosemite and Sequoia. Wright, *Wildlife Research and Management in the National Parks*, 77–78.

34 Joseph Grinnell to Arno B. Cammerer, 23 January 1939, Arno B. Cammerer files, MVZ-UC.

35 A. Starker Leopold *et al.*, "Wildlife Management in the National Parks," in *Transactions of the Twenty-eighth North American Wildlife and Natural Resources Conference*, edited by James B. Trerethen, (Washington, D.C.: Wildlife Management Institute, 1963), pp. 39–41, 43. Philosophically and policy-wise, the elk management situation became more complicated when, in 1967–1968, the Park Service terminated elk reduction in Yellowstone. Likely as a gambit to find an acceptable justification in a politically charged situation, the Park Service attempted to base its decision to terminate reduction on the Leopold Report's recommendations—which in fact had urged continued reduction. Starker Leopold, who was the report's principal author (and who also had studied under Grinnell), continued to doubt the wisdom of the Service's new "natural process" elk management policy. In June 1983, a little more than two months before his sudden death, Leopold made perhaps his last written comments on this issue. Seriously questioning the natural process concept of park management as it applied to elk and other grazing animals, he in effect sided with the Park Service biologists of the 1930s, observing that the national parks were "too small in area to relegate to the forces of nature that shaped a continent." National Park Service, United States Department of the Interior, News Release, "National Park Service Director Hartzog Initiates Elk Management Program for Yellowstone National Park," 1 March 1967, with attachment, George B. Hartzog, "Management Program, Northern Yellowstone Elk Herd, Yellowstone National Park," 1 March 1967; A. Starker Leopold to Jack Anderson, 16 March 1971, Hartzog Papers; and A. Starker Leopold to Boyd Evison, 9 June 1983, Leopold Papers. See also A. Starker Leopold, Interview Conducted by Carol Holleuffer, 14 June 1983, Sierra Club Oral History Project, Sierra Club History Committee, typescript, 19–20.

36 The Lamar Valley bison herd, introduced in the early twentieth century, came from two subspecies, both different from the remnant wild herds located in other areas of the park. While the wild herds at times interbred with the introduced Lamar Valley herd, they were almost always left alone and did not receive the intensive management as did those in the Lamar Valley. See Margaret Mary Meagher, *The Bison of Yellowstone National Park*, National Park Service Scientific Monograph Series no. 1 (Washington, D.C.: National Park Service, 1973), 26–37.

37 Also, both Fauna No. 1 and No. 2 recommended reestablishing bison in Glacier National Park, in cooperation with local Indian tribes. The comments on bison are found in Wright, Dixon, and Thompson, *Fauna of the National Parks*, pp. 117, 147; and Wright and Thompson, *Fauna of the National Parks*, pp. 59–60.

38 For carrying capacity figures, see Curtis K. Skinner, *et al.* "History of the Bison in Yellowstone Park" [with supplements] 1952, typescript, various pagination, YELL; M. R. Daum to Theodore C. Joslin, 9 January 1929, YELL; and Meagher, *Bison in Yellowstone*, p. 32.

39 Harlow B. Mills to Ben Thompson, 21 June 1935, Entry 34, RG79; Skinner, "History of the Bison in Yellowstone Park."

40 George Wright and Ben Thompson, *Fauna of the National Parks of the United States*, Fauna Series No. 2, p. 59.

41 Specifically regarding elk, Wright cited the situation in Mount Rainier, where non-native elk from Yellowstone had been transplanted—making it, in his opinion, "impossible ever to realize the restoration of the native Roosevelt elk to the park." George M. Wright to Arno B. Cammerer, 18 January 1935, Central Classified File, RG79.

42 Edmund B. Rogers to the Director, 10 December 1937, YELL; *Annual Report of the Secretary of the Interior for the Fiscal Year Ending June 30, 1939* (Washington, D.C.: GPO, 1939), pp. 280–281; *Annual Report of the Secretary of the Interior* (Washington, D.C.: GPO, 1940), pp. 180–181; Arno B. Cammerer to the Secretary of the Interior, 6 February 1936, YELL.

43 Palmer H. Boeger, *Oklahoma Oasis: From Platt National Park to Chickasaw National Recreation Area* (Muskogee, Oklahoma: Western Heritage Books, 1987), pp. 107, 111–112, 135–137; *Annual Report of the Secretary of the Interior for the Fiscal Year Ending June 30, 1935* (Washington, D.C.: GPO, 1935), 198; John Ise, *Our National Park Policy: A Critical History* (Baltimore: The Johns Hopkins Press, 1961), p. 584.

44 Horace M. Albright, "Our National Parks As Wild Life Sanctuaries," *American Forests and Forest Life* 35 (1929): 507.

45 George M. Wright to the Director, 19 December 1931, Entry 35, RG79.

46 Joseph Grinnell to Arno B. Cammerer, 9 November 1933, Arno B. Cammerer files, MVZ-UC.

47 Skinner, "History of the Bison in Yellowstone Park"; Rudolph L. Grimm, "Report on Antelope Creek Buffalo Pasture," (1937), typescript, YELL.

48 In 1945, Victor Cahalane recalled that the Park Service "practiced very limited control of wolves and coyotes in our Alaska areas from about 1932 to 1939 or 1940." Victor H. Cahalane to Mr. Drury, 14 March 1945, copy from the files of William E. Brown. See also William E. Brown, *A History of the Denali-Mount McKinley Region, Alaska* (Santa Fe, New Mexico: National Park Service, 1991), p. 198.

49 Horace M. Albright, "The National Park Service's Policy on Predatory Mammals," *The Journal of Mammalogy* 12 (1931): 185. Quotes from the 1932 superintendents' conference are found in National Park Service, "Policy on Predators and Notes on Predators" (1939), various pagination, typescript, Central Classified Files, 715, RG79.

50 National Park Service, "Policy on Predators and Notes on Predators."

51 Wright, Dixon, and Thompson, *Fauna of the National Parks*, p. 147.

52 The quote is found in National Park Service, "Policy on Predators and Notes on Predators."

53 Wright and Thompson, *Fauna of the National Parks*, p. 71.

54 Curtis K. Skinner to Dr. Mills, 12 March 1935, YELL.

55 Frank W. Childs, "Report on the Present Status of Wildlife Management in Yellowstone National Park With Suggested Recommendations for Future Treatment," 19 April 1935, YELL. There was also interest among Yellowstone's staff in restoring some of the park's extirpated species. Naturalist Assistant Harlow B. Mills wrote to Ben Thompson in 1935 that,

As a policy I can see no great obstacle in the way of our, at least, attempting the introduction of cougar and wolves into the Park. They were a vital part of the picture at one time, a picture which can never be the same in the Park in their absence. This should be done, I realize, with considerable forethought and care, but I believe that it should be done, nevertheless.

Harlow B. Mills to Ben Thompson, 21 June 1935, Entry 34, RG79. Such interest would have been in accord with the recommendations of Fauna No. 1 that "any native species which has been exterminated from the park area shall be brought back if this can be done. . . ." See Wright, Dixon, and Thompson, *Fauna of the National Parks*, p. 148.

56 Murie, *Ecology of the Coyote*, p. 16; Sumner, "Biological Research and Management," 14.

57 C. A. Henderson to David Canfield, 21 November 1935; and David Canfield to C. A. Henderson, 30 November 1935, Entry 34, RG79. Victor H. Cahalane, "Evolution of Predator Control Policy in the National Parks," *Journal of Wildlife Management* 3 (1939): 236.

58 David Madsen, Memorandum for The Director, 20 May 1939, Entry 36, RG79. See also Susan R. Shrepfer, *The Fight to Save the Redwoods: A History of Environmental Reform* (Madison: The University of Wisconsin Press, 1983), pp. 61–63.

59 Joseph Grinnell to Arno B. Cammerer, 10 April 1939, Central Classified File, RG79.

60 Horace M. Albright to the Director, National Park Service, 18 October 1937, Central Classified Files, RG79.

61 Murie, *Ecology of the Coyote*, pp. 146–148.

62 Thomas Dunlap, in *Saving America's Wildlife*, p. 75, indicates that some Park Service officials "wanted to fire" Murie. Alston Chase, in *Playing God in Yellowstone: The Destruction of America's First National Park* (Boston: The Atlantic Monthly Press, 1986) pp. 126–128, describes the "fierce Park Service resistance" which Murie faced during the coyote controversy. Sumner, in "Biological Research and Management," p. 15, recalled that, following the coyote study, "Murie's findings, and his personal concepts of ecological management of park resources, continued to be unpopular in various administrative circles." Given, however, that Murie was very soon assigned to a similar study of wolves in Mount McKinley National Park, it is clear that he had support in high places, very likely from Director Cammerer himself.

63 Horace M. Albright to A. B. Cammerer, 11 January 1939, Central Classified Files, RG79; *Annual Report of the Secretary of the Interior* (1939), p. 282.

64 Murie, *Wolves of Mount McKinley*, xiii-xv; Albright to Cammerer, 11 January 1939. Murie's wolf study is discussed in Brown, *A History of the Denali-Mount McKinley Region, Alaska*, p. 198.

65 Madsen to the Director, 20 May 1939.

66 Wright, Dixon, and Thompson, *Fauna of the National Parks*, p. 63.

67 David H. Madsen, "A National Park Service Fish Policy," (ca. early 1930s), typescript, Entry 36, RG79; and Madsen, "Outline of a General Policy of Handling the Fish Problem in the National Parks," 10 May 1932, typescript, Central Classified File, RG79. The records do not indicate whether Madsen was first detailed to the Park Service in 1928 or in the early 1930s.

68 John D. Varley, "Record of Egg Shipments from Yellowstone Fishes, 1914-1955," Yellowstone National Park, Information Paper No. 36, May 1979, YELL.

69 David H. Madsen, "Report on Fish Cultural Activities," 5 April 1935, Central Classified File, RG79.

70 David H. Madsen to Arno B. Cammerer, 6 October, 1933, Central Classified File, RG79.

71 Wright, Dixon, and Thompson, *Fauna of the National Parks*, pp. 148, 63.

72 Arno B. Cammerer, Office Order No. 323, 13 April 1936, Entry 35, RG79.

73 Cammerer, Office Order No. 323, 13 April 1936; *Annual Report of the Secretary of the Interior* (1936), p. 124.

74 John D. Varley, "A History of Fish Stocking Activities in Yellowstone National Park Between 1881–1980," Yellowstone National Park Information Paper, no. 35, 1 January 1981, typescript, 9, 13, 17, 19, 21, 26, 52–53, YELL. The stocking of Mammoth Beaver Ponds took place in 1936, quite possibly in the months after the park had received the new fish policy, issued by Cammerer in mid-April of that year. In the case of McBride Lake, also in the Yellowstone drainage, exotic rainbow trout were introduced in 1936, where previously only native cutthroat trout existed. Varley, "History of Fish Stocking," 17.

75 Varley, "Record of Egg Shipments"; *Annual Report of the Secretary of the Interior for the Year Ending June 30, 1937* (Washington, D.C.: GPO, 1937), p. 44. As another example of fish production and shipment during the 1930s, the collection of approximately sixty million trout eggs in one year from several unspecified national parks, with about half of them being shipped to various states, is mentioned by Cammerer in *Annual Report of the Secretary of the Interior* (1936), p. 124.

76 Carl P. Russell, "Opportunities of the Wildlife Technician in National Parks." Paper presented at the North American Wildlife Federation conference, St. Louis, Missouri, 1 March 1937, typescript, HFLA. Victor H. Cahalane, "Thoughts on National Park Service-Bureau of Fisheries Agreement," draft, 4 August 1939, Entry 36, RG79. Cahalane accepted that the Service would continue its dependency on other agencies for fish culture work. And Director Cammerer had reported in 1937, the year after the new fish policy was issued, that cooperation was closer "than ever before" between the Service and the Bureau of Fisheries and state game departments. Cooperation became even closer in 1940, with the transfer of the biologists to the Bureau of Biological Survey and its subsequent merger with the Bureau of Fisheries. Cahalane, "Thoughts on National Park Service-Bureau of Fisheries Agreement"; *Annual Report of the Secretary of the Interior* (1937), p. 44.

77 John C. Paige, *The Civilian Conservation Corps and the National Park Service, 1933–1942: An Administrative History* (Washington, D.C.: National Park Service, 1985), appendix A, p. 162. The National Park Service Act authorized the Service to "sell or dispose of timber in those cases where . . . the cutting of such timber is required in order to control the attacks of insects or diseases or otherwise conserve the scenery. . . ." Hillory A. Tolson, *Laws Relating to the National Park Service, the National Parks and Monuments* (Washington, D.C.: U.S. Department of the Interior, 1933), p. 10.

78 *Annual Report of the Secretary of the Interior* (Washington, D.C.: GPO, 1933), p. 157.

79 John D. Coffman, "John D. Coffman and His Contribution to Forestry in the National Park Service," n.d., pp. 36–39, typescript, HFLA. Because of the CCC's heavy emphasis on forestry, Coffman was also given the huge responsibility for overseeing CCC operations within the national parks. In 1936, however, the director consolidated oversight of these operations with the Service's state parks assistance program (also funded by the CCC). Assistant Director Conrad L. Wirth supervised this expanded office combining all CCC-related national and state park work, leaving Coffman free to concentrate on directing forestry management in the parks, which continued to rely on CCC manpower and money. See Coffman, "John D. Coffman and His Contribution to Forestry," p. 44; Conrad L. Wirth, *Park, Politics, and the People* (Norman: University of Oklahoma Press, 1980), p. 118; and Paige, *Civilian Conservation Corps*, pp. 39–40, 48.

80 "A Forestry Policy for the National Parks," approved by Horace M. Albright, 6 May 1931, typescript, Entry 18, RG79.

81 Stephen J. Pyne, *Fire in America: A Cultural History of Wildland and Rural Fire* (Princeton: Princeton University Press, 1982), p. 300.

82 *Fauna of the National Parks*, p. 33.

83 U.S. Office of National Parks, Buildings and Reservations, "Instructions for Superintendents of Eastern National Park ECW Camps and CW Projects Concerning Roadside Clean-up, Fire Hazard Reduction, Brush Disposal," Chapter IX, 3, Supplement No. 7 to *Forest Truck Trail Handbook* (Washington, D.C.: U.S. Forest Service, 1935); George M. Wright to the Director, 28 February 1934, Central Classified File, RG79.

84 Adolph Murie, memorandum for Ben H. Thompson, 2 August 1935, Entry 34, RG79.

85 L. F. Cook, memorandum for the Chief Forester, Reply to Dr. Murie's report on the Glacier National Park Cleanup Project, 28 August 1935, Entry 34, RG79.

86 L. F. Cook, memorandum for the Chief Forester, Re: Criticism of Forestry Recommendations by Other Technicians, 28 August 1935, Entry 34, RG79. Personal communication with Bruce Fladmark, Glacier National Park, August 1991, regarding clearing in the McDonald Creek area.

87 Murie to Thompson, 2 August 1935.

88 Cook to Chief Forester, 28 August 1935. In Cammerer's 1939 annual report, the director discusses fire prevention and fire protection work undertaken with CCC funds and enrollees. *Annual Report of the Secretary of the Interior* (Washington, D.C.: GPO, 1939), pp. 272–275.

89 *Annual Report of the Secretary of the Interior* (1933), 180–181.

90 George M. Wright to Arno B. Cammerer, 1 August 1935, Entry 35, RG79; Victor H. Cahalane to A. E. Demaray, 23 September, 1935, Entry 34, RG79. For comments on CCC involvement in insect and disease control see Paige, *Civilian Conservation Corps*, pp. 101–103.

91 Adolph Murie to George M. Wright, 26 March 1935, Entry 34, RG79. Similar statements regarding insect control are found in biologist Harlow B. Mills' letter to Ben Thompson, 21 June 1935.

92 Cook to Chief Forester, 28 August 1935; and *Annual Report of the Secretary of the Interior* (1939), pp. 272–274. For similar comments made earlier by Cammerer, see *Annual Report of the Secretary of the Interior* (1937), pp. 42–43.

93 Sumner, "Biological Research and Management," p. 13.

94 The utilitarian aspects of the National Park Service Act and the act's ramifications for national park management are discussed in Richard West Sellars, "The Roots of National Park Management: Evolving Perceptions of the Park Service's Mandate," *Journal of Forestry* 90 (1992): 16–19; and Richard West Sellars, "Science or Scenery? A Conflict of Values in the National Parks," *Wilderness* 52 (1989): 29–38.

95 Russ Olsen, *Administrative History: Organizational Structures of the National Park Service, 1917 to 1985* (Washington, D.C.: National Park Service, 1985), p. 63. Under Coffman, the Park Service also provided considerable training in forest protection, including techniques in fire, insect, and disease control. In many parks, rangers, park naturalists, and maintenance staffs all received this training. John W. Henneberger, "To Protect and Preserve: A History of the National Park Ranger," 1965, typescript, unpublished manuscript, copy courtesy of the author, p. 307.

96 Tom Ela, interview with the author, 26 January 1989; Arthur Wilcox, interview with the author, 17 March 1992.

97 As an example of the growing strength of the forestry programs, a list of one hundred thirty-seven professionally trained foresters in the National Park Service by 1952, shows most of them in key positions. Robert N. McIntyre, "A Brief History of Forestry in the National Park Service," March, 1952, Appendix A, typescript, BL. About eight wildlife biologists were transferred back into the Park Service around the end of World War II, yet Sumner later recalled that Fauna No. 1 itself became "forgotten." Moreover, the number of biologists did not increase until the 1960s. Sumner, "Biological Research and Management," 16–17, 19.

RICHARD A. SKOK

FOREST SCIENCE AT UNIVERSITIES IN THE UNITED STATES

INTRODUCTION

Scholarly inquiry organized along disciplines is a central mission of United States universities. Such inquiry or "research" provides an integral part of university educational programs and at the same time furthers our scientific base in a resolute yet often unforeseeable manner. Forestry research relies on many scientific disciplines because of the integrative nature of forestry. Thus, forestry research at United States universities has emanated from diverse sources but primarily owes its identifiable, substantive history to the organized efforts of the faculties of forestry schools and the agricultural experiment stations of those universities.

Forestry research encompasses the breadth of studies directed at increasing our understanding of both forests and forestry for the full array of uses and benefits our forest resources offer. The need for multidisciplinary approaches has increased as the complexity of the issues that forestry addresses has grown. The application of science from basic disciplines to forestry needs over the past century has characterized much of the research done at universities. This in no way diminishes the value of these efforts but rather focuses attention on the problem-solving strategies that forestry faculties have utilized.

This report focuses principally on the forest science contributions that the schools of forestry and the agricultural experiment stations at American universities accomplished during the twentieth century. Much of this information reflects the emergence of major research activity since the 1950s. Prior to that time, forestry faculties were small in numbers, carried heavy instructional responsibilities, and generally lacked significant research funding and facilities.

THE UNIVERSITY MISSION

Undergraduate programs in forestry typically provided the initial impetus for establishment of forestry faculties at universities. Beginning in 1898 at Cornell University, forestry programs designed to educate professionals for resource management in the United States grew to degree programs accredited by the Society of American Foresters being offered at forty-seven schools of forestry by 1995.[1]

Developing the human resources through graduate programs essential for the conduct of forestry research may well be the most important contribution of universities to the nurturing of forest science over the past century. Graduate programs in forest science emerged during the twentieth century at many of these institutions and immensely benefited from the linkage to active research. During the early years of this century, many of these were graduate programs authorized through the agriculture colleges in which forestry and related fields most often were located. As forestry programs grew after the mid-1900s, some gained greater independence while retaining close linkages to related fields in agriculture. Forest pathology, forest entomology, and forest soils are examples of graduate fields in forest science that generally have retained their traditional alliances to departments within agricultural colleges.

Research is a recognized mission of university faculties. It forms an essential component of graduate education and a recognized means to maintaining faculty capabilities and involvement in relevant science as well as practical issues of the times. Broadly, university forestry research has had a mission of both building basic knowledge in forest science as well as solving problems of importance to the state and national forestry communities.

Forestry programs at institutions that are recognized graduate-research universities of this nation have been best positioned to make substantial contributions to forest science. In general, they have had better access to: 1) the supporting disciplines important to comprehensive graduate and research programming; 2) the facilities and equipment that have become increasingly crucial to forest science; and 3) the sustainable core forestry faculty resources essential to both the research and instructional missions.

Where forestry programs were associated with land grant universities, the extension service system provided a means of moving research results from the academic community to those in need of this knowledge within a state. The Smith-Lever Act of 1914 provided the federal authorization for the establishment of this service in each state. Matching state funds were required for the federal formula funds distributed annually to the states. More recently the national Forest and Rangeland Renewable Resources Extension Act of 1978 has proven a catalyst to enhance the funding of outreach in the natural resources program area. Slightly more than five hundred full-time equivalent extension personnel nationally were funded in forestry and related fields in 1985.[2] This outreach mission is another important contribution to forest science made by U.S. universities.

THE UNIVERSITY SETTING FOR FOREST SCIENCE

Beginning in the latter part of the nineteenth century, agricultural programs at universities often served as "incubators" for the emergence of forestry research at universities. The land grant university established in each state under the Morrill Act of 1862 most often served as location for such programs. These universities were created to assure that the benefits of the "agricultural and mechanical arts" were accessible to the broadest possible segment of the nation's people.

Whether it was research to address increasing concerns such as disease and insect epidemics, or the deforestation of large forest areas through exploitation or in failed attempts to convert forests to other land uses, or the farm woodlot as an integral part of many farming operations, or simply the desire to seek expanded economic opportunities from the local timber resource, the outcomes were similar. Limited attention and resources were focused on research programs at universities in those states where such forestry issues were of concern to leaders with vision and commitment.

Contributing to emergence of a small but focused research effort on forestry issues at the universities was the federal Hatch Act of 1877. This established the present system of agricultural experiment stations at land grant universities. Formula funding to the individual states authorized by this act required matching nonfederal monies. Forestry research, considered as one of the agricultural sciences, fell under the purview of these stations. Despite this, directors of these stations were generally reluctant to allocate many of the funds they controlled to forestry matters. Illustrative of this, less than 2.5 percent of Hatch funds were allocated to forestry and related research as late as fiscal year 1960.[3] For earlier years the share was even less.[4] This lack of funding fueled a long smoldering concern among the expanding forestry school community in the 1950s. Efforts were begun to seek federal funding directed specifically to forestry research needs.[5]

This lack of research funds severely limited university forestry scientists until the advent of the McIntire-Stennis Cooperative Forestry Research Act. Passed by Congress in 1962, it authorized federal formula funding to the states to be used at public universities that had graduate degree programs in the forest sciences. There are currently sixty-two member institutions participating in the McIntire-Stennis Research Program. Matching non-federal funds were required. With this impetus, the number of forestry scientists at eligible institutions nearly doubled to 860 by 1984 from the initial appropriation in 1964.[6]

FUNDING FOREST SCIENCE AT UNIVERSITIES

Non-federal funding has been the most important category of support for forestry research over the years. It has accounted for approximately two-thirds of the forestry research dollars at universities in recent years. State funds have been the largest component of non-federal support for forestry research over the years at universities.[7] A significant portion of these have been provided through agricultural experiment station appropriations. These state funds, along with the federal funds through the McIntire-Stennis program and to a lesser degree through Hatch funding, have provided the backbone of support for forestry research over the years. They have been considered the "hard funding" that could be counted on for long term research support essential to many aspects of forest science.

Competitive grant funding, principally through research programs in the U.S. Department of Agriculture (USDA), grew significantly in the past decade as a source of support for university forestry research. These funding sources are characterized as having targeted research subjects and with grants awarded for three years or less on the basis of peer review competition. Such "soft funding" approaches have had increasing appeal from appropriating authorities.

Universities have received important research support over the years through cooperative agreements with federal and state agencies. The Forest Service has been a principal source of such agreements with up to ten percent of their appropriated research funding utilized for collaborative research at universities in recent years.[8] Such agreements have generally targeted research needs of such agencies where they have not had personnel or facilities available to address specific problems. They at times have been an important mechanism to focus on development of advanced degree graduates in emerging scientific fields to meet future research agency needs.

Forestry research has also received support at selected universities through the establishment by the Department of the Interior (USDI) of cooperative fish and wildlife research units and national park research units. These units provide both on campus federal scientists with faculty status and opportunities to access funding for research by faculty and graduate students on designated priority research issues of these agencies. All fifty-six USDI cooperative research units, including those associated with forestry research programs at universities, were combined into the National Biological Service in 1993.[9]

Funding through foundations, industry, and other private sources also has contributed to the overall university research effort in the forest sciences. Historically, these have been of particular importance to the private universities with forestry programs. More recently, public forestry schools have availed themselves of such sources as their parent institutions have become more aggressive in the private development funding arena.

Forestry research cooperatives have been used as a means for combining both private and public funding for targeted research initiatives. Tree improvement and growth and yield cooperatives have been among the most common of these undertakings.[10] A number of universities have participated in forestry research cooperatives and generally have provided the leadership for the partnerships. This approach combined the scientific and technical talents of multiple organizations by focusing them on the specific research needs of forest resource managers for whom such cooperatives were organized.

This brief exploration of funding for forest science at universities indicates a complex and continually evolving network of sources that have been used. Despite this, evidence exits that the funding base has been considerably short of what would be necessary to adequately address the pressing forestry issues through research and actually has been declining in real dollars as well over the past decade.[11]

This plant physiologist works with a microtome to cut slices from test samples of chemically treated trees, in preparation for microscope study. The research involved experiments with chemicals that cause standing trees to shed their bark, in an effort to lower the cost of finished forest products. The State University of New York College of Forestry at Syracuse and a group of northeastern forest industries collaborated on the project.
(AMERICAN FOREST PRODUCTS INDUSTRIES PHOTO,
IN THE FOREST HISTORY SOCIETY COLLECTIONS)

RESEARCH PROGRAM PLANNING, COORDINATION, REPORTING, AND REVIEW

Forestry research support through significant public funding has presupposed accountability. Forest Service and university reliance on federal research funding appropriated through the USDA has had oversight through systematic research reporting as well as planning and coordination efforts. University forestry programs have been full partners in these efforts through a regional and national network representing the participating research organizations.

The most comprehensive of these efforts occurred over the period from 1976 to 1978 in response to a request from the USDA Agricultural Policy Advisory Committee for review of the conduct and content of forest and associated rangeland research in the United States. The resultant regional/national plan evolved from participation of nearly one thousand research users and scientists.[12] The plan served as a basis for forestry research program direction at both the national and local level during the following decade. The most recent widely distributed forestry research plan was a collaborative effort involving the forestry schools, the Forest Service, and Cooperative State Research, Education and Extension Service (CSREES). It presented both a national and regional research program for the decade of the 1990s.[13]

A Cooperative Forestry Research Advisory Council established by the McIntire-Stennis Act regularly advises the secretary of agriculture on the plans and accomplishments of this program. The council is comprised of representatives from forest industries, public forestry agencies, non-governmental groups, and forestry schools.

Individually, university forestry research programs that receive funding from the McIntire-Stennis or Hatch Act receive periodic comprehensive reviews by a visiting team of scientists and program managers. Institutional program reviews are provided to enhance the quality and responsiveness of research activities as well as to avoid unnecessary duplication of effort. These reviews are voluntary and, when requested, are organized by CSREES. Nearly all university programs receiving such funds participate voluntarily in this review activity.

All university forestry research programs receiving federal funds from programs administered through the USDA must report annually by project the progress of work done. The information is recorded by project in the Current Research Information System (CRIS) and is accessible to those who wish to obtain the reports. CRIS maintains information on the entire system of agricultural and food sciences administered through the USDA. Normally, all forestry research projects at a forestry school regardless of the source of the support funding are reported in CRIS annually. Additionally, the forestry schools and colleges have in collaboration with CSREES periodically published summaries of research progress at universities.[14]

CONTRIBUTIONS TO FOREST SCIENCE

Identifying university contributions to forest science in a brief manner is challenging because of the many facets and complexities that contribute to scientific outcomes. Tracing the network of people, ideas, and studies that produce scientific results of significance to forestry is not unlike piecing together a jigsaw puzzle. The selected examples discussed below are not intended to be an exhaustive list.

In general, much of the research conducted in the nation's university forestry programs has provided knowledge useful in answering problems and addressing opportunities surrounding the use and protection of forests ranging from local to global in nature. To do so has meant commitment both to long term studies as well as maintaining the flexibility to respond to changing issues with limited resources.

Tree Improvement and Genetics Research

Research and application of genetics to tree improvement provides one of the best examples where science and forestry have been clearly focused on long term goals associated with forest tree productivity. Historically, the utilization of timber from forests of the United States was almost exclusively dependent on the trees from natural forests. Realizing that successful methods for increasing productivity of annual crops and animals in agriculture might be applied to trees, some farsighted forest scientists were encouraged to pursue genetic selection as a silvicultural approach. University scientists at several institutions have made important contributions both to the science of forest genetics, its application to tree improvement, and in the education of individuals who have been crucial over the long term in conducting many aspects of the tree improvement programs in this nation.

Nothing better illustrates this contribution than the southern pine tree improvement work. The history of these efforts are traced in a very readable manner by Zobel and Sprague.[15] The challenges were many at the outset. Among these were the: 1) lack of basic scientific knowledge about the biology of forest trees; 2) long time periods that would be required to yield uncertain results given the nature of tree growth and maturation; and 3) need for sustained support and funding that would be required to be successful in tree improvement work. Universities played a pivotal role in developing and operating the partnerships needed to confront these barriers.

These partnerships took the form of tree improvement cooperatives. They typically involved forest industry firms, state forestry agencies, and universities. Individual companies provided funding and, in some cases, land or personnel to support the work of the cooperatives. A university provided the direction for research and applications operations of the cooperative as well as the central research personnel including faculty, staff, and graduate assistants. In addition, the facilities for office needs and conduct of much of the research as well as some funding were provided by the university.[16]

Early cooperative efforts for southern pine tree improvement began in an organized and sustained manner in the 1950s. Three cooperatives with particular significance in the development of southern pine tree improvement were those located at North Carolina State University, Texas A&M University, and the University of Florida.

The first of these cooperatives was established at Texas A&M University in 1951. Professor Bruce Zobel was hired to serve as its head. Zobel's work in tree improvement has spanned more than forty years. He began at Texas A&M, but moved to North Carolina State University in 1956 to head its newly begun tree improvement cooperative program. He is considered by many as the "father of southern pine tree improvement."[17]

Work of these three university-based cooperatives emphasized seed selection from parent trees with desired traits. Rate of growth, form, wood density, and resistance to forest pests were among traits considered. Seed orchards were established, and the long process of testing these seed sources followed. By 1987, these three cooperatives accounted for 8,350 acres of seed orchards with genetically-improved forest trees, producing 1.3 billion seedlings annually and resulting in the planting of 1.8 million acres annually.[18] Although not all of this tree improvement impact was for southern pine, the largest share was. Significant gains in tree and stand volume for loblolly and slash pine are demonstrated in various studies over the years.[19] The economic viability of southern tree improvement has also been shown in a number of analyses.[20] Thus, the contributions of these university-based cooperatives and their research and application to southern pine forest productivity is clear.

The various cooperatives served as the fertile training ground for both the science and application of tree improvement work for a large number of graduate students over the past forty years. As an example, Professor Zobel cites his work as either major adviser or a member of the graduate committee for more than one hundred seventy graduate students from 1951 to 1990.[21]

Both students from the United States as well as those from other countries participated in tree improvement graduate studies and research. Many of the latter subsequently returned to their homelands, often developing nations, and utilized the scientific knowledge gained at American universities to further forest tree productivity in their native countries. The long term compounding effects on forest science of this educational role of the universities is not readily measured but has profoundly influenced tree improvement globally and will continue to do so into the foreseeable future.

Biotechnology has emerged in the past decade as a potent opportunity for advancement in the field of tree genetics. Although traditional tree improvement efforts have contributed significantly to the nation's timber productivity, tissue culture and genetic engineering are being utilized by forest scientists at several universities. The expectation is that these techniques will decrease the time required to achieve desired genetic responses as well as increase the range of response that can be more readily achieved.[22]

Structural Particleboard Research

Forest products utilization research has provided the technology for extending the commercial timber resource base, thereby providing for consumer needs while insuring jobs and economic security for many communities nationwide. This role is an important one for research as the nation has faced a stable or declining forest land base and at the same time an increasing demand for products and services from these lands.

Structural particleboard is an excellent example of how a concentrated and collaborative effort among private and public organizations can lead to the development of a new and important wood product industry in North America utilizing what had been a relatively low-valued wood resource. University research efforts played an important role in this endeavor which essentially spanned three decades beginning in 1950.[23]

The oriented strandboard (OSB) segment of the structural panel industry has seen spectacular development in recent years. OSB is a substitute for plywood in construction. From 1976 to 1992 North American production of OSB increased nearly twelve-fold reaching eleven billion square feet (³/₈ in.).[24] Several reasons can be cited for this impressive growth. The availability of low-valued timber sources, e.g. aspen, suitable for OSB along with the increasing cost of veneer logs used for plywood, particularly in the West, provided favorable economic conditions for OSB. Coupled with this was the timely emergence of the technology and research basic to overcoming obstacles to both OSB production and marketing.

In a review of twelve hundred research articles identified from the scientific literature in English over a thirty plus year period ending in the early 1980s and deemed to have contributed to the development of structural particleboard panels, university research in North America accounted for approximately twenty percent of these publications.[25] Universities were found to have been particularly important scientific contributors to research on raw materials and the mechanical/physical properties related to the production of particleboard. Forest products industry representatives considered public sector structural particleboard research to have been particularly beneficial in developing the basic scientific knowledge needed, in providing means for transfer of research results as well as in encouraging new technologies and research approaches, and in development of the necessary human resources complement essential to the innovation and application required for structural board industry success.[26]

Haygreen, et al.[27] identified U.S. universities and their particular research contributions over an extended period as follows:

Washington State University has been a leader in particleboard research and in technology transfer efforts since 1960. They sponsored an annual symposium starting in 1967 which brought together researchers, equipment suppliers, and industry representatives from around the world. Waferboard research at the University of Minnesota has focused on raw material and process variables. At Purdue University, technology was perfected to make thick NSP (new structural panels) with a steep density profile. At Michigan State University, important fundamental research provided an understanding of the relations of furnish type to board properties. At Oregon State, efforts were directed at improving durability, evaluating test methods, and improving resin systems. At Auburn, research evaluated the properties of a variety of veneer and particleboard composites and studied their mechanical properties. At the University of California, pioneering research was done on fundamentals of bonding reactive wood surfaces without synthetic resins.

The growth of research activity from 1950 through 1985 shows a strong chronological correlation with the development of North American OSB industry capacity.[28] Estimated internal rates of return for structural particleboard research range from nineteen to twenty-two percent over an assumed twenty-one year lifetime of the investment.[29] This case of public and private collaborative forest products research and development efforts clearly illustrates an exceptionally productive outcome in which several U.S. universities were important players.

Forest-wildlife Interactions

Forests serve as an important habitat for a large number and variety of wildlife in the United States. Forest ecosystems meet the requirements of different wildlife species not only through the variation in tree species but through the differing structure of forest stands due to densities and age. As a result, changes in a forest either through natural causes or through deliberate human intervention often can impact the population levels of a given wildlife species or the type of wildlife species found in a "changed" forest habitat. Understanding the impacts associated with changes in forest conditions has been of increasing interest to university scientists both in wildlife and forestry. Studies originally focused most heavily on forest game species such as whitetail deer and ruffed grouse. More recently they have accelerated on non-game species as concerns over threatened and endangered species have become more widely recognized.

Ruffed grouse and its forest environs have been the subject of long term studies that have received consistent support by a small group of university researchers and those interested in grouse as a game bird. Aspen forests have been found to be a particularly important habitat for ruffed grouse. Much of the research and the education on the relationship between ruffed grouse and aspen forests results from work done by the late Professor Gordon Gullion and others at the University of Minnesota over a thirty year period.[30]

Aspen is widely distributed in North America being found in Alaska, thirty-nine of the forty-eight contiguous states of the United States, and all of the provinces of Canada.[31] In the United States, the Lake States accounted for the greatest concentration of aspen forests with nearly twelve million acres of timberland in 1987.[32] Aspen, once viewed as an undesirable species because it had little commercial value, has become a very important component of timber utilization in the this region. Pulpwood production of aspen more than doubled to reach in excess of four million cords during the period from 1962 to 1986 across the northern tier of states in the eastern United States and has continued to grow.[33]

The ruffed grouse association with aspen forests meant that studies focused on understanding the ecology of the grouse have necessarily involved aspen forests as a prime habitat component. Aspen is both an important food source for grouse and critical as a protective cover.[34] An important outcome of such studies has been the development of guides for silvicultural practices designed to benefit the ruffed grouse population as well as the importance of aspen as a commercial timber source for many mills in rural communities. Studies also have been done at Minnesota[35] and Pennsylvania State University[36] to ascertain the impact of nongame wildlife on habitat management for ruffed grouse.

Aspen clearcuts of twenty acres or less and with the maintenance of patches of several age classes of differing male aspen clones within a relatively close distance of one another were found to be significant in their positive effect on the density of ruffed grouse.[37] Aspen is an intolerant species. It was established in the Lake States on a broad scale through the extensive logging and fires in the latter 1800s and early 1900s. Thus, clearcuts are an important harvesting practice to perpetuate such a forest type. Such small scale clearcuts, important to grouse management while consistent with timber objectives, were proposed by Gullion at a time when clearcutting had become of significant public concern. Silvicultural guidelines were developed based on these recommendations and have been adopted by a number of forestry organizations. Properly implemented they have helped alleviate the concerns of the public over extensive clearcuts, provided desired ruffed grouse habitat, and enabled the regeneration of aspen on productive sites.

Adapting New Technologies to Forestry

To meet efficiently and effectively the needs of forestry through research often has meant borrowing new basic technology developed in other scientific fields. Universities have been active contributors in such efforts with many faculty and most forestry schools involved to one degree or another over time.

The use of aerial photography, remote sensing techniques, and satellite imagery to address, in more cost-efficient manners, the extensive and often remote character of forest resources is well known. Applications have ranged over a broad spectrum of problems from land use classification to early detection of insect and disease infestations. Much of this basic technology was developed initially for military uses and adapted to forestry uses through research and development as it became available for non-military use. Laboratories associated with forestry programs at several forestry schools have been particularly involved in this line of application. Among these universities are Purdue, Wisconsin-Madison, California-Berkeley, Minnesota, Michigan, and Georgia. More recently, the linkage of computers to remote sensing techniques for geographic information systems applications in resource management reflects further extension of this principle.

The complex nature of forest resource systems involving biological, physical and socioeconomic interactions has led to the adaptation of computer modeling to enhance our understanding of these relationships and the likely outcomes under various scenarios.[38] Contributions have ranged from studies of timber supply and demand to hydrologic patterns of forest watersheds to modeling of long term outcomes on ecosystems from varying forest practices. In the modeling of such systems, the need for basic knowledge upon which models are built has become increasingly apparent. The capability of computers has generally outpaced our understanding for model building but in turn has motivated efforts to achieve the greater basic scientific knowledge needed.

Forest scientists were involved early in adapting computer modeling to needs of forestry. Carter, reporting forestry school progress, noted:

> Simulation models for pine and Douglas fir plantations have been developed in Georgia, Virginia, California and Washington. While scientists in Missouri, Wisconsin and Indiana have developed mathematical models of hardwood forests. In Michigan and Minnesota, scientists have developed economic analysis techniques which greatly assist the forest landowner in optimizing the return from his investment. Most of these programs consider not only timber production, but wildlife, recreation, water yield, aesthetic and other values as well. . . .[39]

These early efforts have been vastly strengthened as our basic understanding of the underlying forest systems and the capabilities of computing have improved.

FOREST SCIENCE FOR THE PROFESSION

As forestry emerged as a profession to protect and manage the forest resources of this country a century ago, it was necessary to borrow heavily from the central European experience and knowledge on this subject. The knowledge base regarding American forest characteristics and conditions was very limited. Among the results of this was an absence of teaching materials available to forestry faculty to meet the educational requirements of students preparing for work in America as professional foresters. This presented a daunting handicap in the early years of professional forestry education in the United States. It also served as a strong motivation to undertake research to begin to fill this void.[40]

*One of Yale University's earliest Forest School classes—the Class of 1905.
Students and professors, including Dean Henry S. Graves (standing,
back row at far left) take a break from their field camp exercises at the
Vantine House, Milford, Pennsylvania.*
(FOREST HISTORY SOCIETY PHOTO)

Forestry faculty were small in number in the new forestry schools and
often had limited or no research experience. Some, however, made important contributions in the form of textbooks they authored. Evaluating, syn-
thesizing, and integrating existing and new scientific knowledge relevant to
forestry along with that gained through professional experience relevant to
forestry in America was crucially important to improving professional forestry education in the nation.[41] These texts also provided the foundation for
those students who were to pursue graduate education in specialized fields
important to forestry. They represented the new generation of scientists and
teachers crucial to the growth of forestry in this nation in the period from
the 1920s through the 1960s.

Texts on the subject of forest measurements and mensuration have a long lineage of distinguished faculty including Henry Graves and H. H. Chapman of Yale University whose works spanned a half century and are continued today by their successors. Silvicultural texts with periodic edition updates spanned over years from 1921 and were classics of their time. These were authored by authorities beginning with Hawley of Yale and continued by Baker of the University of California and Westfeld at the University of Missouri.

In the field of forest economics and policy, faculty members again were leaders in harnessing these disciplines of the social sciences in their application to forestry. Vaux and Josephson observe,

> A major contribution to the training of forest economists (and of others) was the publication of a number of college level textbooks on forest economics and forest policy. Those by Worrell, Duerr and Gregory have been widely used and have added much to the definition of the field of forest economics.[42]

All these authors were distinguished faculty at well recognized U. S. forestry schools.

These examples as well as those that one could trace in fields such as forest entomology, forest pathology, forest products, wood utilization, dendrology, and forest management indicate the pervasive and long term nature of this type of contribution in forest science to the profession and practice of forestry. Particular publishers fostered this effort over the years. The McGraw-Hill forestry series, begun in 1931 and continuing since, has covered a wide range of forestry topics and prominently features faculty authors. Other important publishers who have contributed a significant lineage of forestry texts over the years include the Ronald Press, Macmillan, and John Wiley and Sons.

EDUCATION OF FOREST SCIENTISTS

Forestry education in the United States has shown a distinct move toward greater emphasis on graduate degree programs over the past half century. Such programs have been an important source of forest scientists for both the public and private organizations of the nation engaged in research in forestry and forest products.

Enactment and funding of the McIntire-Stennis Cooperative Forestry Research Program in the early 1960s was an especially important stimulus to forestry graduate programs. The legislation included among its objectives enhancement of graduate education in the forest sciences as well as research. As noted by USDA Assistant Secretary for Science and Education Bentley on the occasion of the 25th anniversary of this act,

> Through the McIntire-Stennis [Act] , thousands of young men and women have received advanced degrees . . . who will continue the efforts to both use and protect our natural resources. . . . [this] ensures a continuous infusion of intellectual resources and future scientists into the forestry research community.[43]

Doctoral degrees granted through forestry school programs are the clearest available measure of this contribution to forest science by universities. During the period 1900 to 1960, a total of 575 doctoral degrees in forestry and related fields were granted in the United States through these programs.[44] In 1990 alone 192 such degrees were granted.[45]

With the increasing complexity of forestry problems addressed by research as well as the growing sophistication of the science required, scientists from basic disciplines have been increasingly recruited to forestry research organizations. These recruits have supplemented the degrees awarded through programs offered by forestry schools, particularly over the past twenty years. This trend will likely become even more evident in the years ahead. It is also reflected in the increasing number of graduate students in forestry graduate programs who have not had previous exposure to academic studies in forestry.[46] Both changes are consistent with a part of the findings of the Committee on Forestry Research established by the National Research Council in 1989.[47] They already have been reflected in growing diversity in both the education content and research agendas of most university forestry programs.

NOTES

1 Anon., "Institutions with SAF-Accredited Curricula," *Journal of Forestry* 93 (1995): 47.

2 National Research Council, *Forestry Research: a Mandate for Change* (Washington, D.C.; National Academy Press, 1990), p.23.

3 R. H. Westfeld, "Opportunities for Research and Graduate Education in Forestry," *Journal of Forestry* 61 (1963): 420.

4 Frank H. Kaufert and William H. Cummings, *Forestry and Related Research in North America* (Bethesda, MD: Society of American Foresters, 1955), pp. 64, 65.

5 R. H. Westfeld, "Opportunities for Research and Graduate Education in Forestry."

6 "A Quarter Century of Progress: the McIntire-Stennis Cooperative Forestry Research Program, 1962–87" (Washington, D.C.: National Association of Professional Forestry Schools and Cooperative State Research Service, USDA, 1986).

7 National Research Council, *Forestry Research: a Mandate for Change*, p. 20.

8 John H. Ohman, "Current research: U.S. Forest Service," *Progress and Promise: A Commemoration of the McIntire-Stennis Cooperative Forestry Research Program* (Washington, D.C.: National Association of Professional Forestry Schools and Colleges and Cooperative State Research Service, USDA, 1986), pp.14–15.

9 James A. Allen and Virginia Burkett, "The National Biological Service," *Journal of Forestry* 93 (1995): 15–17.

10 The University of Maine established a Cooperative Forestry Research Unit (CFRU) in 1976 that has operated continually since. An advisory committee drawn from CFRU cooperators sets research priorities and reviews proposals. The CFRU has addressed a broad range of forestry issues over the period of its existence. Annual Reports are available from this cooperative. Centers have been a similar means of organizing research and outreach programs across academic units at universities. The University of Washington's Center for International Trade in Forest Products is a relatively recent example that has attracted substantial private and public funding support directed at global forestry trade issues that have become of increasing importance.

11 Ronald i. Giese, "Forestry Research: an Imperiled System," *Journal of Forestry* 86 (1988): 15–22; National Research Council, *Forestry Research: A Mandate for Change*, pp. 55–58.

12 USDA, Association of State College and University Forestry Research Organizations, and National Association of State Universities and Land-Grant Colleges, *Program of Research for Forests and Associated Rangelands: A Collection of Major Documents Produced by the Regional/National Planning Process*, (1978).

13 USDA Forest Service, National Association of Professional Forestry Schools and Colleges and USDA Cooperative State Research Service, *Forests for America's Future: A Research Program for the 1990's* (1989).

14 Examples are: 1) *A Quarter Century of Progress: The McIntire-Stennis Cooperative Forestry Research Program, 1962–1987* (National Association of Professional Forestry Schools and Colleges and Cooperative State Research Service, USDA, 1986); 2) *Forestry Research Progress in 1974 and 1975* (Association of State College and University Forestry Research Organizations, 1976); 3) *Forestry Research Progress in 1968: McIntire-Stennis Cooperative Forestry Research Program* (Cooperative State Research Service, USDA).

15 Bruce J. Zobel and Jerry R. Sprague, *A Forestry Revolution: The History of Tree Improvement on the Southern United States* (Durham: Carolina Academic Press and the Forest History Society, 1993); See also their chapter in this work.

16 Zobel and Sprague, *A Forestry Revolution*, p. 106.

17 A. E. Squillace, "Tree Improvement Accomplishments in the South," *Proceedings of the Twentieth Southern Tree Improvement Conference* (Charleston, SC: 1989), p. 11; Zobel was recognized by the Society of American Foresters for his contributions to the biological sciences that advance forestry through his work in southern pine tree improvement with the Barrington Moore Memorial Award in 1968. Other university scientists who also have made major research contributions to southern pine tree improvement and have been recognized for their work with this award are: Claud Brown, 1976, Texas A&M University and Johannes P. Van Buijtenen, 1994, Texas A&M University. In 1981, Jonathan W. Wright of Michigan State University was similarly honored for his work with tree improvement in the northern states.

18 Squillace, "Tree Improvement Accomplishments in the South," p. 10.

19 Squillace, "Tree Improvement Accomplishments in the South," pp. 11–13.

20 Bruce J. Zobel and J. Talbert, *Applied Forest Tree Improvement* (New York: Wiley and Sons, 1984), pp. 451–453.

21 Zobel and Sprague, *A Forestry Revolution,* p. 108.

22 Gregory N. Brown, "Future Expectations: Academic Perspectives," *Progress and Promise: A Commemoration of the 25th Anniversary of the McIntire-Stennis Cooperative Forestry Research Program* (Washington, D.C.: National Association of Professional Forestry Schools and Colleges and Cooperative State Research Service, USDA, 1986).

23 John Haygreen, Hans Gregersen, Andrew Hyun, and Peter Ince, "Innovation and Productivity Change in the Structural Panel Industry," *Forest Products Journal* 35 (1985): 32–38.

24 Henry Spelter, "Capacity, Production, and Manufacturing of Wood-based Panels in North America," General Technical Report FPL-GTR-82, (Madison, WI: USDA Forest Service, Forest Products Laboratory, 1994), p. 13.

25 Haygreen, et al., pp. 34–35.

26 Haygreen, et al., p. 36.

27 Haygreen, et al., pp. 35–36.

28 Spelter, p. 5.

29 David N. Bengston, "Economic Impacts of Structural Particleboard Research," *Forest Science* 30 (1984): 685–697.

30 Gordon W. Gullion, "Management of Aspen for Ruffed Grouse and Other Wildlife—An Update." *Proceedings of the Aspen Symposium* (North Central Forest Experiment Station, Forest Service-USDA. General Tech. Report. NC-140, 1990), pp. 135–143.

31 Gordon W. Gullion, "Management of Aspen for Ruffed Grouse," p. 135.

32 John S. Spencer, Jr., Earl C. Leatherberry, and Neal P. Kingsley, "The Lake States' Aspen Resource Revisited," *Proceedings of the Aspen Symposium*, p. 244.

33 John A. Youngquist, and Henry Spelter, "Aspen Wood Products Utilization: Impact of the Lake States Composites Industry," *Proceedings of the Aspen Symposium.*

34 Gordon W. Gullion, "Management of Aspen for Ruffed Grouse," pp. 135, 137.

35 Gordon W. Gullion, "Management of Aspen for Ruffed Grouse," pp. 139–140.

36 Richard H. Yanner, "Nongame Response to Ruffed Grouse Habitat Management in Pennsylvania," *Proceedings of the Aspen Symposium,* pp. 145–153.

37 Gordon W. Gullion, "Management of Aspen for Ruffed Grouse," pp. 136–137.

38 Professor Harold E. Burkhart, Virginia Polytechnic Institute State University, received the Society of American Foresters' highest recognition for accomplishment in the biological sciences to advance forestry, the Barrington Moore Memorial Award, in 1991 for development of improved methods for modeling forest stand dynamics and growth and yield. *Journal of Forestry* 89 (1991), p. 44.

39 Mason C. Carter, "Renewing the Timber Supply," *Forestry Research Progress in 1974 and 1975.* (Association of State College and University Forestry Research Organizations), p.7.

40 Richard A. Skok, "Forestry Education in the United States," in *The Literature of Forestry and Agroforestry*, edited by Peter McDonald and James Lassoie, (Ithaca: Cornell University Press, 1996) p. 95.

41 Arthur B. Meyer, "The Literature of American Forestry," *American Forestry Six Decades of Growth*, edited by Henry Clepper and Arthur B. Meyer. (Bethesda, MD: Society of American Foresters, 1960), pp.54–55.

42 Henry J. Vaux and H. R. Josephson, "Development and Accomplishments of Research Programs," *Forest Resource Economics and Policy Research: Strategic Directions for the Future,* edited by Paul V. Ellefson, (Westview Press, 1989), p. 16.

43 Orville G. Bentley, "An Overview of Forestry Research," *Progress and Promise: A Commemoration of the 25th Anniversary of the McIntire-Stennis Cooperative Forestry Research Program*, p. l.

44 Samuel T. Dana and Evert W. Johnson, *Forestry Education in America Today and Tomorrow* (Bethesda, Md.: Society of American Foresters, 1963) p. 171.

45 Skok, "Forestry Education in the United States."

46 Donald P. Duncan, et al., "Forestry Education and the Profession's Future," *Journal of Forestry* 87 (1989): 31–37.

47 National Research Council, *Forestry Research: a Mandate for Change*, pp. 7, 8.

GEORGE R. STAEBLER

INDUSTRIAL FOREST RESEARCH AND FOREST SCIENCE

Organized forest research conducted by private industry is comparatively new. In the past, landowners who kept their harvested land for succeeding crops of trees learned from extant forestry knowledge how best to regenerate and manage a new forest. From colonial times through the nineteenth century, however, forestry proved generally an extractive industry. It's not surprising, then, that owner-financed research was confined primarily to the product, rather than forest renewal.

By the middle of the twentieth century, that situation experienced change—albeit slowly. In 1955, the Society of American Foresters reported that fifty large U.S. landowners spent $1.5 million on forest research in 1953. Most of that research had developed in the previous ten years. In contrast, the entire wood products industry spent $26.5 million the same year on forest products and utilization research.

Organized, private industrial forest research has depended largely on a handful of companies. International Paper, Champion International, Union Camp, Weyerhaeuser, Westvaco and, in its day, Crown Zellerbach, have all supported research organizations with adequate scientific staffing and infrastructure. In addition, many smaller companies and forest landowners employ foresters who are responsible for research. Those foresters may conduct research themselves, but more often they are expected only to stay abreast of current research developments through professional contacts or participation in research cooperatives. The research organizations of only a few large companies, therefore, have greatly influenced the private forest research of today. The development of forest research by one of those companies—Weyerhaeuser—may serve as an example of private industry's contribution to forest science and forestry.

THE WEYERHAEUSER EXAMPLE

Forest research had been a line item in the Weyerhaeuser Company's annual budget for twenty-four years when the organization, founded in 1900, reached a turning point in 1966. That year, senior managers asked themselves, "Should the company initiate a program of greatly intensified forest management on its millions of acres of U.S. forestland?" Immediately the results of company research—begun in 1940 with a handful of permanent, thinned and unthinned sample plots—took on added significance and utility. Furthermore, when the senior managers judged the answer to be "yes," demands for information, data, and more research became insistent. Forest research took on increased stature as part of an effort that became widely known as High Yield Forestry.

This turning point in the company's conduct of formal forest research helps illustrate some principles of private forest research that certainly have been true at Weyerhaeuser and may apply elsewhere:

> Forest research data from a variety of disciplines almost always must be developed and accumulated before the information becomes fully useful. Weyerhaeuser's forest research organization has been well equipped to respond at crucial junctures in the company's evolution because it had developed a foundation of varied research. This research, some of which may not have had apparent immediate applications, was continued even through economic downswings.

> In an environment driven by the bottom line, the success and continuity of private industrial forest research can hinge on the vision, expertise, and dedication of a few individuals. Their contributions can occasionally involve "bootlegging" promising research.

> Close interaction between researchers and operational foresters improves problem solving and ensures rapid technology transfer.

These ideas, clearly evident in the company's approach to research prior to and during 1966, persisted at least into the 1990s as Weyerhaeuser and the forest products industry faced new challenges. Just as Weyerhaeuser forest research was well equipped to respond to the demands of High Yield Forestry then, so the research organization provided the necessary vision and expertise to lead the company's 1992 development of "Weyerhaeuser Forestry" and cutting-edge environmental forestry techniques. The private forest research organization's role in paving the way for contributions as diverse as yield improvement and biodiversity management has been demonstrated in Weyerhaeuser forest research from its very beginning.

THE FORMATIVE YEARS OF WEYERHAEUSER FOREST RESEARCH

On May 1, 1942, Weyerhaeuser hired Donald G. McKeever to establish a formal forest research program for the company. Previously affiliated with the U.S. Forest Service's Northern Rocky Mountain Forest and Range Experiment Station, McKeever had prepared a plan for company forest research that so impressed C. Davis Weyerhaeuser, manager of reforestation and land, and Managing Forester Bill Price, they hired him, explicitly sanctioning the establishment of Weyerhaeuser forest research.

Previous interest had existed and research occurred at lower organizational levels of the company, however, prior to McKeever's arrival. In 1940, for instance, Weyerhaeuser branch foresters established a partial-cutting trial in ninety-year-old western hemlock in Grays Harbor County, Washington. Sample plots were installed to assess the results. In 1941, the company dedicated the Clemons Tree Farm, later designated No. 1 in the American Tree Farm System. Forest researchers were assigned to help manage the tree farm, and, a year later, McKeever's staff took over the Clemons studies.

The new research organization worked from Centralia in southwestern Washington, a location central to at least a million acres of the company's forestland. Its purpose, according to the fledgling research organization's 1942 annual report, was to "assist in providing facts, figures and standards requested by management and operating foresters to synthesize a complete and practical science of producing, protecting, improving and harvesting forest products for maximum continuous profits within the limits to which the business can be adjusted. Research in marketing, distribution and utilization will be conducted by other departments." That broad statement of purpose has withstood the test of time.

The 1942 report noted that the forest research staff's priorities were regeneration studies and protection from fire. Early regeneration research aimed at promoting natural regeneration through studies of:

Seed production.
Natural seed fall patterns.
Scarification techniques.
Germination.
Seed blocks.

The study of seed coatings and rodent baiting also addressed the need to protect tree seeds prior to germination. Other efforts emphasized protecting growing trees from insects and disease, although forest pathology didn't become a distinct research area at Weyerhaeuser until 1956, with the hiring of Dr. Keith R. Shea.

Growth and yield research began in 1944. That year, researchers established permanent sample plots in both plantations and young natural stands on the St. Helens Tree Farm in southwestern Washington. These plots became particularly significant later because they provided almost two decades of key data for yield tables produced in 1962—yield tables on which Weyerhaeuser based its decision to pursue High Yield Forestry.

Seed-eating insects are captured using the cylindrical cartons.
After the bugs have eaten their way out of the cones in the cartons,
they crawl toward the light—and into this entomologist's test tube.
Once captured, the bugs are identified and studied.
(WEYERHAEUSER TIMBER COMPANY PHOTO, 1959)

Still, in 1944, the Weyerhaeuser forest research staff's entire field-travel needs were served by one company vehicle, plus personal autos—an indication not only of the staff's limited budget, but also of the modest value the company placed on the research.

Personnel changes in 1948 provide evidence of Weyerhaeuser forest research's growing maturity. Paul Lauterbach filled the role of Don McKeever, who died tragically in a woods accident. A research staff member since 1945, Lauterbach was managing stand improvement studies when he became head of the research staff. The same year, Longview, Washington, Branch Forester Edwin F. Heacox was promoted to managing forester. Heacox believed that the company's forest research activities, centered in Centralia, should no longer be regarded simply as training for operational foresters. He recommended that forestry specialists on the job be directed to conduct a well-balanced program of forest research and to regard themselves as career researchers. This attitude prompted a different philosophy toward current studies and encouraged the recruiting of new forest specialists during the prime years of their professional lives. Ultimately, the personnel changes of 1948 helped Weyerhaeuser's top management build a compact, flexible research organization staffed by young, vigorous specialists with the best academic training and field experience. This team, including new project leaders recruited in the areas of regeneration, silviculture, and forest management, possessed a talent for unselfish cooperation with each other and with other professionals.

In 1955, William H. Cummings signed on as supervisor of forest research. His charter directed him to expand company research, covering more completely the full range of forest science. Cummings organized Weyerhaeuser forest research into:

Forest sciences, including tree physiology, forest soils, and genetics.
Forest protection, including forest entomology, pathology, and wildlife biology.
Forest management, including regeneration, silviculture, and forest management.

Specialists in many of these fields were already working on Cummings' research staff. He quickly recruited scientists in tree physiology, pathology, and wildlife biology to create a remarkably well-rounded research program lacking only a genetics expert. The company had provided a new building in 1954 to house the research organization and a group of old-growth cruisers who had also occupied rented quarters in Centralia up to this time. In 1956, a chemistry lab, growth chambers, greenhouses, an insectary, offices, and specialized laboratories for tree physiology, pathology, and wildlife studies augmented the Centralia Research Center's excellent facilities.

Dr. George S. Allen replaced Cummings in 1961. A former dean of the Faculty of Forestry at the University of British Columbia, Allen was the first in the company to be given the title of director of forestry research. He moved quickly to complete the research organization by hiring a trained forest geneticist. The addition marked the beginning of a remarkable program of tree improvement in the Pacific Northwest. Tree improvement for loblolly pine had already been under way in North Carolina since 1956.

A LANDMARK YEAR: 1966

During the tenures of Cummings and Allen, Weyerhaeuser forest research came of age as a fully industry-supported, peer-recognized, scientific organization. Recognized scientists filled the company's nine specialist positions. Six held doctorates, and a seventh would earn one later. Their research programs received adequate scientific, technical, and staff support. Professional interchange between Weyerhaeuser researchers and those in other organizations was strong, a testament to the forestry community's high regard for the company's research.

Then came the company's landmark decision: should it initiate a program of greatly intensified forest management on its private forestlands? Answering that question objectively required an appraisal of how much merchantable wood could be produced under intensive management—and what that management should entail. Weyerhaeuser managers turned to their forest research staff for the answer. Never were twenty-four years of research results and the knowledge and experience of a dedicated group of scientists so immediately productive. In fact, in the preceding months, a request from Heacox had already started Allen's team working to synthesize the cumulative knowledge gained from the company's studies. Now, Allen assembled his project leaders to formulate a response to management's charge.

The research staff focused on the most immediate need: yield tables for managed stands of Douglas-fir. In 1962, the staff had constructed variable-density yield tables for natural stands, using data from plot measurements begun in 1944. The forest research team decided the 1962 yield tables could be adjusted to reflect the effects of thinning and fertilization. They drew on related research to make those adjustments, creating defensible managed-stand yield tables. They presented expected yields for Douglas-fir, by site quality, assuming:

Plantations or natural stands of a given initial density.
Stocking control (precommercial thinning).
Fertilization.
Commercial thinning at five-year intervals.

The yield tables included projected stand characteristics—number of trees, diameter distribution, tree heights, and wood volumes—at five-year intervals for each specified thinning and fertilizer regime. The projections assumed precise and successful regeneration, control of competing vegetation, and prevention of animal damage, as well as the exclusion of catastrophes such as wildfire.

Known as "the target forest," these yield tables were analyzed to determine financial rotations, eventual forest value, and return on investment. That analysis illustrated the benefits implicit in greatly intensifying the management of Weyerhaeuser's Douglas-fir forests.

Silviculturist George R. Staebler was promoted to director of forestry research in mid-1966 when Allen resigned for health reasons. Over that summer, a Timberlands team including Staebler shared the "target forest" conclusions throughout the company's forestry and timberlands management organizations. In October 1966, Weyerhaeuser's board of directors voted to proceed with the recommended intensified forestry on the company's forestlands.

On the occasion of Weyerhaeuser's original land purchase, founder Frederick Weyerhaeuser reportedly had said, "This is not for us nor for our children, but for our grandchildren." In fact, Frederick's great-grandson, George H. Weyerhaeuser, led the company in 1966 when it made the momentous decision to proceed with an effort named High Yield Forestry. Harry Morgan, Jr., corporate vice president with responsibilities for timberlands and raw materials, as well as forestry research, reorganized the timberlands structure to best implement High Yield Forestry. By this time, the fledgling forest research group begun more than two decades before had grown from its formative years into a mature industrial forest research organization with significant impact on corporate forestry and policy.

Weyerhaeuser forest research faced even more change, though. The implementation of High Yield Forestry demanded more and more support data and additional study. The further expansion of Weyerhaeuser forest research can be traced through the development of individual research areas, including:

Silviculture.
Growth and yield.
Regeneration.
Tree improvement.
Soil survey.
Environmental forest science.

SILVICULTURE RESEARCH

Weyerhaeuser forest research began with a few measured plots in an experimentally thinned stand. Although a cornerstone of forestry from the field's very beginning, thinning had been more written about than practiced. Weyerhaeuser researchers performed additional studies of operational thinning in 1965. This research included time-and-motion studies of thinning operations and subsequent mill studies of the logs produced. The promising results influenced the decision to include thinning in the managed-stand yield tables upon which Weyerhaeuser High Yield Forestry was based. Hence, successful implementation of High Yield Forestry depended partly on the yield gains from thinning.

Over the years, company researchers conducted many other small-scale thinning trials, known as Pilot Management Studies, in Weyerhaeuser's Pacific Northwest forests. Careful measurements and assessments of results have provided a wealth of useful data. Many similarly designed studies also took place for loblolly pine, providing the basis for large commercial-thinning programs in all of Weyerhaeuser's southern U.S. regions.

Weyerhaeuser forest researchers were responsible for the design and initiation of a large, cooperative Douglas-fir thinning experiment known as the Levels of Growing Stock (LOGS) study. The project involves five cooperative participants and nine installations of a replicated nine-treatment experiment. It permits evaluation of repeated thinning, controlled by the amount of growing stock in the residual stands. The LOGS experiment starts with stands under 40 feet tall and extends through 60 feet of height growth on the tallest trees—that is, from approximately 20 to 40 years, depending on site quality. Several of the studies, including the two on Weyerhaeuser property, are complete. The results include a wealth of information useful particularly to people building growth and yield models. A number of USDA Forest Service experiment station papers have already been published on the LOGS study, and several more are expected. Studies with similar designs have also been initiated in loblolly pine stands.

Weyerhaeuser has also performed other silvicultural studies at its research centers in the northwestern and southern United States. Though thinning research predominates, the studies also include research on stocking control, herbicides, and disease control (notably root rot).

Foresters have always assumed that combinations of silvicultural practices lead to synergism: that the total effect of two or more simultaneous practices is greater than the sum of the effects of treatments applied singly. Identification of select families in the tree improvement program has brought synergism's potential salutary effects increasingly to the fore. Prior to that, however, Weyerhaeuser researchers installed several well-designed, small-scale thinning/fertilizer studies over the years. A progression of similar studies (with continuously improved designs) are now in place in family block plantations where researchers can apply the full range of silvicultural practices individually and in combination to stands of known genetic heritage. Test variables include planting stock, initial spacing, vegetation control, fertilizer, stocking control, thinning, and pruning. The results are measured in value of product per unit area of forest, incorporating yield, product quality, piece size, and other value measures.

The major supply of timber in the next century will come from plantations. The yield tables used as the basis for Weyerhaeuser's Target Forest in the mid-1960s, however, were natural-stand yield tables, only assumed to apply to plantations. In 1978, the company installed a massive plantation forecasting study to provide meaningful data for more refined yield tables. The plantation study incorporates thinning and fertilizer treatments to assess possible synergistic effects.

Since 1985, researchers have also installed additional western conifer plantation studies beginning with planting. They established five-acre plantations of densities from one hundred to twelve hundred trees per acre. The stands, intended for future stand treatment studies, will permit investigation of silvicultural control of tree quality. A surprising early result: after only seven to ten years—well before crown closure—the trees in the denser stands are larger in both height and diameter. The reasons for this unexpected phenomenon are unknown, but the implications may be of tremendous importance in the intensive management of western forests. The same experiment is being repeated using seedlings from the better Douglas-fir families identified in the tree-improvement program. More than forty five-acre test beds have been established for all western conifer species, as well as for red alder and bigleaf maple. Already, early treatment results indicate that timing of the first thinning proves particularly critical to tree and wood quality.

Increased concern for stem quality in recent years has also generated pruning studies, leading to extensive operational pruning of both Douglas-fir and loblolly pine. Carefully designed mill studies of loblolly logs harvested from long-term study plots in Louisiana are helping researchers determine the effect of pruning on sawmill recovery of clear lumber. The results of the studies, a cooperative effort with Louisiana State University, provided excellent data on what to expect from trees grown under High Yield Forestry prescriptions. Pruned Douglas-fir logs have also provided similar, although much less extensive, data.

Weyerhaeuser's forest research organization also participates in the University of Washington Stand Management Cooperative. This well-managed, extensive thinning/nutrient study involves private forest products companies as well as federal and state forest managers in Washington and Oregon. Weyerhaeuser forest research depends heavily on the results of this important regional study for the synergistic effects of thinning and fertilization.

GROWTH AND YIELD RESEARCH

A 1930 publication familiarly known as Bulletin 201 first systematically quantified the enormous productive potential of Douglas-fir forests. Richard E. McArdle of the Pacific Northwest Forest and Range Experiment Station wrote Bulletin 201, formally titled *The Yield of Douglas-fir in the Pacific Northwest.* Many foresters consider it among the most important influences on the Northwest forest industry. Before Bulletin 201, most of the Pacific Northwest industry followed the "cut-out-and-get-out" practices carried over from the nineteenth century by "lumber barons" in the Northeast, the Lake States, and the South. McArdle's publication encouraged the shift to true sustained yield in the Douglas-fir forests of the Pacific Northwest.

Bulletin 201 presented "normal" yield tables giving the volume and other characteristics of fully stocked natural stands. "Normal" in this case connoted full stocking beginning at an early age—usually ten years. Many, perhaps most, natural stands, however, were not fully stocked. The normal yield tables could be adjusted to reflect different stocking levels, but neither research nor experience really corroborated those adjustments.

This shortcoming helped convince early researchers like Weyerhaeuser's McKeever to make their own empirical tables that would reflect average conditions actually found in their forests. At the time, Weyerhaeuser owned vast harvested acreage that was naturally and variably restocked. McKeever adjusted the normal yield tables in Bulletin 201 to forecast the yield of initially understocked stands. His tables reflected the stands' "trend toward normality," based on studies by the Pacific Northwest Forest and Range Experiment Station. McKeever's work showed that understocked Weyerhaeuser stands would yield more than might otherwise be expected.

Designed to correct the shortcomings of the Bulletin 201 tables, the company's growth and yield studies gained momentum following installation of the first permanent sample plots in 1944. By 1950, researchers had systematically established growth plots on all of the company's Douglas-fir tree farms to sample the full range of stand conditions. This work provided 201 growth plots and 275 five-year measurements that supplied data for the important set of yield tables constructed in 1962.

The new variable-density yield tables showed volume and other statistics for stands ranging from grossly understocked to fully stocked. These tables reflected stand development up to 120 years of age. Furthermore, while the yield tables in Bulletin 201 reflected only net volumes, Weyerhaeuser's 1962 yield tables also included gross production—that is, net volume plus the volume of the accumulated mortality up to a given age. It was widely believed that thinning could offset and help capture volumes lost to mortality in naturally developing stands. Gross production became particularly important when combined with two other elements: stocking-level measurements and site class.

Stocking levels were quantified with a new measure: growing stock index (GSI). Forest Management Group Leader James E. King had developed the GSI as a doctoral study at the University of Washington. The novel concept was based on quantifying the basal area trend for stands of different densities and then identifying the curves that passed through class medians at an index age. Resulting yield tables then could be arranged by GSI as well as by site class.

As used by McArdle in Bulletin 201, site class was defined as the height of dominant trees at one hundred years of age. It became readily apparent, however, that one hundred-year-old trees would not be common in managed forests. A younger index, such as fifty years, would be more useful for constructing the new site curves needed for empirical yield tables that applied to stands not fully stocked. Contributions from soils research helped make the resulting site curves far superior to those used in Bulletin 201, which seriously underestimated site quality at younger ages. Forestry professionals widely accepted the new site curves, published as a Weyerhaeuser Forestry Paper.

This work helped answer the company's questions about how much its forests could produce if intensively managed. The new variable-density, gross- and net-yield tables, constructed from Weyerhaeuser plot data using improved site curves, formed the basis for a technically defensible approximation of managed-stand yields from sites on company forestlands. Those yield approximations provided justification for embarking on a specified management program. The "intensive management" practices quantified in the new tables included:

Planting for fully stocked regeneration.
Fertilization.
Both commercial and pre-commercial thinning (also known as
    stocking control).

Genetic tree improvement was not considered as a potential source of yield gain in 1966 because no data were yet available.

These new yield tables used a high GSI (full, though not maximum, stocking). They specified the number of planted trees that needed to survive to achieve full stocking. They also assumed thinning gains equal to the difference between the net and gross yields tabulated in the empirical tables.

Estimates of the potential fertilization gains were derived from forest soils studies. Though these fertilizer studies had not been designed to quantify gains in volume yield, they had included measurements of diameter and height growth and the gains related to the kind and amount of fertilizer applied. Since Weyerhaeuser's new yield tables were constructed from equations using diameter and height, they made it possible to postulate a fertilizer/gross-yield table.

Yield was adjusted further upward to incorporate improvement in tree form anticipated through careful, periodic thinnings. Stand characteristics (number, diameter, height, and volume of trees) and the thinning regime required to capture full gross yield were derived using an approach Staebler had described in *Forest Science* in 1960, "Theoretical Derivations of Numerical Thinning Schedules for Douglas-fir."

That paper's opening sentence—"Next to the ax and good judgment, yield tables are probably the forester's most important management tool"—proved prophetic for Weyerhaeuser. Economic analysis of the yield tables developed by company researchers and the derived fiscal forecasts prompted management enthusiastically to embrace intensive forest management.

High Yield Forestry became a household word to foresters nationwide. Researchers fielded immediate requests for similar calculations on loblolly and ponderosa pine and western hemlock. The growth and yield scientists made the best possible approximations for these species, although without nearly as much supporting data. As better species yield tables were developed, they prepared more defensible target forest tables.

Yield tables are never finished because new silvicultural treatments are constantly being perfected and judged by their effect on stand increment and yield. Further, much work is needed to incorporate previously unconsidered stand, tree, and wood characteristics into yield tables. Cubic-volume and board-foot measures alone are not enough. Weyerhaeuser research currently emphasizes quality measures such as straightness, taper, knot size and distribution, juvenile wood, and specific gravity.

Since such calculations require enormous amounts of data, the history of yield tables must also consider equipment advances: namely computers. The very long-term nature of silvicultural studies frequently dictates study designs that may be less than 100 percent effective—or even studies that are foregone completely. Computers, however, allow researchers to use early results from well-designed, long-term studies to build models with usable results, a task impossible only a few years before.

REGENERATION

Regeneration has always ranked among the Weyerhaeuser forest research staff's first priorities. Increased understanding of natural regeneration prompted the company to begin helicopter seeding in 1949. The pioneering effort developed into an aerial seeding program covering 12,000 to 19,000 acres per year. Planting and soils research begun earlier in the decade was also applied to seedlings hand-planted on an additional 4,000 to 9,000 acres annually. By 1958–1959, the company was supplementing aerial seeding by planting nearly ten million trees a year.

Although a solid foundation of planting research and practice had been built by 1966, much work remained to meet the new demands of High Yield Forestry. The Target Forest established presumably achievable goals, but reaching those goals depended on immediate and successful regeneration.

Years of research and experience suggested planting would be the only reliable and consistent means of establishing the required stand. First, though, Weyerhaeuser needed company-owned and operated nurseries to meet planting-stock requirements. Both company foresters and scientists believed in the possibility and the necessity of better nurseries producing better planting stock.

The sandy soil types most suitable for forest tree seedling and seedling transplant nurseries are rare in Oregon and especially in glaciated western Washington. Nonetheless, a Weyerhaeuser soils scientist found them in southwestern Thurston County, Washington. The area soon became the site of the first Weyerhaeuser Douglas-fir nursery. Beginning in 1967, company foresters and research scientists worked in a remarkably close team to develop the nurseries and advance nursery practices.

Common industry practice had considered the nurseryman's job completed with delivery of bundled seedlings to the loading dock. At that point, they became the responsibility of the forester in charge of tree planting. When plantations failed (not an unusual event), the tree planters blamed the nurserymen and the nurserymen faulted the tree planters. Weyerhaeuser's research and operational team focused everyone on one goal: healthy, established tree seedlings. Nursery staff immediately became concerned with the delivery of seedlings to the planting site and how those seedlings were planted. Simultaneously, the forester and tree planter took a new professional interest in the nursery.

Transpiration studies of seedlings conventionally bundled for transport revealed that it was at least as important to protect seedling tops from wind and weather as it was to protect roots. As a result, refrigerated vans delivered seedlings to planting sites. The researchers also studied a variety of other nursery practices, including:

Optimum seed-bed planting densities.
Undercutting and root development.
Freeze damage control through sprinkler irrigation.
Production of superior transplant stock.

These studies brought about improvements resulting in hardier seedlings with the increased survival rate needed to efficiently establish the target forest.

Few developments in nursery research have proven more consequential than an understanding of the effects of lifting date and storage on seedlings' eventual success in the field. There is a time for lifting and a time for planting; storage is required when the times do not match. Until the early 1970s, cold storage at 2° F caused problems, principally mold and effects on seedling dormancy. Research helped solve those problems with freezer storage at -2° to 0° F. Later the same decade, further refinement of the studies showed that managed cold storage could tailor seedling dormancy to a planting time.

Company foresters also quickly exploited another planting stock improvement: containerized or "plug" seedlings. Again, research and operations worked closely to iron out the problems of commercial-level production. Ultimately, nursery-produced, cold-storage seedlings and transplants better met the need plugs had been expected to fill.

Nonetheless, plugs proved important in Douglas-fir rooted-cutting programs. Douglas-fir has long been considered a species very difficult to root. Weyerhaeuser forest research physiologists found the plugs useful for speeding the rooting process and facilitating transplanting of the rooted cuttings. This important technique permits large-scale family plantings from only a few full-sib seeds that have been produced in controlled crosses among parents proven superior in progeny tests.

Nursery research also increased knowledge of other tree species. Ponderosa pine, for instance, was once almost universally managed by the selection system. Perpetuation of the forest depended on natural regeneration. Weyerhaeuser's Target Ponderosa Pine Forest, however, was based on plantation forestry requiring a pioneering effort in artificial regeneration. So the company established a nursery at Malin, Oregon, to supply the expected seedling needs of its Klamath Tree Farm. Weyerhaeuser regeneration scientists led by James Dick participated to ensure a successful nursery operation.

Once planting stock was available at Klamath Falls, however, a major problem arose: establishing plantations in what is essentially the tension zone for the east-of-the-Cascade Range pine type. In this high-elevation, low-rainfall region consistently successful plantings required more understanding of the ecological relationships among temperature, precipitation, soils, nursery seedling characteristics, competing vegetation, and animal damage.

Weyerhaeuser hired an Oregon State University scientist to tackle the problem. He worked with the company's research scientists and local foresters to modify the Scholander "pressure bomb" into an instrument that accurately measured the transpiration potential of seedlings before and after planting, as well as diurnal and seasonal variation. This highly engineered, sophisticated tool proved very useful in regeneration and nursery research.

Highly successful plantations increasingly became the norm following harvest of all Weyerhaeuser forests. The plantations actually began with harvest methods tailored to specific sites and designed to favor regeneration: To prevent soil damage, tractor skidding had been abandoned in favor of high-lead logging and, later, more forgiving forms of ground skidding. Broadcast burning or scarification helped prepare sites, after which highly skilled crews planted the seedlings.

Once established, the company's plantations had to conform to certain specifications to ensure proper size and density for the stocking control and first commercial thinnings called for in the managed-stand yield tables. Weyerhaeuser established a checkpoint for Douglas-fir known as the CASH standard: Correct Age, Stocking, and (breast-high) Height. The standard differed only by site quality.

Animal-damage prevention and vegetation control also became standard practice. In western Washington, deer, elk, mountain beaver, rabbits, mice, and probably other animals threatened or often destroyed Douglas-fir plantations. On the Klamath Tree Farm, deer and pocket gophers were of most concern, though porcupines also damaged many sapling-sized ponderosa pines. Weyerhaeuser hired wildlife biologist Dr. W. H. Lawrence in 1956 to pursue actively animal damage control, addressing problems and seeking any wildlife management opportunities. Potent pesticides and closely controlled and monitored application methods were very successful in mountain beaver control. Later chemical bans led to trapping techniques.

Weyerhaeuser chemists developed a big-game repellent made first from putrefied fish (spent salmon) and later from eggs. When applied to nursery stock before lifting, the repellent effectively controlled deer and elk damage. The company licensed the product to a Minnesota manufacturing firm, and Weyerhaeuser foresters used it for several years. Foresters also made good progress controlling depredation from black-tailed deer in western Washington and Oregon by adjusting hunting seasons, including either-sex seasons, in high deer-population zones. Since the setting of hunting seasons requires biological knowledge and involves both sportsmen and government agencies, the effort required extensive communication and education—a technique that became known as biopolitics.

High Yield Forestry requires regeneration of all harvested land within one year. That prescription is meaningless, though, unless the plantations survive successfully. Long-term plantation success required recognition of the necessity to manage every link in the chain, from seedling-viability improvements, through site selection and preparation, to control of competing vegetation, fertilization, and prevention of damage or loss from disease, fire, or insects. Over time, the research organization reflected the disciplines needed to address each of these links in the chain of success, and joint efforts between forest research and operations staff succeeded in developing prescriptions for each, fulfilling the regeneration obligation of High Yield Forestry.

TREE-IMPROVEMENT RESEARCH

Early introductions of tree improvement to large-scale industrial forestlands came through cooperatives organized by Texas A&M University and North Carolina State University. The North Carolina Pulp Company, acquired by Weyerhaeuser in 1957, was a charter member of the very successful North Carolina State University/Industry Cooperative Tree Improvement Program. Similarly, the company's first involvement with Douglas-fir tree improvement was the 1957 establishment of a small seed orchard at Camp McDonald in southwestern Washington as part of a cooperative sponsored by the Industrial Forestry Association.

The company's first geneticist, Dr. Robert C. Campbell, signed on in 1962. He initiated an aggressive, wide-ranging program of tree improvement fully backed by research. Under his guidance, Weyerhaeuser established seed orchards in Washington and Oregon. Four blocks in the Washington orchard and two in Oregon produced seed for carefully designated geographic areas of company ownership.

The establishment of seed orchards began an aggressive Douglas-fir breeding program. Led by Dr. John H. Rediske, the tree-improvement research staff conceived and documented a detailed, thirty-year plan for orchard production of improved seed. They estimated costs and returns and calculated net present value of the program to the company. The 1971 proposal, known as "The Potential for Tree Improvement," projected the potential returns of as-yet-undeveloped techniques, such as early flower induction, that would greatly reduce the time between generations.

The clear analysis of problems, opportunities, rewards, and risks presented in "The Potential for Tree Improvement" captured senior management's support. The great impact of "The Potential for Tree Improvement" came from its extraordinarily incisive thinking, characteristic of

Controlled pollination is practiced on the Klamath Falls Tree Farm of the Weyerhaeuser Company in Oregon. Pine conelets are protected from open pollination by a plastic bag, while a research scientist injects pollen from the Apache pine of the Southwest. The goal is to produce a hybrid tree with better root vigor to penetrate and survive on local pumice soils.

(WEYERHAEUSER TIMBER COMPANY PHOTO)

Weyerhaeuser's tree-improvement program from the beginning. This imaginative viewpoint and management support of the related forest research effort have made it a central focus for the future of the company's forests and forestry.

The benefits of tree improvement were originally assessed exclusively as gains in tree volume and, hence, yield. In 1986, company forest products analysis led to greater emphasis on value that includes wood quality. Wood specific-gravity and stem form (the latter particularly in Washington state's

Douglas-fir families) now receive increased emphasis in tree-improvement efforts.

In 1991, Weyerhaeuser researchers analyzed twelve-year-old progeny tests of first-generation families. Concurrent wide-adaptability studies showed that low-elevation families could be moved throughout the states of their origin (Washington and Oregon) without loss from climatic or other ecological factors. These results are useful in the design of progeny-test trials of second-generation families. The results also, of course, prove of great importance to the regeneration of harvested land on the company's widely dispersed ownership.

With perfection of early flower induction and other seed orchard management techniques, Weyerhaeuser's second-generation orchard in Medford, Oregon, produced significant quantities of seed in 1995, three to nine years after initial grafting. Rooted cuttings made from seedlings that originated from controlled pollination in the Medford orchard will enable speedier, more efficient progeny testing and earlier selection for the third-generation orchard. The company is directing current research efforts toward controlled pollination and volume production of seed of full-sib families. (See also the Southern Forestry Research section.) The ability to multiply each seed through rooted cuttings and (if successful) somatic embryogenesis emphasizes the enormous opportunity of mass, controlled pollination.

Weyerhaeuser's support of forest research, the research organization's contributions to genetics and tree improvement, and a proactive, well-conceived and planned operational tree-improvement program have, indeed, made "The Potential for Tree Improvement" a self-fulfilling prophecy.

SOIL SURVEYS

Weyerhaeuser initiated forest soils research in 1951 with the hiring of a forest soils scientist. The hiring was justified, if for no other reason, in the words of Managing Forester Edwin F. Heacox: "because Weyerhaeuser has two million acres of the stuff; we ought to know something about it." Since then, in fact, forest soils research has greatly advanced Weyerhaeuser forest management, including the contributions of fertilizer studies to High Yield Forestry.

Soil mapping of the entire company ownership provided another major contribution. Dr. E. C. Steinbrenner, who joined the research staff in 1952, worked closely with soil specialists from Washington State University. They jointly developed a survey technique that used aerial photography to identify land forms. When combined with other physiographic characteristics, that information could be used to delineate soil-type boundaries. The techniques proved efficient, economical, and accurate.

The researchers completed a full-scale pilot test in 1959 on almost a quarter of a million acres covering Washington's Snoqualmie Tree Farm. By the time of its completion the following decade, the soil mapping had covered almost three million acres in the southern U.S. as well as several million acres in the Northwest.

The value of the soil survey lay in the forest-management interpretations of soil series. Site indices for the tree species adapted to a given soil were determined from prediction equations based on measurable physical soil characteristics and profiles. Researchers collected data for the equations when they made measurements for the new site index curves described earlier under Growth and Yield Research.

Recently, growing environmental concern about the ecological impacts of forest management has increased the importance of the soil survey to company foresters. Accurate soils information is useful for assessing:

The effects of harvesting and road construction on erosion and stream sedimentation.
The likelihood of mass failures on harvested and unharvested land.
The need for and design of riparian zones.

Fortunately, the Weyerhaeuser soil survey already provides much of the necessary information for managing these issues. With updated soil-management guidelines, refined operability-risk rating tools, and integration with the company's Geographical Information Systems, that foundation of data can help ensure not only plantation productivity but also the environmental responsibility implicit in Weyerhaeuser's commitment to stewardship of its lands.

ENVIRONMENTAL FOREST SCIENCE

Concerns for the environment as a societal issue developed in the late 1960s, just as High Yield Forestry was getting under way, and congealed on the first Earth Day, April 22, 1970. Forests as an environmental centerpiece, and the practice of forestry as a major agent in altering that centerpiece, became the center of attention—and attack. As a large industrial-forest landowner, Weyerhaeuser recognized that forest stewardship entailed more than repeated growth and harvest of trees. Ironically, the company was lionized in

1941 for initiating the American Tree Farm movement and ensuring sustainable wood fiber production—and, thirty years later, vilified for focusing so heavily on tree farming!

The passage of the Clean Water Act in 1972 marked an even more imperative turning point. After company study and analysis of the new climate in which the industry found itself, the forest research staff took on the task of defining an active Weyerhaeuser environmental program. Dr. W. H. Lawrence marshaled the effort. As a result, in 1974 Lawrence took charge of a new Weyerhaeuser research group: forest environmental sciences. The new program included:

Wildlife management.
Fisheries management.
Hydrology.
Vegetation management.

Lawrence recruited fisheries specialists and hydrologists to augment existing expertise.

One of the new group's first tasks was a study of Washington state's Deschutes River watershed. Weyerhaeuser, a principal landowner in the river's headwaters region, was harvesting the timber in 1974. The watershed is particularly sensitive because the river flows into Capitol Lake before entering Puget Sound. Weyerhaeuser's forest research and operations organizations worked together to develop a comprehensive plan for the river basin. Studies of water flows, water quality, and sedimentation provided preharvest benchmarks. Hydrological input and data from the company's basic soil survey led to engineering changes in road design. Foresters also modified their harvest practices based on fisheries and wildlife studies. The plan and its results were well received by government agencies and environmental organizations concerned with land use in the upper reaches of a river with highly visible lower reaches.

Weyerhaeuser's environmental sciences staff contributed a storehouse of information for Washington's landmark Timber, Fish and Wildlife Agreement, developed in the mid-1980s and signed in 1987. The accord, developed by consensus among various forest stakeholders, succeeded in reducing forest harvest delays caused by conflict and increased regulation under the Washington Forest Practices Act. Both the development and application of the agreement depended on up-to-date technical expertise.

Today, environmental-science researchers focus on watershed analysis, Habitat Management Plans (HMPs), and Habitat Conservation Plans (HCPs) to help the company respond to environmental laws, especially the Endangered Species Act. This focus reflects the company's belief that comprehensive forest and wildlife management offers more efficiency and greater likelihood of success than sequential responses to individual environmental issues. Ecologists, botanists, and specialized wildlife biologists joined the research team to help develop ways to manage the forest for wood production while protecting wildlife and other forest resources. In 1994, the team completed a formal HCP to protect northern spotted owls living on and around 209,000 acres of the company's Millicoma Tree Farm near Coos Bay, Oregon. This was the first HCP for the northern spotted owl in Oregon to win approval from the U.S. Fish and Wildlife Service. A multi-species HCP is also under way for one hundred thousand acres in southwestern Washington, and the company is applying lessons from these efforts to its other forestlands across the country.

The development of the environmental sciences group from its start until the mid-1990s complements, parallels, and even leads Weyerhaeuser's changing approach to forest management. Just as forest research's contributions to High Yield Forestry set the tone at Weyerhaeuser for several decades, environmental forest research has assumed that role for Weyerhaeuser Forestry—since 1992 the formal name for the company's brand of forest management. The term is meant to convey concern for wildlife, fish, water quality, biodiversity, and forest aesthetics, as well as tree growth and management for wood production. Under the sponsorship of Charles W. Bingham, executive vice president of timberlands and corporate affairs, forest environmental research became central to Weyerhaeuser efforts to adjust its forest-management practices to meet environmental regulations and the demands of society. Just as forest research prepared the company for High Yield Forestry, it has spurred the development of Weyerhaeuser Forestry—though probably with less advance recognition of where the developments were heading. As the company and the industry continue to feel out a precarious path toward the future, forest research leads the way by providing new technical data that allows the charting of new directions.

SOUTHERN FOREST RESEARCH

Weyerhaeuser moved into southern pinery in 1956 with a land-acquisition program based in Columbus, Mississippi. In 1957, the company acquired the North Carolina Pulp Company, along with more than half a million acres of North Carolina forest that provided raw material for the Plymouth pulp mill and scattered small sawmills. Soon after, Weyerhaeuser built a new North Carolina pulp mill at New Bern. This expanded mill capacity placed more demand on the company's forests, which in turn called for new silvicultural practices to increase growth and sustained-yield capacity. The need for forest research, already multiplied by High Yield Forestry, increased further. The forest research organization, then led by Vice President of Research and New Business Development R. D. Pauley and Director of Forestry Research Staebler, responded in 1969 by setting up a Southern Forestry Research Center in New Bern. Dr. Norman E. Johnson, who had served as a company entomologist before becoming a professor at Cornell University, returned to Weyerhaeuser to organize the new effort.

Johnson established the company's first field station at Columbus, Mississippi, to provide research support for the management of the half million Weyerhaeuser acres in Mississippi and Alabama. The field station placed scientists directly within an operating area to work on applied projects mutually identified as priorities by research and operations foresters. This field-station concept subsequently spread throughout Weyerhaeuser, helping to strengthen the partnership and communication between operations and research.

Also in 1969, the company acquired Dierks Forests, Inc., gaining 1.8 million acres of primarily loblolly pine in Oklahoma and Arkansas, as well as a huge manufacturing capacity. The southern operation's center of gravity shifted to Hot Springs, Arkansas. Shortly thereafter, the Southern Forestry Research Center moved from New Bern into what had once been the Dierks wood products research laboratory. The New Bern and Columbus facilities became regional field stations.

The Southern Forestry Research Center become a near replicate in organization and forestry disciplines of the Western Forestry Research Center in Centralia, Washington. Mensurationists, biometricians, and silviculturists strengthened and refined the managed-stand yield tables made in the West in 1967. Those tables, since computerized to provide instantaneous information, form the basis for rigorous harvest and management decisions in the loblolly pine forests.

As with Douglas-fir High Yield Forestry, quick, reliable regeneration became of utmost importance. Each field station addressed the regional implications of similar, interrelated concerns:

Planting stock and practices
Soil and soil/site impacts on stand performance

171

Management of hardwood forests
Control of competing vegetation
Fertilization
Growth and yield model development
Genetic improvements
Water quality and wildlife habitat management
Optimization of overall forest value and long-term productivity

For instance, researchers in North Carolina had to study and understand drainage, bedding, fertilization, and planting as both single and interacting practices. In particular, they had to solve drainage problems in the North Carolina pocosins to ensure successful loblolly plantations on sites previously occupied by low-value, slow-growing pond pine. Water-level studies led to water management rather than simple drainage. Newly planted trees could not survive with "wet feet," but neither could the water level be allowed to drop so low that tree roots couldn't reach it during dry seasons.

Despite the well known phosphorous deficiency of pocosin sites, determining fertilizer requirements required extensive research. Prescriptions and economical application techniques were developed. In addition, the scientists defined how much vegetation control was required to unlock loblolly's enormous early height growth potential.

The Columbus field station focused on successful regeneration on all sites under all conditions. Most cases required site preparation to achieve the company's goals for established plantations. Since good site preparation depends on soil characteristics, the company's soil survey proved useful in developing the necessary prescriptions.

Since the Mississippi/Alabama forests had more hardwood than other southern regions, the Columbus field station also emphasized hardwood research. Field station staff conducted an in-depth survey of the region's hardwood potential and initiated regeneration studies with Mississippi State University. Hardwood silviculture has grown in importance as company policy shifts to retaining more and better riparian zones along streams.

Mississippi State University also cooperated with Weyerhaeuser's Columbus field station in a study of the effects of intensive pine plantation management on wildlife. This issue would eventually become critical. The study's most significant finding, however, held that intensive forest practices are entirely compatible with, or even needed to maintain, stable deer and wild turkey populations.

In Arkansas and Oklahoma, studies of ripping (or subsoiling) techniques and machinery resulted in important gains in regeneration success. The region's characteristically impervious, rocky, droughty soils and steep slopes tend to make planting difficult and survival rates poor. Ripping along the contours of hillsides provided a place to plant seedlings, as well as catchment basins that held water where it was most useful. Understanding derived from the company's soil survey again contributed to successful application of ripping.

Tree improvement became a major component of Weyerhaeuser's southern forest research. In 1972, the southern research team hired a geneticist, Dr. W. T. Gladstone, who later became manager of southern forestry research. Gladstone accelerated internal tree-improvement efforts and reinforced the company's role in the already-successful North Carolina State University cooperative program. Extensive, carefully designed provenance tests proved that loblolly pine of North Carolina parentage significantly outperformed local strains in Arkansas and Oklahoma. Further research showed that any maladaptation risks could be offset by careful planning and site control. As a result, appropriate company land throughout the South now supports North Carolina trees.

Plant physiologists in the Hot Springs laboratory perfected early loblolly flowering to shorten the time between generations in tree-breeding programs—as had been proposed in the Potential for Tree Improvement in 1971. The company has established a third-generation seed orchard in Georgia to capitalize on the significant gains that resulted.

Weyerhaeuser also established fifty-acre, half-sib family plantations in North Carolina using seed from first- and second-generation orchards. Although casual observation reveals the expected family differences, careful assessment of eventual harvest yields will provide the first quantified measure of loblolly tree improvement benefits, vindicating, it is to be hoped, a long commitment. In the meantime, a 1986 appraisal of trials to date—including a twenty-two-year-old full-sib progeny test—concluded that first-generation improved seed will easily provide the projected fifteen percent gains.

A decision to adopt even-aged management on the Arkansas and Oklahoma loblolly forestland raised serious questions about post-harvest erosion and water quality. To obtain the necessary scientific input, Weyerhaeuser hydrologists undertook studies on fifteen watersheds ranging from four to fifteen acres in size. They equipped streams emerging from the watersheds with weirs to measure runoff and to permit sampling for sediments and dissolved nutrients. Foresters then harvested the watersheds using selective

cutting and clearcutting, with and without ripping. The results were compared with untreated controls. Their measurements indicated slight increases
in runoff and some increase in dissolved nutrients during the first year after harvest, with quick and complete returns to preharvest levels. This research, performed cooperatively with Oklahoma State University, the University of Arkansas at Monticello, and the U.S. Forest Service, provided
much scientific data on stream flows in a part of the country dependent on
clean water from managed-forest watersheds.

Today the southern U.S. forest, particularly that composed of loblolly
pine, is well along as an established and developing third forest—the forest
following the harvested second growth that developed naturally subsequent
to the removal of virgin growth earlier in the century. Southern forest research certainly has contributed significantly to the productive third forest
on Weyerhaeuser land. That forest includes genetically improved loblolly
pine in highly successful plantations receiving advanced silvicultural treatment and, ultimately, harvested with efficient machinery and methods. An
even better-quality fourth forest can be expected.

STRATEGIC FOREST RESEARCH

Attention to return on investment dictates that industrial forest research
emphasizes applied research, technology transfer, and technical service.
Nonetheless, Weyerhaeuser's research program has not neglected fundamental forest research, often identified as exploratory research or strategic biology. Though the company has contracted a fair amount of such research to
third parties, particularly universities, much has been performed internally,
too.

Tissue-culture research, for example, has been identified as exploratory
research since its inception in 1972. "The Potential for Tree Improvement"
described tissue culture as a potentially useful means of shortening generation time. In 1973, soon after the Institute of Paper Chemistry successfully
propagated multiple whole plants from a single aspen seed embryo,
Weyerhaeuser began its own work by hiring a University of Georgia scientist versed in tissue-culture techniques. A few years later, the company provided a five-year, $1 million tissue-culture grant to the Oregon Graduate
Center and made other grants to the Institute of Paper Chemistry to help
continue tissue-culture work. In 1979, the company transferred its tissue-
culture studies to new laboratories at the Weyerhaeuser Technology Center
in Federal Way, Washington, where the technology was perfected. Though

tissue culture has not played a role in the accelerated tree-breeding program because of high costs, researchers are still pursuing it as a means of multiplying highly selected tree families and individuals.

Multiplying individual trees by rooted cuttings was also identified as exploratory research until the methods were proven practical. Other research efforts similarly received the exploratory label. The line between exploratory and applied research, never clearly drawn, has often been ignored—not unlike the division between research and operations.

FINAL THOUGHTS

Whether because of the zeal of early professional foresters working in an environment where they had to make their own waves, or because their company truly perceived its future in capturing the full productive capacity of its extensive forestlands—or both—forest research has played a substantial role in Weyerhaeuser's present corporate health and its position as a major player in industrial forestry and in the development of forest science. Two factors set the company's brand of industrial forest research apart from more conventional research:

A nearly indistinguishable line between research and field foresters and their respective problem-solving efforts.

Operational implementation of promising research results before those results are experimentally verified by standard probability measures.

From the beginning, Weyerhaeuser forest researchers and field foresters have worked closely to locate, establish, and tend experiments of all kinds. Joint meetings permit efficient communication of research results and help identify problems amenable to research solutions. Researchers frequently see immediate results from the application of their efforts—a reward often as appealing to the scientist as formal publication. Researchers also recognize that technical service need not be rigorously separated from conventional academic research, or even clearly defined, in the industrial setting. Technical service is essentially another means of communicating research results.

Further, sound business decisions may be based on 50/50—or 51/49— chances of success. That working hypothesis has meant that company researchers often see their findings applied before statistical evaluation is even available. Instead of depending solely on small, carefully measured test plots, operational tests of fertilization, for example, often include "skips" where no fertilizer is applied so that effects will be immediately visible.

175

Although these factors set Weyerhaeuser's industrial forest research apart from more conventional research, the company's research organization nonetheless has developed into a truly professional scientific organization. Sabbaticals granted to a number of company researchers enabled them to participate in overseas, operational, or university assignments. Titles such as senior science advisor, denoting a scientist unburdened by administrative duties, were bestowed first on Staebler and later on others, including Roy Stonecypher and Lawrence. In 1993, Weyerhaeuser recognized forest research's contributions to the company by assigning vice presidential status to Dr. Rex McCullough, director of forest research. By the end of 1994, eight other Weyerhaeuser vice presidents had been forest researchers earlier in their careers. They include Dr. Norman E. Johnson, senior vice president of technology, responsible for corporate strategic research and information technology.

Regardless of the titles they may have achieved, though, every scientist who has ever been part of the Weyerhaeuser research staff contributed greatly both to the research organization and to the company. Many of those named here were the company's first specialists in a given discipline or played a part in pivotal periods for the research organization. Many more important individuals contributed, or are still contributing, to the critical mass of knowledge, experience, and determination required to shape a successful, professional scientific organization with significant impact on the company and on forest research.

The contributions of Weyerhaeuser's industrial forest researchers have been recognized outside the company. At least twenty company researchers have become university professors. Research staff members are frequently sought for their industry experience and their ability to lend a business perspective to forestry curricula. In addition, several currently employed Weyerhaeuser forest scientists teach as adjunct professors at nearby forestry schools. The research staff also participates in numerous collaborative research efforts with universities, government agencies, industry associations, and private research firms.

The forest research organization also sponsored two of the company's research and development science symposiums. The first, "Forest Plantations: The Shape of the Future," took place from April 30 to May 3, 1979, during the tenure of Dr. J. Laurence Kulp, vice president of research and development. Scientists from around the world participated and presented papers intended to clarify the future of forestry and the forest industry when it becomes completely dependent on man-made forests. The second forestry symposium took place in August 1994. It addressed "Forest Potentials: Productivity and Value." International scientists again participated by addressing the theoretical basis for believing that forest productivity can be improved much beyond the current state of the art. These significant symposiums helped demonstrate the role of private forest research in the advancement of forestry science and the profession.

Before the inauguration of High Yield Forestry in the 1960s, Weyerhaeuser's annual forest research budget was less than a quarter of a million dollars: the approximate cost of planting a harvested township of Douglas-fir. The research staff members were reminded to ask themselves if their efforts were worth a township-sized plantation. By 1979, the forest research budget under Director of Forestry Research Conor Boyd approached $10 million annually. After several years of reductions in the early to mid-1980s, during the tenure of Director Mark R. Lembersky and thereafter, the budget for 1995 again neared $10 million.

Industrial forest research is not easy. The bulk of it is long term, as far as the effects on current annual profits. It requires a distinctly visionary view— not unlike the industry itself, which accepts a future of managing repeated forest growth and harvest in thirty- to seventy-five-year cycles. Companies that possess the required vision substantially ease the researcher's path. Hence, forest research at Weyerhaeuser has flourished, contributing significantly not only to the company itself but also to the forestry profession and its underlying sciences.

Supervisors/Directors of Weyerhaeuser Forest Research

Donald G. McKeever	1942–1948
Paul Lauterbach	1948–1955
William H. Cummings	1955–1961
George S. Allen	1961–1966
George R. Staebler	1966–1979
Conor Boyd	1979–1981
Mark R. Lembersky	1981–1984
William H. Lawrence (acting)	1984
Rex McCullough	1984–present

PART TWO

Research Fields

DAVID BRUCE

STATISTICAL METHODS IN FORESTRY RESEARCH

INTRODUCTION

In the original sense of the word, "Statistics" was the science of State-craft; to the political arithmetician of the eighteenth century, its function was to be the eyes and ears of the central government. It could tell the Prince how many able-bodied men he might mobilize, how many would be needed for the essentials of civil life; how numerous or how wealthy, were sectarian minorities who might resent some contemplated change in the laws of property, or of marriage; what was the taxable capacity of a province, his own, or of his neighbors.[1]

"Efforts to obtain information about every member of the target population through a census go back to at least the third millennium BC in Babylonia, China, and Egypt. The purpose was usually to provide a basis for taxation; or proscription."[2] In 1662, John Graunt first demonstrated the value of reducing such masses of data to meaningful tables. When not predicting the return of a comet, Edmond Halley improved on Graunt's tables of life expectancy. In the next hundred years, wealthy gambling patrons of mathematicians stimulated the production of models of chance events and calculations of probabilities. At the same time, astronomers dealt with uncertainties in their measurements of the positions of stars and of the movement of planets. Carl F. Gauss published the method of least squares in 1808, but because it required laborious calculations, its early use was confined mostly to astronomy.[3]

181

Many natural phenomena were studied in the eighteenth and nineteenth centuries with descriptive statistics and sometimes with graphs used to show relations between two variables. The keen observations of the investigators and their practical knowledge of the subject matter provided penetrating insights into their subjects of study. Such study methods are still needed to provide a sound basis for asking the important questions that can be convincingly answered by well-designed objective quantitative studies supporting valid inductive inferences.

Modern statistical theory was developed not to meet the needs of actuarial science, gambling, and astronomy but for the needs of medicine, genetics, agronomy, and other blossoming sciences in England. The methods of exact measurement and statistical analysis in biological settings described by Francis Galton in 1889 in "Natural Inheritance" inspired biologist Walter F. R. Weldon and applied mathematician and philosopher Karl Pearson (at University College, London, from 1891) to study this area. Pearson and Weldon edited the still-prominent journal *Biometrika* starting in 1901. Pearson also contributed many mathematical refinements, including chi-square tests and correlation coefficients.[4] In 1908, William S. Gosset (using the pseudonym "Student") published the t-test.

In England in 1912, Ronald A. Fisher, who later profoundly influenced modern forestry statistical methods, published his first paper on statistics on use of maximum likelihood. This was followed in the next twelve years by a series of papers that laid the foundations of statistics as a separate scientific discipline with a great potential for application in all sciences.[5] Before this, mathematicians had regarded deviations as nuisances in fitting exact distributions. But now, biologists found that the deviations in samples not only helped to describe the great variation encountered in nature but also to distinguish between differences attributable to chance and those from other causes. A major advance in statistics was the development of small sample theory.

A few American foresters interacted with Fisher before World War II and introduced modern statistical methods into forestry research in the United States. This interaction resulted from a series of individual decisions and actions; the failure of any one could have disrupted the entire sequence. It seems a matter of chance, not any kind of planning, that it was the United States Forest Service research organization that led the way to the introduction of modern statistical methods in forest research before World War II. Had this not happened, the introduction might have been delayed for twenty or more years. It is a fascinating story, with many loose ends, that includes a game of musical chairs and describes the development of a solid foundation for the research needed today in America's forests. This introduction was influenced through time, by changes in knowledge of statistical methods, by improvements in computing devices, and by refinements in research planning.

Because there is no complete record of what happened and those who know parts of the story do not agree on all details, I will try, as best I can, to set the record straight. I will not cover developments in the past twenty-five years, which omits the great proliferation of electronic computers and the general inclusion of statistical training as a requirement for advanced science degrees at universities.

EARLY FORESTRY IN THE UNITED STATES

A congressional bill drafted in 1874 led to Franklin P. Hough's four "Reports on Forestry" and to his appointment in 1876 as forestry agent and then, in 1881, as chief of the new Division of Forestry in the U.S. Department of Agriculture (USDA). The bill provided for presidential appointment of a person of approved scientific attainments to report on forestry, someone who also knew statistical methods and was familiar with forests. Hough's reports include descriptive statistics showing the extent, content, and problems of America's forests, their potential as a basis for commerce, and much more.[6]

The third chief of the Division, appointed in 1886, was Bernhard E. Fernow, a leader in the American Forestry Association, who had been educated as a forester in Germany. He set up research programs, among which were tests of timber strength, and he initiated cooperative forestry projects with the states, including tree planting in the Great Plains. He prepared many bulletins, circulars, and other publications selling forestry to the American public. He and several others advocated the setting aside of forest reserves, which started in 1891, but he did not have the opportunity to participate in their administration. His vision of intensive European-style management of public forests has never been realized.[7]

When Gifford Pinchot became chief of the Division in 1898, he continued the research, cooperation, and education efforts. In 1905, the Division of Forestry, its name changed to "Forest Service" (USFS), was assigned responsibility for the administration of the forest reserves, which two years later were renamed, "national forests." This change in assignment resulted in the rapid growth of a decentralized administrative branch of the Forest Service. Although in many ways subordinated to the administrative branch, the investigative function of the Forest Service continued. In 1902 one-third of the division's budget went to investigations, but this diminished with time.

In 1908 a "forest experiment station" was established at Fort Valley, Arizona, and in 1910 the Forest Products Laboratory (FPL) began at Madison, Wisconsin. Between 1911 and 1917, under Chief Forester Henry S. Graves, the Forest Service established seven more "experiment stations" as headquarters for field investigations. Although some were abandoned during World War I, others, such as Fort Valley, Wind River, and Priest River, still exist as "experimental forests."

In this same period, most of those involved in the later interaction with Fisher were members of the Forest Service. Two exceptions were George W. Snedecor and Francis X. Schumacher. Snedecor had taught mathematics for eight years before becoming an assistant professor of mathematics at Iowa State in 1913. He started teaching statistics courses at Ames in 1915. Schumacher was in the Medical Corps of the Ohio National Guard in 1916 on active duty near the Mexican border.[8] He probably had completed a year or more of college study before this.

In 1915, Earle H. Clapp was appointed assistant forester in charge of research, after two years in the Washington Office (WO) of the Forest Service as assistant to William B. Greeley, who was in charge of the Silviculture Branch. Before this, Clapp had been assistant district forester in Albuquerque, New Mexico. In 1915, Edward N. Munns was in charge of the Converse Experiment Station in southern California. Also, in 1915, two District One foresters from Montana had been recruited by Dean Walter Mulford to become professors in the new forestry school at the University of California in Berkeley. They were David T. Mason, assistant district forester in silviculture, and my father, Donald Bruce, supervisor of the Flathead National Forest.[9]

From 1915 on, the forestry school in Berkeley, California, taught a different kind of mensuration than most other forestry schools. At California, mensuration included the mathematical basis for the practical side of the art. Bruce had a classical education, including four years of Latin and five of Greek, before he majored in mathematics and science at Yale. Because his father, who taught modern languages in Easthampton, Massachusetts, died in June 1906, the month Don graduated, he returned home, where he taught mathematics and physics for two years at his preparatory school.[10]

After this, he studied forestry at Yale and, in 1910, entered the USFS in District One. He was assigned to timber sales early and wrote the first of two dozen articles on mensuration in 1912.

In 1917, many foresters joined the war effort. Two forestry engineer regiments went to France with the American Expeditionary Forces to operate sawmills and procure lumber and fuelwood for cantonments and trench warfare. Of those named, Graves, Greeley, Mason, and Bruce went to France with the 10th or 20th Engineers, and Schumacher went as a lieutenant in the infantry. Most of the others had special wartime assignments.

After 1919, many changes occurred. Colonel Graves returned to his job as chief of the USFS, but illness led to his retirement and the appointment of Colonel Greeley as chief in 1920. Major Mason decided against teaching as a career and started working for the Internal Revenue Service on timber taxation and as a forestry consultant in Portland, Oregon. Captain Bruce returned to Berkeley to teach mensuration, logging, and forest improvements. Captain Schumacher, who had received the Distinguished Service Cross in Belgium, earned a B.S. degree in forestry in 1921 at the University of Michigan. Munns moved to the San Francisco District Office after completing his wartime assignment. In 1920, Joseph A. Kittredge, Jr., was appointed head of the Office of Experiment Stations in the Forest Service's Washington Office to administer the remaining forest experiment stations still at field locations.[11]

THE INTRODUCTION OF FORESTERS TO STATISTICAL METHODS

Professional forestry training in the United States started at Cornell in 1898 under Fernow, where it lasted but five years. Forestry education at Yale started in 1900 and at several state agricultural colleges shortly thereafter. Most curricula included the practical side of mensuration but little of the underlying theory. The statistics courses of the day were taught in the mathematics department on the other side of the campus, mostly for mathematics majors. With computing done by hand, books of tables to assist computation were common. In 1910, only half of the engineering schools in the United States taught the use of the slide rule, although practical models had been devised more than a century earlier.[12]

Foresters were first introduced to modern statistical methods by a few mensurationists who learned the methods soon after they were developed. In the USFS these mensurationists often had the job title "silviculturist." Later the USFS had "statisticians" and ultimately "biometricians," which reflects the current scope of the field of work. In forestry, the latter two seem synonymous. A mensurationist handles measurement of diameter, height, and volume of standing and felled trees and also estimates of stand volume and stand growth. When nationwide forest inventories started in 1929, the field began to split into "inventory" and "growth and yield," both still requiring basic tree volume determination. Although several histories of forest inventory work have been written, only Timothy G. Gregoire focused on the statistics involved.[13] My experience has been mostly in forest measurements and growth and yield.

In 1924, Kittredge published an article on "Use of Statistical Methods in Forest Research." He said that forest research works with measurements, and that statistical methods facilitate the analysis and interpretation of masses of numerical data. Calculating averages and drawing curves through points on graph paper are examples of the simple statistical methods applied almost daily in forest research. Most of the methods, including those that might be most useful, were unknown or little known to foresters. The amount of calculation in application is large, and the data to be collected in any project should be carefully considered to decide if the necessary and laborious computations are justified.[14] Kittredge brought out not only the general lack of familiarity by foresters with statistical methods but also the primitive computing capabilities of the period.

In 1925 Bruce said:

> Today, there are perhaps a dozen foresters in America . . . who have any real grasp of the statistical method and most of these feel keenly their inexperience . . . the task ahead involves first of all the adoption of this new technique by the profession. The schools must teach it, research men must become expert in its use, and the profession as a whole must become sufficiently familiar with the fundamental conceptions involved to understand the work of the investigator and to interpret and apply his results.[15]

SOME DEVELOPMENTS AFTER WORLD WAR I

In 1919 Fisher considered offers of two posts: one at Galton Laboratory, University of London, working under Pearson, the other at the Rothamsted Experimental Station near London. He chose the latter because he felt that opportunities for independent research would be greater there. Also, he was more interested in finding ways to use statistical methods in practical problems than in developing elegant theories. He stayed at Rothamsted until 1933, when he accepted the Galton Chair of Eugenics as Pearson's successor. His 1919 appointment initially required him to analyze further and extract hidden truths from the mass of data relating to old experiments in fertilizing various crops—"raking over the muck heap," as he described it. The first step in resolving some of the problems he encountered was adaptation of analysis of variance to analysis of replicated experiments; the second was introduction of randomization, replication, and local control. This work led to new ideas about design and analysis of experiments.[16]

Fisher was outspoken about many things. After the first edition of his *Statistical Methods for Research Workers* appeared in 1925, some mathematical statisticians criticized Fisher's failure to give precise derivations. He had not persuaded them that practical application of the statistical method was more important than its mathematical demonstration. Fisher rebutted this viewpoint:

> I want to insist on the important moral that responsibility for the teaching of statistical methods in our universities must be entrusted, certainly to highly trained mathematicians, but only to such mathematicians as have had sufficiently prolonged experience of practical research, and of responsibility for drawing conclusions from actual data, upon which practical action is to be taken. Mathematical acuteness alone is not enough. My revered teacher, Professor Whitehead of Cambridge, used to say in one of his courses: 'The essence of applied mathematics is to know what to ignore.' And when I read current publications in mathematical statistics I am continually and forcibly reminded of the wisdom of this remark.[17]

In 1921 the U.S. Department of Agriculture sponsored a school in Washington, D.C., that offered advanced training for its scientific workers.[18] The USDA Graduate School was unofficial, and training took place outside departmental hours; the school was self-supporting through fees collected from students. Universities accepted its credits, and it cooperated with universities in Washington. The courses first offered included statistical methods and seven other scientific courses. After 1925, Fisher's *Statistical Methods for Research Workers* served as the text book. One of the first students in statistics at the Graduate School was Mordecai Ezekial, who later taught the school's courses in statistics and published *Methods of Correlation Analysis* in 1930. Another eminent instructor, W. Edwards Deming, joined the teaching staff in 1930 and from 1933 through 1953 was head of the USDA Graduate School Department of Mathematics and Statistics. He had started working for the USDA in 1928, transferred to the Census Bureau in 1939, and resigned in 1946 to become a consultant and professor at New York University until 1975. He gained enduring fame by introducing quality management to Japanese industry after World War II.

In 1921, after graduating from the University of Michigan, Schumacher joined the staff of the forestry school at Berkeley, California. One of his duties consisted of assisting Bruce with a study of redwoods. Schumacher reported, "In the evenings we played chess, or he would discourse on statistics—my first introduction to the subject."[19] Schumacher also took statistics courses from the Mathematics Department at the university.

In 1924 Snedecor cooperated with Henry A. Wallace at Iowa State College in conducting Saturday seminars in statistics, including machine calculation of correlation coefficients and regression lines. Wallace borrowed card-handling business machines from an insurance company and brought them to Ames for the seminars. The Iowa State bulletin, "Correlation and Machine Calculations," by Snedecor and Wallace (1925) attained worldwide distribution. "In 1927 Snedecor and A. E. Brandt became directors of a newly created Mathematics Statistical Service, to provide a campus-wide statistical consulting and computation service. It was the forerunner of the Statistical Laboratory, organized in 1933 which was the first organization in the US to recognize statistics as a separate science."[20] In 1931 Iowa State awarded the first M.S. in statistics to Gertrude M. Cox. The combination of the laboratory and the degree program proved unique among universities, and Ames became the Mecca of many outstanding statisticians.

In the summer of 1931, Fisher accepted Snedecor's invitation to lead a statistical training session at Ames. Fisher's daughter described this in the biography she wrote. He gave three lectures a week either on genetics or on theory of statistics. He also held biweekly seminars, supervised research projects by several students, and attended three special weekend meetings that drew speakers and auditors from surrounding states. Accustomed to English summers, Fisher suffered from the heat, partly because Prohibition had not yet been repealed, and beer was not readily available.[21]

In the early 1920s, the headquarters functions of some of the USFS field experiment stations were transferred to the cities in which there were District Offices. When Munns moved from San Francisco to the Washington Office in 1923 to take charge of the Office of Experiment Stations, the organizational directory [1924] listed the Forest Products Lab, six experiment stations, and three experimental forests. While in San Francisco, Munns had known personally the staff of the forest school in Berkeley. Roy A. Chapman once told me that Munns was the first person in the Washington Office to promote the use of Fisher's new statistical methods.

Clapp was Munns' immediate superior. Clarence L. Forsling described Clapp's research management philosophy: "He used to say, 'There are three essentials to do well in research: one, you must work on a significant problem; two, you must have adequate funds; and three, you must have quality personnel.' [Also] 'Be sure that when you find young ambitious persons around, and you are confident that they are above average, get them into research—that's the kind of personnel we want.'"[22] Apparently, Clapp also believed in giving such people full freedom to do the work they personally viewed as important.

Edward N. Munns, an early proponent of statistical methods, headed the Office of Experiment Stations for the U. S. Forest Service.
(FOREST HISTORY SOCIETY PHOTO)

USFS—WASHINGTON OFFICE 1924-1931

When Bruce decided to move east in 1924, he asked Dave Mason, then in Washington, D.C., to help him find the best job opportunity. A few days later, Mason happened to have lunch at the Cosmos Club with three prospective employers. The best offer came from Bill Greeley, but Mason outlined a total of five possible jobs.[23] As a result, Bruce decided to work half-time with the Forest Service and half-time with Dave Mason. In 1929, I visited my father's two offices: the Forest Service in the Atlantic Building and his consulting office in the Albee Building. His meticulous records apparently avoided any charges of conflict of interest, but this arrangement probably could not exist today.

When Bruce left Berkeley the game of musical chairs began. Schumacher replaced him and continued the development of a rigorous approach to mensuration. After Bruce arrived in the Washington Office in July 1924 as head of Forest Measurements, one of the first things he and Munns did was promote the purchase of a Hollerith machine to facilitate computations. Punching, sorting, and tabulating cards was a gigantic improvement over computing with hand-operated mechanical calculators. Edward I. Kotok remembered using the Hollerith machine many years after its purchase.[24]

In Washington preparation of volume and yield tables for southern pines awaited completion. To assist with this, Lester H. Reineke transferred to Washington from the Appalachian Forest Experiment Station, Asheville, North Carolina. As well as handling this major job, which culminated in the printing of USDA Miscellaneous Publication 50, *Volume, Yield, and Stand Tables for Second-Growth Southern Pines*,[25] Bruce and Reineke devised many ways to make data compilation and analysis easier. They added several *Journal of Agricultural Research* articles on mensuration to the two dozen that Bruce had previously published elsewhere.

Other transfers to the USFS Forest Measurements group in the Washington Office included V. A. Clements from California in the spring of 1928 and Bert R. Lexen from Arizona in 1929. By this time a decision had been made to assign promising young mensurationists to the Washington Office for training for a year or more, with the objective of having at least one well-trained mensurationist at each experiment station. Reineke moved to the California Station in 1929 or 1930 and stayed there until after Clements returned from Washington in 1931. Then Reineke moved to the Northeastern Station where he was "mensurationist" until World War II, when he went to the Forest Products Lab as an engineer. Lexen returned to the

Southwestern Station in 1931. It is not known if others went to the Washington Office on details not listed in the USFS organizational directories. In 1934 there were mensurationists, or people specially trained in mensuration, at ten of the eleven experiment stations.

Schumacher reported an important development proposed by Bruce in this period:

> The first idea of a text in forest mensuration came to him (Don) in 1927 while he had been 'idled' for three days in a Florida town by floods. During this time he wrote pages of notes. These he sent me for review with the suggestion that we pool our efforts toward a complete text on forest mensuration. Between 1927 and 1950 we corresponded irregularly, particularly concerning the materials in the three editions of the work. The good reception of our text-book over these 30 years is primarily due to Don's facility in writing and teaching the subject.[26]

In 1927 Schumacher had no plans for moving east, and it is not known if either worked much on the book before 1930. The writing of the book was certainly affected by Schumacher's resignation from the forestry school in California and his move to the Washington Office to become head of Forest Measurements in 1930. At the end of the year Bruce resigned and went to work full time with Dave Mason. Bruce continued the consulting work he had been doing; planning and demonstrating good forest management on large private timber land holdings.[27] In 1931, he and Mason presented a study on sustained yield to Hoover's Timber Conservation Board. I believe he felt that when completed, the forest mensuration text book would serve to summarize his major contribution to forestry up to that time and that future developments in mensuration and statistics were in good hands. Indeed, Schumacher enthusiastically introduced Fisher's new statistical methods to two generations of foresters.

In 1935, *Forest Mensuration* by Bruce and Schumacher was published.[28] Its writing probably consumed many evenings and weekends. In the preface, they acknowledge continuous encouragement during the whole course of the work by Munns, and the searching review of the manuscript and many constructive criticisms by Chapman.

USFS—WASHINGTON OFFICE 1931–1938

After Schumacher went to the Washington Office in 1930, he studied statistics under Deming in the USDA Graduate School and enrolled in George

Washington University for courses in mathematics. After he started statistical training sessions for foresters, his trainees generally began with the elementary agricultural statistics course based on Fisher's book at the USDA Graduate School, and they took other related courses at the Graduate School and local universities.

Schumacher's first trainee was Verne L. (Les) Harper, who had been selected by Munns in 1930. Harper had been working for the Southern Forest Experiment Station at Starke, Florida, since he graduated from the University of California in 1927. In December 1930 he took to the Washington Office his data on a naval stores study. He enrolled in the USDA Graduate School course in statistics, and Schumacher helped him with the analysis of his data. When Harper returned to Lake City, Florida, late in 1931, a new staff member there was James G. (Ted) Osborne, who had just graduated from the University of California. After Harper left Washington, Chapman moved to the Washington Office from the Southern Station to start a detail that lasted two or three years to analyze data he brought with him. He also studied mathematics and statistics at the Graduate School and a local university.

Soon after Harper returned to Florida in 1931 he was put in charge of the research there.[29] He instituted the requirement that a written study plan be prepared and approved before work started on any study. This ensured that most of the procedures, needed supplies, and labor were planned ahead and also that no major disruption in the study would occur should the initial investigator be unable to complete it. Study plan review also made it possible for a qualified statistician to help iron out problems of quantification and analysis, and avoid collection of masses of data from which no meaningful information could be extracted. As Harper was promoted to greater responsibilities in forest management research, first in 1935 in New Orleans, Louisiana, and then in 1937 in the Washington Office as division chief of Forest Management Research, written study plans and problem analyses were required in the units he supervised. When he became deputy chief of research in 1951 he applied the requirement for written plans to all USFS research.

Fisher would have endorsed these requirements because he advocated involving statisticians in research planning to avoid collecting data that was hard to analyze or contained little information. He said that while laborious calculations might increase the information yield of inferior data by possibly five percent, well designed data collection could get ten to twelve times the information at the same cost. "To consult the statistician after an experiment is finished is often merely to ask him to conduct a post mortem examination. He can perhaps say what the experiment died of. To utilize . . . [a statistician's] . . . experience he must be induced to use his imagination, and to foresee in advance the difficulties and uncertainties with which, if they are not foreseen, [the] investigations will be beset".[30]

While Chapman was still in Washington, Schumacher tried a new format for training. He reduced the training period to nine months and increased the number of trainees. Charles A. Connaughton and George M. Jemison attended the 1932–33 training session and mention it in their oral histories.[31] Munns had recruited both of them, and they were detailed to the Washington Office to work on their two sets of field data, relating to forest fires. They took training courses at the USDA Graduate School, one based on Fisher's textbook, and the other on correlation analysis. Their fellow trainees were G. Luther Schnur, Gordon D. Merrick, and E. M. Hornibrook.

Statistical training sessions at the experiment stations started in the early thirties, as did individual statistical advice by those trained in the Washington Office. After he returned from Schumacher's training, Jemison provided the earliest formal training in Missoula in 1933. He taught six or eight staff members every morning for two or three weeks, including Director Lyle Watts.

In 1935, Fisher published *Design of Experiments*.[32] This added much new information to the ideas compressed into a single chapter in his earlier book.

My record of Schumacher's other training sessions remains incomplete. It appears that after two more sessions, the period of training was again shortened. In December 1935, Philip A. Briegleb, along with several from other stations, went to the Washington Office from Portland for a newly instituted three-month training session. I believe that this session started the format followed later — morning lectures and afternoons to study or analyze data. In the fall of 1936, J. E. Lodewick went to the session, and in 1937 W. G. Morris spent several months in the Washington Office working on statistical methods and experimental design.[33] According to Jemison, C. Allen Bickford, Jesse H. Buell, and George H. Hepting each attended one of Schumacher's early training sessions, and I am sure that Osborne did.

In the summer of 1936 Schumacher, Lexen, and Besse B. Day (Forest Survey, California), and possibly other Forest Service employees, attended the second statistical seminar given by Fisher at Iowa State College. From mid-June to late July, Fisher gave three lectures each week either on design of experiments or on the theory of statistics. Day noted that Fisher helped her then with analysis and design of several difficult forestry problems.[34] Later that summer, Fisher traveled by train from Ames to Colorado, Michigan, and New York, ending up in Washington, D. C. Schumacher and several other foresters went by train with him to Asheville, North Carolina, where Fisher lectured in the morning and attended seminars some afternoons. This meeting was attended by about forty foresters, mostly from the USFS; seven came from other government agencies or universities.

The afternoon seminar subjects included a large hypothetical even-aged ponderosa pine study with cutting cycles of five, ten, twenty, and forty years installed in five geographic regions. They discussed questions about how to ensure uniformity of treatment among regions and among generations of foresters. Other subjects discussed included how to sample rainfall in a 2,000-acre watershed with an 1,800-foot range in elevation, and how to determine sampling error with systematic inventory designs. Studies were described that identified variation in fire-caused mortality of southern pines, affected by tree diameter, stand density, vegetation type, age and height of stand, and topography, and variation in gum yields of southern pines as influenced by tree diameter, growth rate, and depth and height of streaks.[35]

Statistics course participants at the Asheville, North Carolina, headquarters of the Appalachian Forest Experiment Station. R. A. Fisher is standing in the front row, center. C. L. Forsling (left) and F. X. Schumacher (right) join him.
(FOREST HISTORY SOCIETY PHOTO)

Finally two studies using Latin-square designs were described. One had been installed in Florida three years earlier. These may have been the first such study designs attempted in forests in America. Both needed missing plot techniques in their analyses: the 5x5 because a fire escaped and burned a plot scheduled for no-burning, and the 4x4 because the stand on one corner plot was much different from the other fifteen, and no shifting of boundaries could find a uniform square area. But what western foresters remembered best was a map of the plots. They were amazed that a Latin-square could be installed in such steep country—until they started laughing when they learned that the contour interval was one-tenth of a foot.

After the Asheville meeting, Schumacher and Osborne took Fisher by car through Georgia to north Florida. By the end of August Fisher was in Massachusetts, where he received an honorary degree from Harvard. After this he went to Princeton and Washington, D.C., and lectured at both places. Finally, he went to San Francisco, where he stayed a month while commuting to Berkeley to lecture. Phil Briegleb went to Berkeley for a week to attend some of these lectures, as probably did other western foresters unable to go to Asheville. The intensive training at Ames for a few and the lectures and seminars for forty foresters at Asheville did much to spread statistical knowledge in American forestry research and to encourage follow-up study.

Several years later Schumacher summarized his previous work and the Asheville meeting:

> The statistical method is indispensible [*sic*] in experimental forestry. Ever since the establishment of the eleven regional forest and range experiment stations about 1926, programs of in-service training in statistical method have been provided by the Forest Service and by the Graduate School of the Department of Agriculture. Perhaps the most notable of these was the course of lectures at Asheville, N.C., by Professor R. A. Fisher in 1936 on the design of experiments, attended by a group of forty foresters from the experiment stations and forestry schools. Fisher's appetite for practical problems, his ready comprehension of intricate detail, and his helpful advice on modes of solution were a stimulating experience to men somewhat accustomed to consider research in forestry as something unique.[36]

Two years after the Asheville meeting, while in India, Fisher said that biological research throughout the world bristled with problems to which only exact tests . . . could be applied. In a few years, more such tests were in general use among experimentalists. "The period which followed has shown the somewhat ludicrous spectacle of entomologists, foresters, plant physiologists and others with no trace of mathematical pretensions, applying mathematical refinements freely and with understanding in their daily work. . . . [These methods] . . . most official statisticians could not understand, and . . . too many teachers in mathematical departments were unable to expound."[37]

In 1937, Schumacher accepted an invitation to move to Duke University where he taught for twenty-four years. He was replaced in the Washington Office by Osborne, who had been his student at Berkeley. The following year Schumacher had this to say about developments during the past few years:

> The mensurationist has . . . the . . . conviction that inductive reasoning from observational data, involving . . . uncertainty, deserves better than rough and approximate . . . [methods]. He has sought and found in modern mathematical statistics improvement in computational technique. This . . . led to . . . the contemplation of defects in observational procedure. . . . He learned that in . . . biology and agriculture, the practical requirements of research had, through the work notably of R. A. Fisher and his associates, moulded mathematical structures . . . [for] observational programs. Finally he learned to devise observational programs in forestry investigations suited to unambiguous mathematical analysis and to exact tests of hypotheses. [38]

He also said the principal problem facing inventory using a systematic design was estimation of a valid sampling error, that use of volume tables should recognize both sampling errors within stands and the differences in tree form and volume between stands, and that yield investigations using measures of degree of stocking provided better estimates of growth than normal yield tables. He concluded that the specialist in each of the divisions of forestry research had learned to design experiments in his own field so as to maximize the amount of information derived, consistent with cost and values.

USFS—WASHINGTON OFFICE 1938–1942

According to Osborne,[39] his twelve-week statistical seminar took place almost every year from 1939 through 1955 with time out for the war. The seminars continued the pattern started by Schumacher in December 1935, with morning lectures and afternoon individual study. Trainees at Osborne's seminars received Graduate School credits for the course.

In 1939, the USDA Graduate School started its first correspondence course at the request of one of the "bureaus," a course based on Snedecor's 1937 book, *Statistical Methods*.[40] The bureau probably was the Forest Service, because Osborne was a friend of Deming, and some of Osborne's trainees were required to complete the correspondence course before going to the Washington Office seminar. This marked the beginning of a large variety of correspondence courses, which continue today, given by the Graduate School.

In 1940, Carl E. Ostrom attended Osborne's training course, along with eight to twelve trainees from other stations. In early 1941, William G. O'Regan helped Ted teach the statistics course. In 1942, Ted took leave to join the Air Force and work on statistical analysis of bombing results.

AFTER WORLD WAR II

In 1946 Osborne returned to the Washington Office and resumed his statistical seminars. Lewis R. Grosenbaugh was one of the trainees in the 1947 course. After Osborne taught the session that I attended early in 1948 and the one Robert W. Harris went to that fall, he asked Lew to teach the 1949 course, with some lectures given by Chapman. Later, Chapman and Austin A. Hasel each taught one training session. The usual group of trainees seems to have been one person from each station, about eight to ten students. These lengthy details must have taken a big bite out of each station's travel budget. I believe twelve sessions took place, so Osborne trained about one hundred foresters in statistical methods.

In the summer of 1946, Fisher taught for six weeks at the Statistics Institute of the University of North Carolina, which Gertrude Cox directed. He then held two meetings in the mountains of North Carolina: a week on the application of statistics to biological problems and the second on the development of modern statistics as a science. It is likely that Schumacher and Osborne were included in the twenty-seven invited to the first meeting. It appears that this was near the height of Fisher's influence on American statistical thought. His daughter reports, "When he returned to the United States in 1946, he was welcomed by the younger statisticians as a great originator and authority certainly, but as a foreigner whose ways were not always their ways, nor his thoughts their thoughts."[41] The new school of statistical thought led by Jerzy Neyman and Egon S. Pearson supplemented but did not replace most of Fisher's developments. About the only exception was the acceptance of their confidence limits instead of Fisher's incompleted work on fiducial inference.

In 1946, Osborne recruited Floyd A. Johnson, who had studied statistics under William G. Cochran at Iowa State, to join the Pacific Northwest Station in Portland. In 1947, Al Bickford moved to Northeastern Station headquarters in Philadelphia to become their statistician. At other stations, mensurationists trained by Schumacher were still available to give statistical advice. About 1952, a drive started to get university trained biometricians at all stations. This was made possible by the great expansion in such training after the war. Ted reported that by 1958 all but two stations had biometricians, and he implied that by 1960 the roster was complete.

In 1949, after he had taught Osborne's course in Washington, Grosenbaugh held a statistical training course at the Southern Station in New Orleans. This consisted of lectures and discussions each day for a week or more. There may have been other similar sessions elsewhere, but the only other group study reported to me was by Walt Smith. He, Herb Fleischer, and four others at the Forest Products Lab took a correspondence course in 1943 from Georgetown University based on Snedecor's book. I helped a similar group in Gulfport in 1949 with a retread of my statistical correspondence course from the USDA Graduate School, and I presume similar study occurred at other stations.

In 1952, Fisher and Frank Yates were at Blue Ridge, North Carolina, from June 6 until July 25. This forty-day session was broken into a series of fourteen conferences; with one or more attended by 196 persons. Fisher's daughter reported, "Fisher and Yates were . . . having a good time as they exposed the folly of purely theoretical approach to mathematical statistics; they incited each other like a couple of school boys to scorn the misguided ideas of pundits who had never done an experiment in their lives but were accepted as the modern authorities on statistics."[42]

I was told that Osborne was the person who decided to end the Washington Office statistical training in 1955. If so, it may be because he felt that with biometricians at each station this training could be done locally. But Station Director Robert A. Cowlin had this to say: "The cost in research time and money of the . . . program was not always justified by the results . . . It was a heavy dose for . . . those who did not have academic training in statistics . . . and for some who had previous training, a considerable part . . . was repetitious."[43] I suspect that the high cost, coupled with an increasing supply of research-oriented foresters with graduate training in statistics, and a diminishing number of older foresters without similar college or Washington training, were responsible for a decision approved by all to end the Washington Office seminars.

After 1955 Osborne did many things to promote statistics in the USFS. He convened three meetings of station statisticians: one in Washington, D.C. in 1957, another in Athens, Georgia, a year later, and a final one in Washington probably in 1960. At the first meeting, the subject matter included proposed training sessions at the stations and also Osborne's proposed research planning factor manual. In February 1957, Harper wrote to all stations, forwarding notes on the meeting, and requesting comments on a uniform statistical training program that would be conducted at all stations.[44] The second meeting continued these subjects, but broadened the consideration to other responsibilities of station biometricians, and how the Washington Office could support these. The final discussions centered on problems with computers; both acquisition and contracting for services. Computer problems lasted a long time; while George Furnival was biometrician in the Washington Office in 1966, he tried to help the stations. He told me that the computers favored by National Forest Administration were quite unfriendly to research.

Another major job Osborne undertook was amplifying and revising his lecture notes for use by station biometricians. He finished this task in 1961 and sent an 855-page mimeographed handbook to each station, to forestry schools that had requested it, and to some of his trainees.[45] He also prepared a two-volume handbook of research planning factors, comprised of systematic descriptions of research studies, their objectives, design, kind of measurements taken, and the variability observed in the study. He went to each station, inspecting their statistical services and needs and arranged advanced training sessions, such as a refresher course for biometricians at North Carolina State University, and a session in California on multivariate regression applied to log grade studies.

After Osborne retired from the USFS in 1961 to take the chair vacated at Duke University when Schumacher retired, questions about statistics that he had tried to settle continually re-emerged. Possibly the most important was the question of what comprised the major job of the biometrician. Station biometricians who had experience in biology before they received statistical training felt that helping scientists design and analyze their studies and programs had highest priority. Those starting with training in mathematical statistics felt they should study uses of statistics in forestry research. I believe Fisher would have sided with the former, but those who designed position classification reviews rated independent research much higher than what they considered to be merely a service job.

DECENTRALIZED STATISTICAL TRAINING AFTER 1955

After 1955, group training ranged from major efforts at three stations to little or none at others. In 1955 at the Pacific Northwest Station, Floyd Johnson prepared a 132-page mimeographed syllabus for a seminar in applied statistics.[46] That year he gave a two-week seminar to about twenty trainees at Wind River, Washington, and he repeated this in Portland in 1956 and 1957. Not satisfied with the apparent results of this kind of training, Johnson prepared two correspondence courses each with thirty lessons of about ten pages apiece. He reviewed each completed lesson and then discussed the responses with the trainee, making sure any missed concepts were clarified. From 1958 to 1976, when he retired, several hundred foresters and other scientists had started these courses, with about a third completing them.

In about 1957 at the Central States Station, Chester E. Jensen held a two-week statistical course in Ames, followed a few weeks later by a four-week session. These followed the customary daily pattern and were repeated a year or two later. When he was in Ogden at the Intermountain Station in the 1970s, Chet held several one-week training courses on interactive regression systems. Each was for five or fewer trainees, and over the years a total of fifteen to twenty individuals took the courses.

In 1957 Osborne recruited Bill O'Regan, who had studied statistics at the University of California in Berkeley, to become station biometrician in Berkeley. However, Station Director Jemison decided Bill's services were needed at field locations. So O'Regan started his peripatetic teaching career, holding classes at each field location where the staff needed his help. These rounds lasted several years, before he settled down in Berkeley with shorter trouble-shooting field visits. Soon after he arrived he scheduled morning lectures in Berkeley for the staff there. These were given twice a week for three or more months. This cycle was repeated twice.

In about 1958 and again about two years later, Frank Freese taught ten-week statistical seminars at the Southern Station in New Orleans, each with about a dozen trainees. He had worked at the station as a research forester for eight years, took the USDA Graduate School correspondence courses, but missed out on Osborne's seminar. In 1956 he took a year's leave and went to Iowa State and enrolled in every statistics course he could. When he returned to New Orleans in 1957 he became station biometrician. In 1963 he transferred to the Forest Products Laboratory.

Freese remains best known for the statistical handbooks he prepared. He based the first on the worked examples he wrote down each time he learned a new statistical procedure. Clem Mesavage and others saw these and asked for copies, so in 1956 Frank put them together as a handbook, published by the Southern Station as *Guide Book for Statistical Transients*.[47] Popular demand got a revision in 1959, a reissue in 1963, and in 1967 a remodeling into what is called USDA *Agriculture Handbook 317*. His sampling handbook with a similar genesis became USDA *Agriculture Handbook 232* in 1962. After they became USFS bestsellers, they were translated into Spanish, Japanese, and Turkish, and in 1973 Burgess Publishing combined them to sell in college bookstores. In 1995 a combined version was still on sale in the Oregon State University Bookstore in Corvallis. The Forest Products Lab published his *Linear Regression Methods for Forest Research* in 1964, which became as popular as his earlier handbooks.

After Jacob L. Kovner transferred from the Southeastern to the Rocky Mountain Station, he taught three statistical courses. The first was a two-week elementary course started in 1958, followed a year later by a four-week advanced course. Twelve to fifteen trainees participated in each session, and there were at least three cycles of this training. Later Kovner taught one or more two-week refresher courses for many people with previous statistical training.

At the Northeast Station, Al Bickford gave at least one three-week statistical training session in Philadelphia and two or more in New Haven, some in the early 1960s. At the Lake States Station in 1957 Suren R. (George) Gevorkiantz held an elementary course in statistics and an advanced course was planned for 1958. I found no one with a clear memory of any organized statistical training at the Intermountain Station by Meredith J. Morris, but some correspondence indicated that all stations had held at least one course. Although David O. Yandle and Frank Freese were biometricians at Forest Products Lab after 1959, neither conducted group training there. Yandle remembers a popular one-week course at Forest Products Lab on quality control given by a professor from Rutgers.

At the Southeastern Station, an elementary course was given in 1957 and another in 1958, with an advanced course planned for later that year. Tom Evans assisted by Jerry Clutter taught statistical courses at the Southeastern Station in Asheville, which may have been after 1958. In 1963, after Jerry started teaching at the University of Georgia in Athens, he arranged two three-week statistics courses there for Forest Service employees. An elementary course one year was followed by an advanced course the next. This cycle was repeated four times and several instructors taught at more than one course. These include George M. Furnival, Peter Dress, Warren E. Frayer, Al Bickford, Tom Evans, and others. This seems to be the first university effort to replace the local USFS training efforts, most of which ended at about this time. In the seventies, statistical training courses were started as continuing education opportunities, first at Colorado State University and later at several other universities. These university courses made possible statistical training for scientists at those stations where decentralized training had not become as effective as Osborne had visualized it.

In April 1970, the National Science Foundation sponsored a symposium at Fort Collins, Colorado, on biometric training at universities. There were fifty-five participants, twenty-four from forestry or natural-resource schools, eleven from statistical departments at universities, eight from the USFS, three from other agencies using biometry, and eight graduate students. All had graduate level statistical training, with some experts from forest schools and the USFS. Not surprisingly, they agreed that undergraduate programs in natural-resources biometry should be developed at interested institutions, either in the statistics department or the natural-resources department or both; that equal time be allotted to four areas—general university requirements, mathematics, natural resources, and electives; and that six areas in statistics be covered. Recommendations for graduate programs received less discussion but continued along the same line. A need for satisfying demands for continuing education in biometry was noted.

The report on this meeting stated:

Requirement of the [biometry] courses in natural resources curricula is almost a complete waste unless the faculty of the department(s) of natural resources are able and willing to incorporate modern quantitative science in their courses. . . . It is fully realized that many departments are not in the position to immediately adopt these recommendations. Some are; but, for many, it will require several years of transition and continuing education of faculty to increase their capabilities. It is anticipated that adoption of these recommendations will be realized within the next decade.[48]

CONCLUDING REMARKS

I have tried to name all of those involved in the chain of events that culminated in the in-service statistical training in the USFS. Bruce and Munns were among the first to recognize the need, and Munns actively supported early developments. Schumacher became a world authority on use of statistics in forestry, and developed the training program. Osborne followed through and trained more than one hundred in statistical methods.

As time thinned the ranks of these trained foresters, a younger generation with a wide range of statistical and biological education replaced them. Today, there are a few statisticians who, with better training and adequate biometrical research experience, outclass the previous generation. But also there are many with good mathematical training but inadequate research involvement who, once they gain more practical experience, can become good statistical mentors.

Today it seems there are not enough statistical mentors to go around. I feel it is more productive of valid inductive inferences and good forestry research to involve statisticians in planning studies than it is to ask them to conduct postmortems before or after questionable results are published. The potential benefits of good planning should be balanced against the track record of those trying to find new uses for statistical methods in forest research.

The new generation of statisticians and biologists with statistical training owes its existence to the great expansion of government support for higher education and scientific research after World War II. Specialists active now can answer questions no one knew how to ask fifty years ago. But government support brought with it grantsmanship, necessary to get funds to do worthwhile research. For some investigators, unfortunately, this can tend to make obtaining and maintaining funding more important than sound research results.

Current descriptive studies of the kind typical of the nineteenth century generate more hypotheses than can be tested with available funds. This is due in part to the increasing complexity of the studies and their large demands on money and space. Meeting these demands is justified by the consequences for land managers who fail to answer complex questions in a defensible way. But despite their brilliant concepts, some scientists skip gaily from one subjective study to the next, leaving the landscape littered with hypotheses that need solid reproducible studies to establish their validity. Recognition of the crying need for personal involvement of these concept generators in rigorous testing of such hypotheses was never more critical, especially in the currently popular field of ecological studies. A return to the fundamentals of valid statistical inference based on soundly designed quantitative studies would avoid embarrassment and recriminations later.

Some still seem to believe that the pseudo-replication of ten repetitions of the same anecdote gives nine degrees of freedom for testing its validity, especially when different people tell the story. Another distressing tendency is the rush of some researchers to advocacy, with the same objective as grantsmanship; getting more funds to support research. When scientists become advocates, they abandon objectivity. The resulting subjective pseudoscience may impress the politicians, the press, and the public, but not the scientific community.

Immediately after World War II, few if any foresters had access to mainframe computers. In about 1965, desktop electronic calculators began to replace automatic mechanical computers. These were soon supplemented by batch processing on mainframe computers. Development of computers continues, and personal computers today exceed the capabilities of yesterday's mainframes. This has resulted in a gradual but sustained improvement in analyses of numerical and other quantifiable data. It also has resulted in the proliferation of statistical and mathematical packages. Each of these includes many highly useful programs, but they also contain traps for the unwary. It is now easy to use inappropriate analysis techniques and statistical measures. Also it is tempting to reanalyze data that did not produce an anticipated result. This process stops when a significant difference is found, with no regard for how this data-dredging affects the probability level in statistical tests.

But the good seems in many instances to outweigh the bad. We used to dream about the robust analytical techniques now available. Where once there seemed to be an obsession with testing a single null hypothesis, as if this were all there was to a research study, now it seems more generally recognized that a good research study asks many questions. Most of these can be answered by quantitative comparisons that can be tested to see if observed differences are more likely than not to be due to chance variation alone. Where we used to see research reports that seemed mainly to be discussions of the statistical analysis that somehow overlooked the findings, today there are better presentations of study results supported by comments on the appropriate statistical measures.

In summary, it appears that the centralized statistical training between 1935 and 1955 in the USFS was a major investment needed to improve forest research in the United States. After 1925 the demonstrated improvements in research results using new statistical methods emphasized the need for such training. By 1955, a pool existed of research foresters with a background of field experience and USFS statistical training; most remained with the USFS, but some resigned to join universities or industries. The result was a general improvement in the quality of forestry research that came from the investigators' knowledge of statistical methods, and more importantly from their recognition of problems requiring expert statistical advice. Many trainees later advanced to supervisory positions in research where they could both demand proper experimental design and analysis and also interpret correctly the statistical measures quoted by investigators.

NOTES

1 R. A. Fisher, "Presidential Address, First Indian Statistical Conference, 1938," *Sankhya* 4 (1938): 14–17. I started to work in forestry research in 1937 at the Northeastern Station under Les Reineke, and a few months later in 1938, at the Southern Station I worked for Al Bickford. Much of what I learned about statistics came ten years later, at a twelve-week seminar taught by Ted Osborne and from three USDA Graduate School correspondence courses, one before the seminar and two after. To this first-hand experience I have added fifty interviews with old-timers and some library research.

2 E. L. Lehman, "Statistics, an overview," in *Encyclopedia of Statistical Sciences*, Ed. Samuel Kotz and Norman L. Johnson (New York: Wiley, 1982–1989), Vol. 8, pp. 683–702.

3 H. O. Lancaster, "History of Statistics," in *Encyclopedia of Statistical Sciences*, Vol. 8, pp. 704–711.

4 E. Seneta, "English Biometric School," in *Encyclopedia of Statistical Sciences*, Vol. 2, pp. 511–512. Lancaster, "History of Statistics", p. 707.

5 Lancaster, "History of Statistics," pp. 707–708. J. F. Box, "Ronald Aylmer Fisher," in *Encyclopedia of Statistical Sciences*, Vol. 3, pp. 103–111.

6 Harold K. Steen, *The U.S. Forest Service : A History* (Seattle: University of Washington Press, 1976), pp. 10–11.

7 Steen, *The U.S. Forest Service*, pp. 23-46. Char Miller, "Wooden Politics: Bernhard Fernow and the Quest for a National Policy, 1876–1898," *in The Origins of the National Forests*, ed. Harold K. Steen (Durham: Forest History Society, 1992), pp. 287–300.

8 *Forestry Education at the University of California, The First Fifty Years*, ed: Paul Casamajor, (Berkeley: the California Alumni Foresters, 1965), pp. 47–54.

9 Casamajor, *Forestry Education*, pp. 52–54.

10 Casamajor, *Forestry Education*, pp. 108–109.

11 Steen, *The U.S. Forest Service*, pp. 142–147. I have used USFS directories to identify people's assignments. Before 1920 these were issued quarterly and labeled "Field Program". Until about 1950 they were issued twice a year as "Directory," and after that as an annual "Organizational Directory." Complete sets are rare.

12 William Aspray, *Computing Before Computers*, (Ames: Iowa Univ. Press, 1990), p. 27.

13 Timothy G. Gregoire, "Roots of Forest Inventory in North America," *Proceedings of the 1992 Society of American Foresters National Convention, Richmond VA, October 25–27*, (Bethesda, MD: Society of American Foresters,), pp. 57–66. D. D. Van Hooser, N. D. Cost, and H. G. Lund, "The History of the Forest Survey Program in the United States." in *Forest Resource Inventory and Monitoring and Remote Sensing Technology—Proceeding of IUFRO Centennial Meeting*, (Tokyo: Japan Society Forestry Plan. Press,), pp. 19–27. Douglas S. Powell, William H. McWilliams, and Richard A. Birdsey, "History, Change and the U.S. Forest Inventory," *Journal of Forestry*, 92 (1994): 6–11.

14 Joseph Kittredge, Jr., "Use of Statistical Methods in Forest Research," *Journal of Forestry* 22 (1924): 306–314.

15 Donald Bruce, "Forest Mensuration Today," *Journal of Forestry* 23 (1925): 282–289.

16 Frank Yates, "Appreciation of Sir Ronald Fisher," *Biometrics* 19 (1962): 442–447.

17 Fisher, "Presidential Address," p. 16. Ronald A. Fisher, *Statistical Methods for Research Workers*, Sixth Edition, (London: Oliver and Boyd, 1936), p. 339.

18 Anonymous, "Teaching Statistics at the Department of Agriculture Graduate School in Washington," *Biometrics Bull.* 1 (1945): 33–34. Centennial Committee, U. S. Department of Agriculture, "Century of Service—the First 100 Years of the United States Department of Agriculture," processed copy of typed reports, various paging, ink numbers at bottom of pages, n.d. [USDA was established in 1862], pp. 14–15, 23, 30, 35, 45–46, 90, 101. Nancy R. Mann, "In Memoriam, W. Edwards Deming, 1909–1993," *Journal of American Statistical Association*, 89 (1994): 365–366.

19 Anonymous, obituary "Donald Bruce, '10, 1884–1966," *Yale Forest School News* (April 1967): 18.

20 T. A. Bancroft, "George Waddell Snedecor", in *Encyclopedia of Statistical Sciences*, Vol. 8, pp. 526–527.

21 Joan Fisher Box, *R. A. Fisher—The Life of a Scientist* (New York: John Wiley and Sons, 1978), pp. 313–323.

22 Clarence L. Forsling, "Concerns of a Pioneer Western Researcher," Oral History Interview by Elwood R. Maunder, 1978. Forest History Society, Santa Cruz, CA, pp. 319–320.

23 David T. Mason to Donald Bruce, April 1, 1924, author's file.

24 Edward I. Kotok, "The U.S. Forest Service: Research, State Forestry, and

FAO," Oral History by Amelia R. Fry, 1963, Resources for the Future and the United States Forest Service, pp. 186–196.

25 U.S. Forest Service, *Volume, Yield, and Stand Tables for Second-Growth Southern Pines*, (U.S. Dept. Agr., Misc. Pub. 50, 1929), 202 pp.

26 Anonymous, obituary, "Donald Bruce."

27 Rodney C. Loehr, ed., *Forests for the Future, The Story of Sustained Yield as Told in the Diaries and Papers of David T. Mason, 1907–1950* (Saint Paul, MN: Forest Products History Foundation, Minnesota Historical Society, 1952), pp. 55, 63, 68, 82, 109.

28 Donald Bruce and Francis X. Schumacher, *Forest Mensuration* (New York: McGraw Hill Book Co., 1935), p. 360.

29 Verne Lester Harper, "Research Planning at the Field Level", Oral History Interview by Elwood R. Maunder, 1978, Forest History Society, Santa Cruz, CA, pp. 1–6.

30 Fisher, "Presidential Address," p. 17.

31 Charles A. Connaughton, "Forty-three Years in the Field With the U. S. Forest Service," Oral History Interview by Elwood R. Maunder, 1976, Forest History Society, Durham, NC, pp. 29–31. George M. Jemison, Oral History Interview by Elwood R. Maunder, 1978, Forest History Society, Durham, NC, pp. 156–168.

32 Ronald A. Fisher, *The Design of Experiments* (London: Oliver and Boyd, 1935), p. 252.

33 Robert W. Cowlin, "Federal Forest Research in the Pacific Northwest, The Pacific Northwest Forest and Range Experiment Station," n.d. [1971?] typewritten manuscript, 549 pp., Pacific Northwest Research Station, Portland, OR.

34 Box, *R. A. Fisher*, pp. 427–430.

35 Anonymous, "Proceedings (Afternoon Discussions) Statistical Methods Seminar, Dr. R. A. Fisher," Auspices of the United States Forest Service, Asheville, North Carolina, Aug. 17–26, 1936. 70 pp., typewritten.

36 Francis X. Schumacher, "Statistical Method in Forestry," *Biometrics Bull.* 1 (1945): 29–32.

37 Fisher, "Presidential Address," p. 16.

38 Francis X. Schumacher, "New Concepts in Forest Mensuration," *Journal of Forestry* 36 (1938): 847–849.

39 James G. Osborne to Directors, Aug. 18, 1961, 2 pp. [bound with Handbook].

40 Anonymous, "Teaching Statistics at the Department of Agriculture Graduate School."

41 Box, *R. A. Fisher*, pp. 427–430.

42 Box, *R. A. Fisher*, pp. 427–430.

43 Cowlin, "Federal Forest Research in the Pacific Northwest."

44 V. L. Harper to Directors, Feb. 7, 1957, 8 pp.

45 James G. Osborne, "A Handbook for Teaching Statistical Methods and Experimental Design to Research Foresters, Foreword," USDA Forest Service, n.d. [1961], mimeographed, 601 p.

46 Floyd A. Johnson, "Syllabus for a Seminar in Applied Statistics, Pacific Northwest Forest and Range Experiment Station," Jan. 1955, rev. Feb. 1956, 132 pp., mimeographed. Floyd A. Johnson, "Statistical Methods, Training, Correspondence Course A," n.d., 297 pp., mimeographed. Also, "Course B," n.d., pp. 298–591, mimeographed.

47 Frank Freese, "A Guidebook for Statistical Transients," Southern Forest Experiment Station, n.d., 77 pp., mimeographed. Frank Freese, "Elementary Forest Sampling," Agricultural Handbook 232, 1962, USDA Forest Service, 91 pp. Frank Freese, "Elementary Statistical Methods for Foresters," Agricultural Handbook 317, 1967, USDA Forest Service, 87 pp. Frank Freese, "Linear Regression Methods for Forest Research," USDA Forest Service, Forest Products Laboratory, Madison, Wisc., 136 pp.

48 Warren E. Frayer, editor, "Proceedings of the Symposium on the development and implementation of courses and curricula in natural-history biometry, sponsored by the National Science Foundation, and conducted at Colorado State University, April 20–24, 1970." no page numbers.

BONNIE CHRISTENSEN

FROM DIVINE NATURE TO UMBRELLA SPECIES

The Development of Wildlife Science in the United States

The study of wildlife in North America has never been a simple activity. More has been at stake than accurate knowledge of animals, in large part because the term "wildlife," itself, suggests more than just "animal." "Wildlife" refers to animals in a natural environment, living freely.[1] In North America, these animals belong to the people, not to private land owners, and Americans have had a complicated relationship with *their* wildlife. Wild animals have provided a source of food, wealth, sport, and spiritual fulfillment. Americans have used wildlife to symbolize nature and their own relationship to the natural world. Their ideas and values about wild animals have shaped the development of American wildlife science, which evolved in large part according to pressures from various constituencies who promoted research and management programs designed to meet their own needs and desires.

The demands of these wildlife constituencies have varied considerably over the years, but many of them shared a common interest in "managing" wild animals, often to establish an idealized version of a "natural balance." Wildlife management, the application of scientific information to practical questions of conservation, has been a vital part of the development of wildlife science in America. Management implies control of species and habitats, a control based on certain choices about how to manipulate human beings, wild animals, and the environment. Managers have usually based these choices on the demands of certain groups of people and their perceived value of specific animals. This value has traditionally been determined by the usefulness of particular animals to humans—as food, sport, or aesthetic enjoyment—but, more recently concepts such as scarcity and biodiversity have been added to these considerations. Americans' perceptions of the value of wild animals have shaped the direction of wildlife science and management from the nation's beginning.

In the early national period some Americans valued wild animals because they represented a way to compete with the older nations of Europe. To Thomas Jefferson, for example, they became a symbol of the strength and bounty of the New World. Jefferson hoped to use the study of wild animals to assert the superior qualities of America in order to counter the theories of eighteenth-century European naturalists, such as George Leclerc, Comte de Buffon, who viewed New World animals as degenerate descendants of those in the Old World. Since such conclusions about America's animals suggested that the New World's human inhabitants might also be "less perfect" than those of Europe, Buffon's theory prompted a series of studies on America's wildlife designed to prove its worthiness. One of the assignments Jefferson gave to Meriwether Lewis and William Clark when they started on their famous expedition in 1804 was to study, describe, and collect the wildlife they encountered on their travels.[2] Like so many later studies of American wildlife, the natural history investigations of Lewis and Clark had a motivation beyond simple examination of wild animals. They formed part of a long tradition of American naturalists and scientists who studied wildlife to meet specific needs and satisfy certain constituencies.

American interest in the study of animals continued, in a different form, after the furor over Buffon's claims had died down around 1815.[3] The nineteenth century witnessed a flourishing of natural history studies of wild animals, as well as the professionalization of biology with its emphasis on laboratory studies of animals, and the creation of an explicitly economic approach to wildlife research and management by the end of the century.

Natural historians, the precursors to modern biologists, studied wildlife because they enjoyed observing and identifying animals; they also hoped to find a way of understanding a divine order through the examination of nature's creatures. Natural theology, the belief that one could understand God's plan through the study of nature, proved an important component of American animal studies in the nineteenth century. One of the most famous naturalists in the United States, Louis Agassiz, observed in 1848:

> Animals are worthy of our regard, not merely when considered as to the variety and elegance of their forms, or their adaptation to the supply of our wants; but the Animal Kingdom, as a whole, has a still higher signification. It is the exhibition of the divine thought, as carried out in one department of that grand whole which we call Nature; and considered as such, it teaches us most important lessons.[4]

Nineteenth century Americans collected animal specimens, bought beautifully illustrated books on birds and animals (such as John James Audubon's *Birds of America*), and patronized natural history museums. They took a deep interest in the natural world and wild animals in part because of the explicit ties that naturalists like Agassiz made between the animal world and moral order.

In contrast to the naturalist who examined nature "for evidence of design or to provide recreational enjoyment," the laboratory biologists of the late nineteenth century viewed nature as something that "needed to be probed and manipulated to reveal its secrets." To scientists in the developing field of biology, laboratory work became an essential part of establishing the professionalism of their science. They looked at the natural world, including wild animals, in a new way—in very small sections through microscopes. Instead of examining the whole animal, a scientist with a microscope could study specific parts of animals in great detail. Professional biologists turned to specialties like physiology and embryology; they explored questions of function and moved away from the older naturalist traditions of description and taxonomy.[5] Biologists did not abandon field observations of animals altogether, but their studies often focused on the laboratory, which became the "professional biologist's trademark" and a key to the "hardening" of the science.[6]

The professionalization of biology influenced the way that scientists approached the study of animals, but it also connected wildlife research to other, broader social and economic trends. By the last quarter of the nineteenth century, American leaders began to view science, including the study of wildlife, as a way to advance the nation's economy. The conviction that science could be made to work for the nation led to federal support for land-grant colleges and agricultural research stations, as well as to the creation in 1885 of the Office of Economic Ornithology and Mammalogy (later the Bureau of Biological Survey) in the Department of Agriculture. The new agency provided evidence of federal support for a wildlife science designed to serve the agricultural sector, although eventually its duties extended in important ways to meet the needs of sport hunters and conservationists.

The nation's farmers and ranchers, however, proved the initial constituents of the Office of Economic Ornithology and Mammalogy. Mechanized farming had spread across the nation in the late nineteenth century, and American farmers turned increasingly to the cultivation of single crops to sell on the market. This emphasis on monocultures created conditions ideal for the growth of large insect populations.[7] Economic ornithology and mammalogy sought to find ways to return the animal world to a natural "balance" favorable to farmers. This meant the destruction of insect and rodent pests that hurt farmers' profits by eating crops. Government biologists sought to identify and encourage "friendly" species—robins, owls, or wolves—that helped farmers achieve an economic balance by consuming pests. In analyzing the value of different species, government researchers combined the professional biologist's emphasis on microscope work with the observational methods of naturalists. They used microscopes to identify and count food remains in stomachs and also recorded the interactions of species in the field. Their efforts had little impact on the nation's farmers, however, who preferred the "quick fix" of chemical pesticides to the complicated, and mostly ineffective, process of attempting to rebalance animal populations.[8]

The Bureau of Biological Survey failed in its efforts to re-establish a natural balance in part because it tried to find simple answers to very complicated questions of population dynamics. One difficulty with creating a balance in farmers' fields was that no one at the time "knew what regulated natural populations or how the mechanisms worked." Scientists did not yet have the concepts or research methods they needed to work out these problems.[9] Analyses of stomach contents could show what an animal had eaten, but not the dynamic relationships between predator and prey species. Such examinations could not determine how a prey species responded biologically to the depletion of its numbers through predation or how a "natural balance" could exist in a place as "unnatural" as a vast field of corn. Another difficulty was that the Bureau's definitions of good and bad species sometimes conflicted. Bureau researchers told farmers that predators such as wolves were valuable aids to agriculture because they preyed on rodent pests. But the Bureau also was active in a predator-eradication program to help ranchers kill wolves. Ranchers, who by the mid-1920s contributed up to a quarter of the Bureau's funding, became such a strong constituency that their needs, and their definition of good and bad animals, eventually dominated the Bureau's work.[10]

In addition to farmers and ranchers, the Bureau of Biological Survey served two other important wildlife constituencies in the first decades of the twentieth century: hunters and animal lovers. Both of these groups were part of a larger conservation movement at the turn of the century. Middle-class Americans, who feared a loss of rural values as the nation became increasingly urban and industrial, began "to invest wild places with aesthetic and ethical values" and to place higher value on wild animals, "the embodiment of the wilderness."[11] They promoted nature-study classes in schools and responded enthusiastically to the stories of "realistic" animal writers such as William J. Long and Ernest Thompson Seton.[12] Concerned about extinction, such as that of the passenger pigeon, and the dramatic depletion of the once-vast buffalo herds on the plains, wildlife and wilderness enthusiasts formed organizations such as the National Association of Audubon Societies (1886) and the Sierra Club (1892). These organizations promoted the study and conservation of certain species of wildlife as well as the preservation of their habitats.[13] Sports groups contributed to this movement also, pressuring first state legislatures and then the national government to do more to maintain the nation's game animals.

State governments had a long history of game regulation, but hunting groups like the Boone and Crockett Club, wanted the federal government to add its efforts to game conservation.[14] Hunters argued that game animals —elk, deer, bears—provided an essential link between the nation's ever-growing urban population and its frontier past. Hunting, they stated, took men into the wild and taught them the "hardihood, self-reliance, and resolution" of pioneer heroes such as Daniel Boone.[15] Pressure from these groups, with the active support of President Theodore Roosevelt (one of the founders of the Boone and Crockett Club), encouraged Congress to pass a series of laws in the beginning of the twentieth century to protect wild animals. The first federal game law, the Lacey Act, prohibited commerce in illegal game, and other legislation provided for the establishment of game sanctuaries and migratory bird refuges.[16] Enforcement of the new laws became the responsibility of the Bureau of Biological Survey.

At the turn of the century, the Bureau of Biological Survey thus served an odd dual role directly related to the needs of its different constituencies. The Bureau began a program of predator eradication in 1905 to help ranchers, but it also became the federal government's game conservation agency.

As enforcer of the nation's new wildlife regulations, the Bureau of Biological Survey played a key role in the development of wildlife science and management. Federal game regulations, such as those included in the Migratory Bird Treaty Act of 1918 between the United States and Canada, called for knowledgeable assessments of animal populations. The conditions of this treaty required data on bird populations so that Bureau biologists could establish bag limits and hunting seasons for certain species. To obtain this data, historian Thomas R. Dunlap has pointed out, government researchers had "to develop census techniques, chart migrations, and learn how to predict the level of waterfowl populations." They began a tradition of wildlife management that involved a program of "applied ecological research."[17]

The combination of increasing numbers of hunters and decreasing habitat for wildlife, however, overwhelmed these early attempts at wildlife management. Crises such as the deaths of 14,000 elk in Yellowstone from malnutrition in the winter of 1919–1920 dramatically illustrated the need for more knowledgeable management of the nation's wildlife. As one national parks researcher related, "The idea that species that had been loved and protected for decades could now end up starving on denuded ranges thereby bringing about great public outcry frightened many managers."[18] Government agencies, such as the National Park Service, and private sporting groups, like the Sporting Arms and Ammunitions Manufacturers' Institute (SAAMI), recognized the need for trained biologists to establish wildlife conservation measures based on scientific information.[19] These biologists would come out of a new, research-oriented school of game management that combined theories from animal ecology with conservation-based fieldwork. The new type of management developed from the work of Charles Elton and Aldo Leopold, two scientists who provided the theories and practical techniques that brought these fields together.

Leopold, whose early career in the U.S. Forest Service convinced him of the viability of resource conservation, believed that scientists and managers could work together in the face of game population depletion to restore balance to natural areas that had been disrupted by human activities. In the 1910s and 1920s, Leopold conducted extensive surveys of game populations for both the Forest Service and SAAMI. From these studies he concluded that effective game management required not just control of hunting, but also management of species' environments to maintain proper habitat conditions.[20] That type of management required scientific studies of animals and their habitats. Leopold dismissed most of the earlier attempts to apply basic biology to game management. These efforts, he concluded, had "soon disclosed the fact that science had accumulated more knowledge of how to distinguish one species from another than of the habits, requirements, and inter-relationships of living populations. Until recently," he stated bluntly, "science could tell us, so to speak, more about the length of a duck's bill than about its food, or the status of waterfowl resources, or the factors determining its productivity."[21]

Leopold chided academic biologists who did not contribute practical knowledge to what he saw as the vitally important field of game management. Academic scientists, in turn, dismissed Leopold's work in applying science to management problems. In 1929, Joseph Dixon, who worked at the zoology museum at the University of California, complained to Leopold that "the orthodox Zoology Depts." did not consider game management problems as "sufficiently academic." Of course, he continued, "the questions

Aldo Leopold's 1928 portrait, a few years before he taught game management at the University of Wisconsin, Madison.
(USDA FOREST SERVICE PHOTO)

of parasites, disease and life histories are all biological problems, but there seems to be an ingrained feeling against such problems being handled with a viewpoint of their being applied to present problems [in] game management."[22] Leopold fought to get past this prejudice. He wanted practical scientific data and theory that would help wildlife managers in the "art of making land produce sustained annual crops of wild game for recreational use."[23]

The work of British ecologist Charles Elton, one of the prominent figures in the development of animal ecology, provided wildlife researchers with the kind of theories that Leopold sought. In his 1927 book, *Animal Ecology*, Elton presented four principles for integrating population and community ecology: the niche, the food chain and food cycle, food size, and the "pyramid of numbers." "Niche" was not a new term (it had been used by Joseph Grinnell years earlier), but Elton gave it a functional meaning referring to an animal's "occupation in the biological community." The food chain (later food web) and food size principles acknowledged the significance of jumps in relative body sizes in the successive links of the food chain. Wildlife researchers found the "pyramid of numbers" concept a useful tool for visualizing the relative increase in body size and decrease in population numbers associated with the food chain. Elton's principles provided a guide for field researchers in their studies of the ecological relationships of game animals.[24]

Leopold incorporated many of Elton's concepts into his work of turning wildlife management into a "rigorous applied science." His relationship with Elton also demonstrated the value of cooperation between scientists from different fields. The two men met at a 1931 conference on biological cycles and maintained a steady correspondence over the course of the next two decades. Leopold's biographer called their working relationship a "fortuitous convergence."

> They complemented each other superbly. Elton was abstract, a theorist whose field experience was accurate, but relatively limited. Leopold was, above all, a field man who knew enough theory to give his observations broad application. Elton was laying the foundations of ecology; Leopold was attempting to apply the science even before its foundations were set.[25]

Leopold influenced Elton's thinking on conservation issues, while Elton's ecological ideas contributed to Leopold's developing philosophy of wildlife management.

216

The experience of these two men illustrated the value of communication between the basic and applied sciences, although like many such "convergences," there remained key points of difference. The concept of a balanced nature was at the core of Leopold's view of the natural world. Elton —drawing from his research on animal populations, which included a study of fluctuations of fur-bearing animals using Hudson's Bay Company records —challenged this popular concept. "The balance of nature," he stated flatly in 1930, "does not exist."[26] But the idea of a harmonious nature ideally in balance proved tenacious. In spite of Elton's disavowal of it and a growing critique by other academic ecologists, the "balance of nature" idea, which had shaped much of the early work of the Bureau of Biological Survey, continued to be an influential concept in wildlife ecology and management research in the twentieth century.[27]

Leopold, for one, insisted that in an "undisturbed condition" animal populations lived in a natural balance. He grounded this belief, which influenced several generations of wildlife researchers and managers, in his work on the mule deer of the Kaibab Plateau. The federal government had prohibited hunting on the Kaibab in 1905, while also implementing a program of predator eradication in the area. Unchecked by predation (human and otherwise), the deer population on the Kaibab increased to crisis levels by 1924. Sixty percent of the herd was lost in the winter of 1925–26 through mass starvation and from a belated move to reinstate hunting. Leopold concluded that the population "crash" resulted from "overcontrol" —human action had altered the natural balance, and the result was crisis.[28] One of the lessons Leopold drew from the Kaibab was that management of wildlife in a "balanced" state meant that managers had to look not just at a single species, but also at species interactions, especially those involving predators and their prey. Human beings had disrupted the balance of nature through habitat destruction and predator elimination, but their actions could also help to restore that balance. Under this assumption, wildlife biologists sought to manage species by ensuring the presence of proper "balancing" populations. In the 1940s, for example, the National Park Service attempted to introduce wolves onto Isle Royale as a "natural check" on the island's moose population, which was increasing beyond its habitat capacity.[29]

Leopold's influence extended beyond his work on the Kaibab deer population and his impressive field research. His position in 1933 at the University of Wisconsin as the nation's first Professor of Game Management (later Wildlife Management) helped to legitimize game management as an academic field of study. Leopold's books and articles, especially his 1933 work, *Game Management*, also provided practical guides for wildlife managers and researchers. *Game Management*, the premier work in the field, included suggestions about techniques for capture and artificial propagation of animals as well as advice on how to conduct game surveys and to control diseases in wild populations. Leopold's book was required reading in the 1930s and 1940s for members of the expanding field of game conservation.[30]

Leopold did much to guide the transformation of game management, but the process involved many others. An increasing interest in hunting, encouraged by improvements in transportation, led to renewed activity in the field of wildlife management and research in the 1930s. Herbert Stoddard's extensive research on bobwhite quail in the late 1920s became the first of many "scientific life-history" investigations of game animals. Stoddard, his work sponsored by a private sporting group, sought to determine how best to ensure steady propagation of quail on private game reserves. He approached his study in a methodical and thorough manner; he carefully counted nests, eggs, and chicks, calculated mortality rates, and burned test areas to determine the effects of fire and vegetation re-growth on the quail. Stoddard's work, grounded in "observation and calculation," became a model for later studies done by a new generation of wildlife researchers who worked both for private groups and for the federal and state governments.[31]

The federal government promoted the new research-based management in a variety of ways. In 1935, the head of the Bureau of Biological Survey, J. N. "Ding" Darling, laid the foundations for what became the Cooperative Wildlife Research Units at state universities. Financed through a mixture of state, university, federal, and private funding, these units produced "a new generation of field-trained personnel."[32] The federal government also provided support for wildlife research through the Pittman-Robertson Act of 1937, which placed a ten percent tax on the sale of sporting arms and ammunition and directed the funds toward wildlife work. "P-R" money provided a steady source of funds for state wildlife agencies in their research and management efforts.[33]

By 1940, wildlife managers had moved from being simply game wardens to working as field researchers. They studied animals in the field, charting their activities to determine behavior, migration, nesting patterns, and mortality rates. Drawing from theories of community and population dynamics developed by animal ecologists, and from their own field research, these managers worked out successful techniques and strategies to encourage populations of desirable animals. But, the definition of "desirable" continued to be shaped by interest groups, predominately hunting organizations, who provided the money and political pressure that supported wildlife research and management efforts. Gradually, however, other groups, like tourists to the nation's forests and parks, developed their own conceptions of "valuable" wild animals, especially charismatic mammals. The shift in the 1930s and 1940s to the use of "wildlife" instead of "game" in personnel and group titles provides one indication of the broadening of concern beyond the sporting interests. The federal government reflected this trend in 1940 when Franklin D. Roosevelt reorganized the Bureau of Biological Control into the U.S. Fish and Wildlife Service and moved it from the Department of Agriculture to the Department of the Interior. In spite of these name changes, however, most of the work on wildlife remained focused on game animals because hunters, through "P-R" taxes and hunting licenses, contributed the bulk of the funds available for wildlife agencies.[34]

The U. S. Fish and Wildlife Service used the Grumman Goose
for aerial transects during the 1951 waterfowl survey.
(U. S. FISH AND WILDLIFE SERVICE PHOTO BY REX GARY SCHMIDT)

219

In the years after World War II, wildlife science continued to concentrate primarily on game animals, but important challenges to this traditional focus began to turn the field in a different direction. Some animals came to be valued for their scarcity—not their abundance—and groups besides hunters, ranchers, and farmers gained influence over the direction of wildlife management. Wildlife scientists in this period broadened their focus to non-game endangered species. Basic scientists from a variety of disciplines, who worked to overcome the traditional boundaries between basic and applied wildlife research, joined their efforts and added new perspectives and theories to the field. The game-directed discipline established by Leopold in the 1920s and 1930s had developed by the 1980s into a diverse field in which a variety of different types of scientists worked to establish management plans designed not only to protect certain species of animals but also to maintain ecosystems and biodiversity.

These changes stemmed in large part from a fundamental shift in the postwar years in the way that many Americans lived and in how they viewed animals and the natural world. Animals that had once been valued only for the sport they offered hunters, became important for other reasons. A small group of Americans had always argued for the aesthetic, or even intrinsic, value of wild animals, but in the postwar years their numbers increased dramatically. Attitudes did not change immediately, but as historian Samuel P. Hays has argued, the "massive social and economic transformation in the decades after 1945," had a profound effect on environmental "preferences and values." Educated urban dwellers began to place a "very high value on the presence of wildlife around their homes," and more and more Americans started to identify themselves as wildlife observers and photographers. For Americans, wildlife became a symbol of environmental health and a high standard of living.[35]

But the rising standard of living in the postwar era was tied to practices destructive to the wild animals that Americans were beginning to value. Cheap and abundant food came from extensive use of chemical fertilizers and pesticides; comfortable suburban housing developments encroached on wildlife habitats. Pollution and habitat destruction reduced the populations of many types of wildlife and raised concern among Americans who had begun to incorporate a type of "popular ecology" into their view of the world and wild animals. Endangered species such as the whooping crane attracted national attention in the 1950s, and these animals became symbols of the new environmental concern.[36]

Rachel Carson's 1962 book, *Silent Spring*, dramatically increased public awareness of the effects of chemical toxins on animals and contributed to the creation of new constituencies that demanded different programs of wildlife management and research.[37] Biologists had documented the deleterious impact of pesticides like DDT on nongame birds beginning in the late 1940s, but the public remained almost unaware of the problems before Carson's book. Frightened by the dramatic threat of species extinction, as well as by the danger these pesticides presented to humans, concerned citizens swelled the ranks of environmental groups such as the Audubon Society and Defenders of Wildlife. These national groups, along with various local organizations, mobilized public support for federal programs of environmental protection and research including the National Environmental Policy Act and the Endangered Species Act and its successors.[38]

The endangered species acts of 1966, 1969 and 1973 provided funding and support for research and management of endangered and threatened species that helped to redirect the course of wildlife science. Hailed by scientists as the United States's "most comprehensive and powerful piece of environmental legislation," the Endangered Species Act of 1973 (ESA) proved an important step toward the broadening of wildlife study and management beyond game animals and dramatically appealing species like the bald eagle.[39] Although the ESA promoted the study of single species, its provisions for habitat study and protection "led to an interest in the larger ecosystem of which the species was only one part." It provided citizen groups with a way to challenge environmentally destructive projects and also brought the federal government more fully into the arena of wildlife policy and management.[40] Work on endangered species, often funded by the federal government, contributed to the development and acceptance of important new ideas about population dynamics and the naturalness of change in the management of the nation's wildlife.

By the early 1970s, scientists working with endangered animals realized that the management of small populations posed special problems that required new theories and techniques. Researchers found, for example, that the isolation of populations from each other, often the result of human-created disturbances, undermined the viability of endangered species. Looking for explanations, they turned to the equilibrium theory of island biogeography, developed by scientists in 1967, which "proposed that the number of species inhabiting an island results from a dynamic equilibrium between immigrations and extinctions."[41] Wildlife researchers recognized that nature preserves—national parks, wilderness areas, refuges, even city parks—were for many species essentially "islands in an inhospitable sea

of man-modified vegetation or urban-sprawl."[42] This type of habitat frag-
mentation meant that separate animal populations—what Richard Levins
termed "metapopulations"—could not migrate to, and breed with, other
populations of their species.[43] Part of the work of endangered-species research has thus focused on how to manage metapopulations to ensure that
small populations maintain genetic diversity. Efforts in this area have included the use of DNA "fingerprinting" to determine genetic connections and
the manipulation of the physical landscape to create "corridors" between
habitat patches.[44] From investigations of genetic traits to evaluations of
human-adapted landscapes, work with small, endangered populations led
scientists into increasingly complex and diverse fields of research.

As wildlife scientists working with endangered species recognized the
complexity of the interaction between the "natural" and human worlds, they
also began to introduce new management policies that acknowledged the
dynamic nature of change in the environment. In so doing, researchers finally laid to rest the old idea of the balance of nature in wildlife management. Endangered-species research played an important role in this transition because some endangered species populations, like the whooping crane,
had so few members (fifty-one birds in 1972) that researchers could count
and observe them all. Such intense and complete research revealed erratic
recovery rates that did not match the classic deterministic growth curves
commonly used to predict balanced population interactions. In order to
predict extinction possibilities, researchers had to take a new perspective,
one that accepted "the essentially stochastic quality of population change
over time."[45] Wildlife researchers' acceptance of the "naturalness of change"
and rejection of a mechanistic equilibrium allowed for a new approach to
wildlife management. A 1976 plan to save the Kirtland's warbler, for example, called for the use of prescribed fire after years of fire-prevention
policies to regenerate suitable habitat for the bird. According to ecologist
Daniel B. Botkin, this episode marked "a turning point in the modern perception of the character of nature and the requirements to manage and
maintain nature." The Kirtland's warbler plan showed that sometimes managers had to promote change, not just restore a supposed balance, in order
to preserve a particular species.[46]

In spite of all that scientists and citizens accomplished and learned under the ESA programs, however, the legislation and its policies reflected a
public concern with threatened wildlife that centered on a relatively narrow
number of species. All species did not benefit equally. The Endangered
Species Act of 1973 included plants and invertebrates in its coverage of spe-

cies, but vertebrates have been disproportionately represented on the federal list of endangered and threatened species. Although most "imperiled" species are invertebrates "possessing considerable ecological and scientific value," the public has had little interest in their plight. Because of the ESA's focus on species and the public's preference for certain types of wildlife, concerned scientists have used "large, aesthetically attractive" animals — the ones popularly identified as "wildlife" — to obtain broader ecosystem protection and biodiversity.[47] Scientists have looked to the Northern Spotted Owl, for example, as an "umbrella species" whose protection will potentially save a multitude of species in old-growth forests.[48] Ecologists have been quick to note the irony of having to turn to vertebrates "as indicators of . . . old-growth forest rather than using the forest itself!", even as they acknowledge the importance of the public's preference for certain species in the promotion of environmental research and policies.[49]

Wildlife science in the late twentieth century remained as enmeshed in social, cultural, and economic issues as it had been at the beginning of the century.[50] The growth of powerful constituencies that favored species and habitat protection did not offset the number of equally powerful groups that challenged the conclusions of wildlife scientists and managers when those conclusions threatened personal, especially economic, interests. Jobs, homes in the country, inexpensive drinking water, and better roads could be more valuable to certain people than, for example, the maintenance of every subspecies of mountain lion. Who was to determine the comparative value of lions and timber jobs? The worth of a bear as a hunting trophy or as the subject of nature photography? The value of wildlife remained a contested terrain — one in which science provided some guidelines, but no undisputed conclusions. As researchers acknowledged in a 1989 report on selenium buildup in drainwater associated with irrigation in the San Joaquin Valley: "science cannot judge which is more valuable, the crop or the waterfowl."[51]

Science could not provide conclusive answers to environmental questions, but by the 1970s and 1980s many scientists began to see their role in shaping wildlife and environmental policy as more important than ever. The stakes of extinction, they saw, had increased dramatically. An expanding human population put escalating demands on natural resources including wild animals and their habitats. Some experts predicted hundreds of thousands of extinctions by the middle of the twenty-first century.[52] Other voices, however, dismissed these warnings of doom and demonstrated scientific evidence showing that environmentalists overestimated the damage to wildlife caused by logging, mining, agriculture, and other industries and development. Concerned scientists on all sides of the conservation issue argued that only informed decisions should be included in the debate over how to manage the nation's, and the world's, remaining wild animals. Often, though, these informed decisions contradicted each other.

The accumulation of adequate information about wild animals in their environments has formed a key problem for wildlife scientists.[53] Researchers had long pointed out the difficulty of obtaining data they needed to understand and manage wildlife according to the demands of their various constituents.[54] The vast range of some species, the influence of a multitude of environmental factors on their habits and mortality, and the difficulties of observing them continuously, have been just some of the problems with accumulating data. New techniques of wildlife study have helped researchers to meet some of these problems. Radio telemetry, for example, increased dramatically the ability of researchers to track animals like wolves to determine home range size, movement, mortality, and physiological parameters.[55] But field researchers continued to face problems of "replication, control, and randomization"—the basic components of scientific research.[56] Experiments proved costly, difficult to conduct, and often inconclusive. Testing theories about the impact of hunting regulations on black duck populations, for example, would require the cooperation of wildlife officials from both Canada and the United States to coordinate hunting limits in conjunction with a quantitative study. Such research, however, still could not provide conclusive data on the impact of hunting on black ducks. A variety of factors outside the control of the researchers, such as changes in breeding grounds or severe weather conditions, could affect duck populations even if hunting were completely controlled.[57] Without firm data on how specific human actions might affect certain species, scientists and managers could not offer the type of iron-clad argument for protection or development that would hold up to challenges from opposing interests.

As the politics of wildlife research drove many scientists to question the reliability of their knowledge of wild animals, the sense of environmental crisis created by vanishing species generated an important change in the composition of people working in the field. Environmental concern, and the availability of federal funding, prompted scientists from a variety of disciplines to take a keen interest in questions of applied wildlife research in the 1970s and 1980s. In 1978, University of California at Santa Cruz biologist Michael Soule called on academics and conservationists to come together to meet the challenge of species extinction. He advocated the development of a new field, conservation biology, which would combine "basic science with practical conservation." Conservation biology, he hoped, would bring together basic and applied researchers who would work together to preserve biological diversity.[58]

The decades-long animosity between basic and applied wildlife science has not been easily overcome, however. In spite of some murmurs of cooperation, a great deal of tension remains between the two cultures. Conservation biologists have accused wildlife scientists of focusing their efforts on "single-species, site-specific, problem-oriented research" centered on "species of economic or recreational interest to humans."[59] Traditional wildlife managers, in turn, have charged the conservation biologists with "sacrificing depth for breadth" and for indulging in "data-free analysis." They have criticized the new group for relying too heavily on theory and computer models not based on adequate field data.[60] Some conservation biologists have, indeed, contended that extensive field research is a "luxury that conservation biology may no longer be able to afford." They argue that the impending crisis of extensive extinctions requires the development of "insightful, comprehensive hypotheses," which can be tested with existing information and then applied to management as quickly as possible.[61] This disturbs the more traditional wildlife scientists who consider adequate data bases as essential to the accumulation of reliable knowledge and the formulation of appropriate management policies.

A common concern with certain environmental issues and the realization that their work is vitally tied to human perceptions and social factors, however, joins these groups. Although the new conservation biologists may see their view of biodiversity as more expansive than that of traditional wildlife biologists, both groups have stressed the need to pay attention to "the linkage among physical, biological, and socioeconomic systems, and to the interface between science and policy."[62] Human beings have been intimately associated with the problems that led to the endangerment of wildlife species; humans could not be divorced from the scientific studies and management efforts to save these species. Thomas Allen observed this, noting that, "The study of vanishing wildlife is necessarily the study of . . . [man's perceptions of animals]. What we fear, what we hope, and what we admire in animals will inevitably determine their fate."[63]

Wildlife science is a vital and expanding field. Researchers are working on new cell and molecular biology techniques, studying genetic relationships between animals using DNA "fingerprinting" and employing increasingly sophisticated computer modeling to investigate multi-species interactions. The introduction of several new journals on applied natural-resources science, most notably *Ecological Applications*, indicates the growth in this field of study. But these exciting developments within the field have not moved wildlife science away from the influence of socioeconomic factors. The value that Americans place on wild animals continues to influence the funding and direction of wildlife study and to limit the type of management decisions that researchers can make. Scientists may sometimes find Americans' value judgments on wild animals irrational or frustrating, and they may try to influence those judgments, but they often realize, as did their predecessors, that they have to work within the bounds of what Americans want from, and value about, their wildlife.

NOTES

1 "Wildlife" proves an ambiguous descriptor. As a term it only came into general usage around 1925, and its use as a single word did not begin until 1931 or 1932. Literally interpreted, it means "all living things," from "one-celled plants and animals to the complex species we call birds and mammals." In its popular usage, the definition of wildlife is more restricted—it usually refers only to birds and mammals. Joseph J. Hickey, "Some Historical Phases in Wildlife Conservation," *Wildlife Society Bulletin* 2 (Winter 1974): 167–69. I have used the popular definition of "wildlife," which is all non-domesticated birds and mammals.

2 Paul L. Farber, "Buffon and the Concept of Species," *Journal of the History of Biology* 5 (Fall 1972): 270–72; Charlotte M. Porter, *The Eagle's Nest: Natural History and American Ideas, 1812–1842* (University: The University of Alabama Press, 1986), pp. 7–9; Richard White, "Discovering Nature in North America," *The Journal of American History* 79 (December 1992): 885–887.

3 White, "Discovering Nature in North America," p. 887.

4 Keith R. Benson, "From Museum Research to Laboratory Research: The Transformation of Natural History into Academic Biology," in *The American Development of Biology*, eds. Ronald Rainger, Keith R. Benson, and Jane Maienschein (Philadelphia: University of Pennsylvania Press, 1988), pp. 56–57.

5 Benson, "From Museum Research to Laboratory Research," pp. 72, 76, 70.

6 Gregg Mitman and Richard W. Burkhardt, Jr., "Struggling for Identity: The Study of Animal Behavior in America, 1930–1945," *in The Expansion of American Biology*, eds. Keith R. Benson, Jane Maienschein, and Ronald Rainger (New Brunswick: Rutgers University Press, 1991), p. 165.

7 Thomas R. Dunlap, *DDT: Scientists, Citizens, and Public Policy* (Princeton: Princeton University Press, 1981), p. 18; Frank Graham, Jr., *The Dragon Hunters* (New York: E. P. Dutton, Inc., 1984), p. 4.

8 See, for example, Vernon Bailey, "Birds Known to Eat the Boll Weevil," United States Department of Agriculture, *Division of Biological Control Bulletin* 22 (1905); F. E. L. Beal, "Food of the Bobolink, Blackbirds, and Grackles," United States Department of Agriculture, *Division of Biological Survey* 13 (1900); Sylvester D. Judd, "The Bobwhite and Other Quails of the United States in Their Economic Relations," United States Department of Agriculture, *Division of Biological Survey Bulletin* 32 (1905).

9 Thomas R. Dunlap, *Saving America's Wildlife* (Princeton: Princeton University Press, 1988), p. 53.

10 Dunlap, *Saving America's Wildlife*, p. 39.

11 Roderick Nash, *Wilderness and the American Mind*, revised edition (New Haven: Yale University Press, 1973), p. 145; Lisa Mighetto, *Wild Animals and American Environmental Ethics* (Tucson: University of Arizona Press, 1991), p. 4.

12 Ralph H. Lutts, *The Nature Fakers: Wildlife, Science & Sentiment* (Golden, CO: Fulcrum Publishing, 1990), p. 148.

13 Ronald Tobey, *Saving the Prairies: The Life Cycle of the Founding School of American Plant Ecology, 1895-1955* (Berkeley: University of California Press, 1981), p. 24.

14 Samuel P. Hays, *Conservation and the Gospel of Efficiency: The Progressive Conservation Movement, 1890–1920* (New York: Antheum, 1969, 1959), p. 189.

15 Mighetto, *Wild Animals and American Environmental Ethics*, pp. 34–35.

16 Aldo Leopold, *Game Management* (New York: Charles Scribner's Sons, 1933), p. 13; Hays, *Conservation and the Gospel of Efficiency*, pp. 189–90.

17 Dunlap, *Saving America's Wildlife*, pp. 37–38.

18 Gerald Wright, *Wildlife Research and Management in the National Parks* (Urbana and Chicago: University of Illinois Press, 1992), p. 72.

19 Robert P. McIntosh, *The Background of Ecology: Concept and Theory* (Cambridge: Cambridge University Press, 1991, 1985), p. 167; William L. Robinson and Eric G. Bolen, *Wildlife Ecology and Management*, 2nd Edition (New York: Macmillan Publishing Company, 1989), pp. 3–4.

20 Curt Meine, *Aldo Leopold: His Life and Work* (Madison: The University of Wisconsin Press, 1988), pp. 168, 262.

21 Leopold, *Game Management*, pp. 3–4, 20–21.

22 Quoted in Meine, *Aldo Leopld: His Life and Work*, pp. 266–67.

23 Leopold, *Game Management*, p. 3.

24 Peter Crawford, *Elton's Ecologists: A History of the Bureau of Animal Population* (Chicago: The University of Chicago Press, 1991), p. xiii; McIntosh, *The Background of Ecology*, pp. 92-93.

25 Meine, *Aldo Leopold: His Life and Work*, p. 284.

26 Quoted in McIntosh, *The Background of Ecology*, p. 167.

27 Daniel B. Botkin, *Discordant Harmonies: A New Ecology for the Twenty-first Century* (New York and Oxford: Oxford University Press, 1990), p. 76.

28 Meine, *Aldo Leopold: His Life and Work*, p. 241; Botkin, *Discordant Harmonies*, pp. 76-80.

29 Botkin, *Discordant Harmonies*, pp. 47–48, 36–37, 78.

30 McIntosh, *The Background of Ecology*, p. 168; Meine, *Aldo Leopold: His Life and Work*, pp. 311-12.

31 Dunlap, *Saving America's Wildlife*, p. 72; Leopold, *Game Management*, p. 16.

32 Robinson and Bolen, *Wildlife Ecology and Management*, pp. 449–50.

33 Robinson and Bolen, *Wildlife Ecology and Management*, pp. 4, 446.

34 Meine, *Aldo Leopold: His Life and Work*, pp. 362–365; Robinson and Bolen, *Wildlife Ecology and Management*, p. 4.

35 Samuel P. Hays, *Beauty, Health, and Permanence: Environmental Politics in the United States, 1955–1985* (Cambridge: Cambridge University Press, 1989, 1987), pp. 13, 32, 111.

36 Dunlap, *Saving America's Wildlife*, p. 120.

37 Hays, *Beauty, Health, and Permanence*, pp. 28, 52, 174–75.

38 Hays, *Beauty, Health, and Permanence*, pp. 174–75, 62–63; Robinson and Bolen, *Wildlife Ecology and Management*, pp. 414–416.

39 Gordan H. Orians, "Endangered at What Level?" *Ecological Applications* 3 (May 1993): 206.

40 Hays, *Beauty, Health, and Permanence*, pp. 112–14; Robinson and Bolen, *Wildlife Ecology and Management*, pp. 416–18.

41 Daniel F. Doak and L. Scott Mills, "A Useful Role for Theory in Conservation," *Ecology* 75 (April 1994): 615.

42 M. L. Gorman, *Island Ecology* (London: Chapman and Hill, 1979), p. 65.

43 Doak and Mills, "A Useful Role for Theory in Conservation," p. 619.

44 Nathaniel P. Reed and Dennis Drabelle, *The United States Fish and Wildlife Service* (Boulder, CO: Westview Press, 1984), pp. 112–19; Gary E. Belovsky, John A. Bissonette, Raymond D. Dueser, Thomas C. Edwards, Jr., Christopher M. Luecke, Mark E. Ritchie, Jennifer B. Slade, and Frederick H. Wagner, "Management of Small Populations: Concepts Affecting the Recovery of Endangered Species," *Wildlife Society Bulletin* 22 (Summer 1994): 312–316; Doak and Mills, "A Useful Role for Theory in Conservation," p. 615-16.

45 Botkin, *Discordant Harmonies*, p. 125-27.

46 Botkin, *Discordant Harmonies*, p. 68–70.

47 Stephen R. Kellert, "Social and Perceptual Factors in the Preservation of Animal Species," in *The Preservation of Species: The Value of Biological Diversity*, ed. Bryan G. Norton (Princeton: Princeton University Press, 1986), pp. 60–62.

48 Jerry F. Franklin, "Preserving Biodiversity: Species, Ecosystems, or Landscape," *Ecological Applications* 3 (May 1993): 203.

49 Orians, "Endangered at What Level?", p. 206.

50 Kellert, "Social and Perceptual Factors in the Preservation of Animal Species," p. 56.

51 Peter Montgomery, "Science Friction: Playing Politics with Scientific Research," *Common Cause Magazine* 16 (November/December 1990): 29.

52 Kellert, "Social and Perceptual Factors in the Preservation of Animal Species," p. 51.

53 Reed F. Noss, "Dangerous Simplifications in Conservation Biology," *Bulletin of the Ecological Society of America* 67 (December 1986): 278.

54 See Aldo Leopold, "Why and How Research?" *Thirteenth North American Wildlife and Natural Resources Conference* (1948): 44–48; Charles H. Romesburg, "Wildlife Science: Gaining Reliable Knowledge," *Journal of Wildlife Management* 45 (April 1981): 293–313; William J. Matter and R. William Mannan, "More on Gaining Reliable Knowledge: A Comment," *Journal of Wildlife Management* 53 (October 1989): 1172–76; Donald Ludwig, Ray Hilborn, and Carl Walters, "Uncertainty, Resource Exploitation, and Conservation: Lessons from History," *Science* 260 (2 April 1993): 17, 36.

55 David R. Klein, "The Evolving Role of Research in Management of Wildlife Populations," *Transactions of the XIXth International Union of Game Biologists Congress*, vol. 1 (1990): 12.

56 Ray Hilborn and Donald Ludwig, "The Limits of Applied Ecological Research," *Ecological Applications* 3 (November 1993): 550.

57 James D. Nichols, "Science, Population Ecology, and the Management of the American Black Duck," *Journal of Wildlife Management* 55 (October 1991): 792–93.

58 Ann Gibbons, "Conservation Biology in the Fast Lane," *Science* 255 (January 1992): 20–21.

59 Mari N. Jensen and Paul R. Krausman, "Conservation Biology's Literature: New Wine or Just a New Bottle," *Wildlife Society Bulletin* 21 (Summer 1993): 199.

60 Gibbons, "Conservation Biology in the Fast Lane," p. 20.

61 Peter F. Brussard and Paul R. Ehrlich, "The Challenges of Conservation Biology," *Ecological Applications* 3 (February 1992): 2.

62 S. A. Levin, "Science and Sustainability," *Ecological Applications* 3 (November 1993): 545.

63 Quoted in Kellert, "Social and Perceptual Factors in the Preservation of Animal Species," p. 59.

WILLIAM D. ROWLEY

HISTORICAL CONSIDERATIONS IN THE DEVELOPMENT OF RANGE SCIENCE

Range science is an applied science. It drew upon scientific studies of nature to assist stock growers in the utilization of forage and herbaceous resources. Scientific underpinnings gave credence to range science as it in turn legitimized the professionalization of range management. The term describes practices and methods that deal with the problems of conserving and utilizing grazing lands. For some, the term assumes a strictly regional dimension, describing range grazing practices in the American West from the prairies to the Rocky Mountains and beyond to the Pacific Coast. Despite its regional identification, the development of range science highlighted the differences and close connection in American society between pure and applied science. The former sought knowledge and understanding of nature in generally applicable terms and the latter sought "to achieve an understanding of nature in relation to human endeavor."[1]

By the end of the nineteenth century, several events had encouraged the growth of a field of inquiry and application that was to emerge as range science. In the busy world of American resource use and development, the era of "scientific conservation" arrived with the forest-use and conservation doctrines of Bernhard Fernow and Gifford Pinchot and the bombastic voice of Theodore Roosevelt. In the world of the newly developing American universities the fields of plant science and botany attempted to move beyond the stage of cataloguing and identifying plant species to an understanding of their growth processes and how plants form an ecological community. Most significantly, this involved an attempt to understand the relationship and impact of grazing animals upon vegetational stands, their capacity to sustain growth, and health under grazing pressures.

As the conservation concept of more efficient use of forest resources gained popularity, the movement carried implications for an increasingly misused and deteriorated public western range in the late nineteenth century. A search for remedial measures on the western ranges arose when vast numbers of cattle starved in harsh winters. These disasters occurred under the conditions of free land and open range grazing that prevailed on western public lands. Catastrophic stock losses on the southern plains in the winter of 1885–86, on the northern plains in 1886–87, and in the Great Basin during the White Winter of 1889–90 revealed overstocked ranges and depleted native grasses. Clearly, western graziers needed more rational methods of stock raising, more resource knowledge, and above all, conservation of range resources. A report to the U.S. Department of Agriculture in 1889 declared that "the range system of raising cattle and horses, with its attendant cruelties and losses, is gradually giving way to more humane and thrifty methods." [2]

Overstocking of the ranges became more critical when sheep appeared in great numbers on the western ranges to compete with cattle, which had earlier replaced the vast buffalo herds. Questions abounded of whether these open ranges should be cattle ranges, sheep ranges, or open to gradual settlement by agricultural homesteaders. Western states and territories and Congress itself debated the future of these lands. In some locales of the West the controversies provoked violent range wars. The refusal of Congress to legislate in favor of establishing large ranching empires left much of the western range in turmoil by the end of the century. Still many felt that the situation merely reflected "the natural and unavoidable result of the interaction of productive forces in the development of the resources of the country." [3]

Others saw the destructive side of the process. Agricultural scientists, farmers, and ranchers recognized the depletion of ranges under conditions of overstocking. A supplementary volume to the Census of the United States for 1880 made an in-depth inquiry into the grazing areas west of the Missouri River with the inclusion of Florida, "which, from its peculiar character, is subject to similar conditions." The report acknowledged "the wholly nomadic character of some of the flocks and herds, ranging unrestricted by any tenure of lands, made impracticable an exact count of the animals kept upon these great stock ranges." In addition to surveying the stock resources

of the West, the report included an analysis of forage or range plant resources by the noted professor of agricultural sciences at Yale College, William H. Brewer.

His essay on "Pasture and Forage Plants" detailed the grass, forb ("herbage other than grass"), and browse ("shrubby vegetation") resources of the
western range in a generalized discussion of the varieties of plant types.
These varieties included native and non-native plants, which by 1880 were
flourishing and replacing native plants. He first identified the distinctiveness
of the western range: "The whole aspect of the vegetation of the western
region is unlike that of the agricultural regions east of the Mississippi." He
acknowledged that plants "other than the true grasses furnish a larger portion of the forage of those regions which have rainless summers." More
importantly for the future studies in range science and management, he
noted that, "The native annual species, however, is liable to be easily run out
by heavy stocking." Brewer pointed out that cultivated species such as al-
falfa of the Chilean variety of lucerne has been introduced "to supply the
place of such."

Brewer's survey began to move beyond merely describing the forage
makeup of these ranges when he addressed problems arising from the occupation of cattle and sheep of "the new and natural pastures." Grazing
disturbed the old balance established by nature and a process of change
began in the pasturage in both kind and quantity. Some of the original species diminished and became entirely extinct, while other species diminished
without becoming extinct. "So soon as such a region becomes overstocked,
then the aggregate forage rapidly decreases, but nature in time supplies the
place with other species." Brewer argued that the nature of the change was
determined by the species which composed the original pastures, the climate
of the place, and the kind of animals pastured. When certain of the range
grasses could not reproduce under conditions of intense grazing, they
opened what would later be called an ecological niche for new species to
come in, "which are either less palatable to stock or have some natural pro-
vision by which the seed is protected and self-planted."

He believed especially good illustrations existed of these occurrences in
the valleys and lower ranges of the western Sierra in California. "Originally
the winter and spring herbage of this region was especially rich in variety and
abundant in quantity the country was alike the delight of the botanist and
the paradise of cattle. With settlement and herds the native herbage decreased and European species came in their place." As a result of the overstocking, "pernicious" species often times replaced the native plants.

"Squirrel-grass" or fox-tail was a good example. "This is less relished by stock, and it[s] heads when ripe, break up—the sharply-pointed and barbed seeds and awns doing much damage to stock." By its harmful character the seed is saved from destruction and survives into the next growing season. "So, partly because of its worthlessness when green, partly because it is an enemy to stock when ripe, and partly because of its means for dissemination and self-planting, this pernicious species can hold its own where better kinds are exterminated."

Professor Brewer believed that pastures in the more humid East were capable of great improvement. This did not seem to be the case in the drier regions of the West. He asked ominously what will be the condition in the next few years of those regions so recently overstocked, where the climate is dry and the native vegetation sparse? What species are to take the place of those being reduced or exterminated by overstocking? He could say only that the future would tell. By and large it did appear that the western ranges were free of poisonous weeds or, as Brewer referred to them, "evil species." In fact, some the "naturalized foreign species are worse than the indigenous ones." Evidence suggested as overgrazing continued, the presence of poisonous plants both native and especially foreign would increasingly threaten stock. [4]

These observations by Professor Brewer as early as 1883 recognized a range problem in the West. While his comments reflected California conditions whose ranges had been subjected to European stock grazing since the mission era in the late eighteenth century, they foretold an evolving scenario of disastrous events associated with depleted ranges in the interior West by 1890. What to do about the problem stood open to conjecture. Could investigations and the accumulation of knowledge about western forage plants and grazing practices point to solutions and remedies? The new Department of Agriculture after the Civil War established the Division of Botany in 1868. But botany, like Brewer's essay, was overwhelmingly a descriptive and taxonomic science as Americans struggled to identify and classify the vegetation of their developing West.

Stock graziers by the 1880s called upon the Department of Agriculture and its Division of Botany to turn their attention to the problems of the western range. The new secretary of agriculture from the range state of Nebraska, J. Sterling Morton, noted in his annual report of 1894 that the western ranges supported nearly 150 million head of livestock. He identified western range stock raising as one of the major agricultural enterprises in the United States and called upon Congress to create a separate Division within the USDA to serve it. Congress responded with funds for the creation of the Division of Agrostology in 1895. The new Division directed its attention to grasses and forage plants in the economic service of the western range industry. The noted agrostologist, Professor F. Lamson-Scribner, became chief of the Division whose purpose was to conduct experiments and investigations in order to combat range depletion.

First off, the Division sent botanists on collecting trips into the West. Not only were they to collect specimens, they were to make observations on the condition of the range and its forage cover. Traveling through the West, however, did not substitute for conducting experiments on the ground. In 1897 the Division of Agrostology recommended an experiment station be established near Amarillo, Texas. In March 1898, it set up two experiment sites in west Texas: one near Abilene in Taylor County and another near Channing in Harley County. While no publications came out of the Channing station, bulletins and circulars from the Abilene station offer descriptions of range conditions in Texas in the late decades of the nineteenth century. The investigations revealed a story of depletion of abundant grasses by overgrazing. These studies, according to one source, provided "the first work by the federal government or any state government in the field of range improvement [and] constituted the first range management experiments." Reports from this station reveal that many of the more sophisticated ideas about range improvement that developed in the twentieth century began with the work at Abilene in 1898.[5]

The Division of Agrostology addressed the major questions of rotation, deferred-rotation, range reseeding, and range pitting in these early years. This occurred long before range management defined itself as an organized profession. The experiments showed that range improvements expanded what professionals would call "carrying capacity." By 1901 the Divisions of Botany and Agrostology merged into the Bureau of Plant Industry. Its new director, professor W. J. Spillman of Washington State College, proved more interested in crop agriculture than in range stock problems. With this curtailment of range work, the Bureau of Forestry, to be transformed into the new U.S. Forest Service by 1905 in the Department of Agriculture, addressed range problems and conducted research in the arid and semi-arid West.

In 1891 Congress authorized the president to designate forested public lands as forest reserves. This eventually placed millions of acres of forested and mountain pasture environments under the direct auspices of the Interior Department by presidential proclamation. But no mandate existed defining the purposes of the forest reserves. In a state of limbo, the lands stood off limits to claims under any of the public land acts and apparently to use by the public. Many western communities feared the Interior Department's possession of these lands meant that all of the resources within the reserves, minerals, forage, timber, and water would be inaccessible. Under western pressures Congress devised the Forest Organic Act or Forest Management Act of 1897 that declared forest resources open to use in accordance with rules to insure favorable conditions of water flow and a continuous supply of timber. The act explicitly declared the forests open to "occupancy and use" with specific assurances that mining could occur. It was the under the "occupancy and use" clause that practices of stock grazing went forward on these lands. After 1905 the new U.S. Forest Service in the Department of Agriculture took over the administration of the forest reserves from the Department of the Interior. [6]

As such the Forest Service assumed responsibility for the administration of a significant portion of the western range. Its officers assured stock growers that forest ranges were open to responsible use. While the Forest Service opened the resources to efficient use, it understood the obligation to protect the resource and improve it for use. At the time the two were probably indistinguishable. The goal of efficient utilization required knowledge of range resources and management practices. By 1910 this agency had organized an Office of Range Studies for research and experimentation into the problems of forest ranges. Ideally knowledge derived from these studies would help protect the vegetational resources under conditions of use and, most importantly, for use. Probably most saw research as only a means to achieve more stock numbers on the land. Such efforts by publicly funded agencies began the creation of a body of knowledge that was geared to increasing the "carrying capacity" of the range. Carrying capacity was defined as the numbers of stock a range could support without damaging its ability to support that same number in the following year. Utilitarian goals dominated the scientific endeavor, but studies did seek to understand the nature of the range in terms of scientifically valid procedures. Conceivably, investigations into range problems and plant growth could dictate the reduction of animal numbers on ranges, but usually science appeared as the key to expanding those numbers.

Scientific knowledge about range vegetation purportedly offered itself to tests and verification from multiple sources under various conditions as is the procedure for establishing scientific knowledge. It possessed criteria of exactness and was subject to organization, classification, and replication in terms of experiments. Range and plant researchers dedicated to "science" in the public agencies and in the universities labored under similar standards of scientific achievement.[7] Yet the "science of range management" emerged with two masters in the early part of the twentieth century. The first was the user community that demanded knowledge about how to improve the ranges so as to increase grazing capacity; the second was the standards of the academic community that ideally asked knowledge and understanding of how things worked without the demands of practical application.

It would be convenient to attribute pure science to the universities and utilitarian endeavors to land management agencies, but no such neat division existed. Both sought utilitarian knowledge while desiring to develop the truths of knowledge in scientific terms and standards about the vegetational makeup of the western ranges. The strictly user community outside of the agencies and the universities demanded a science that expanded stock numbers. This community ultimately came to have a great deal of influence over scientific studies at the state agricultural colleges.

Recent studies in science and technology history have addressed the dual nature of range science—its theoretical and applied sides. In their efforts to bridge both needs of society and the standards of legitimate science: "range researchers attempted to develop general knowledge and practical knowledge simultaneously." This brought into sharp relief the often raised question of the relationship of science and society. Without a doubt problems arose from the interaction of stockmen and the range (man and nature). "How should researchers categorize those interactions in such a way that their research contributed both to society and to science?" Should the "occupancy and use" of the range be considered entirely from the viewpoint of the needs and values of stockmen? Or should the problem begin with a thorough understanding of the biological makeup of the range? How could researchers bring both the needs of the stockmen and range into the category of the utilitarian? "As applied scientists, they somehow had to construct a category that incorporated the stockmen and the range simultaneously, but to do that they had to define how to approach the stockmen and the range." [8]

In part, the existence of two institutional structures of research support solved the dichotomy: the state agricultural colleges and the U.S. Forest Service. State agricultural colleges could study the impact of the range on the stock animals and stock growers. The Forest Service studied the impact of the stock animals on the range. In the end however, this resulted in a fundamental difference in conclusions about the ability of the range to support numbers of stock animals. From one point of view the range came to

During a cooperative range inspection trip, Forest Service chief Henry S. Graves joins others along the Upper Porcupine Creek in southwest Montana. Pictured left to right: Nelson Storey, Montana State Game Commission; R. E. Bodley, forest supervisor, Gallatin National Forest; Hans Biering, president Taylor Fork Cattle Company; Forest Ranger Fikas, and Colonel Graves.
(USDA FOREST SERVICE PHOTO BY SMITH RILEY, 1917)

be viewed in terms of its ability to serve a certain kind of use, namely stock grazing; from the other, range health was determined by its natural ecological health and deviation from a natural condition of pristine health. The two approaches served both the needs of society and the abstract goals of science for an extended period. [9]

But the dichotomy appears too neat and even artificial to impose upon the range research efforts at the beginning of the century. Forest Service range science pitched a utilitarian line that saw pristine natural communities in a state of waste and non-utilization. It was in no way dedicated to preserving a state of nature. Constructing an idealized state of original nature might have been its way of understanding what nature was in order to manipulate it for use purposes. To assert that the Forest Service was more interested in ecological conditions and health is to sell the efforts of university science short. At the beginning of the century universities busied themselves with new theories of botany and new ways of understanding the processes of plant growth that laid the foundations of an ecological approach to range matters.

The demand for scientific range resource knowledge oriented toward application came at a time when the world of plant science or botany began to move beyond studies concerned with plant collection, identification, classification, and physiology. The new approaches suggested an attempt to understand the processes of plant development in groups and communities in response to climate, soil, and landscape. An understanding and development of theories relating to the process of plant community development and how dominant forms of vegetation replaced one another in the landscape suggested a move toward "applied botany." Europeans pioneered these theories, but by the turn of the century Professor John Merle Coulter at the University of Chicago and Professor E. E. Bessey at the University of Nebraska produced studies on vegetational process. Their students Henry Chandler Cowles at the University of Chicago and Frederic E. Clements at the University of Nebraska went on to develop a "dynamic ecology" related to the plant world. They identified change as a primary characteristic of the natural world, "and they called for a dynamic, process-oriented science that could explain this change."[10]

Toward this end Clements and Rosco Pound (later to claim fame as a legal scholar) collaborated at the University of Nebraska on a descriptive study in their Ph.D. work entitled The *Phytogeography of Nebraska* (1900). It catalogued species, described plant formations, and correlated their appearance with general features in the environment. The study offered an analysis of environmental factors, but did not attempt to explain causal relationships through experimentation. One source describes it as "a transitional work." Ecology did not provide the dominant perspective in the study of plant distributions, although it was one of many considerations. But by 1905 as an associate professor, Clements published *Research Methods in Ecology*, which declared plant geography "as a small part of a more inclusive and rigorously experimental science of ecology." The work he had done as a graduate student "was reconnaissance, a necessary but rather mundane prelude to more ambitious ecological experimentation."[11] The next step was to understand the processes of growth surrounding the ecology of the plant environment that Clements referred to as a community and others referred to as "phytosociology."[12] This was not a romanticized, idealistic community, but it was an important theoretical concept or tool by which ecologists could examine and understand the growth of plants and their relationship to other plants, species, and most centrally climate.

Ultimately, Clements embraced the idea that the plant community was a "complex organism." As such it experienced stages of growth: infancy, youth, maturity, decline, and rejuvenation—all of which encompassed both change and permanency within the community. To understand this community, according to his *Research Methods in Ecology* (1905), students had to experiment with it, measure it, and count it. This required new quantitative and experimental techniques in field and laboratory. To his botanist colleague in England, Professor A. G. Tansely, he expressed the hope that "botanical teaching in England will soon become a matter of experiment and measurement."[13]

In 1916 Clements published an important work for the methods of applied range science: *Plant Succession: An Analysis of the Development of Vegetation.* This work, plus his *Plant Indicators: The Relation of Plant Communities to Process and Practice* (Carnegie Institute: 1920), launched his studies and scholarship into the practical world of understanding natural processes for concrete utilitarian reasons for benefits to civilization. Since the beginning of the Great War in Europe, Clements saw himself engaged in the discovery of knowledge to advance the cause of agricultural range production through his scientific studies. In his own view his career began in the summer of 1893 with an investigation of the effect of recurring droughts upon the vegetation of Nebraska. Since 1900 he maintained laboratory field work in Colorado near Pike's Peak. In various states botanists, by recommendation of the National Conservation Congress and the work of the Botanical Survey, tried to carry on surveys of vegetation that resulted in classification studies.

In 1913 a more direct attack began upon "the problem of classification and optimum utilization in the West." The focus of the work soon shifted to grazing and by the time of American entrance into the World War I, it was possible to present Secretary of Agriculture David F. Houston with a comprehensive plan of investigation of the western ranges for the purpose of eventually increasing their productivity. "In seeking to carry this into effect," cooperative arrangements were made with several bureaus in the Department of Agriculture and extensive cooperation occurred with experiment stations in twelve western states. From 1913 Clements visited western states repeatedly and especially in the southwest "where the problems are insistent." According to Clements, in an explanatory letter to Secretary of the Interior Franklin K. Lane in 1918, five points of investigation commanded attention: (1) optimum utilization as indicated by the native vegetation; (2) the consequences of overgrazing and the most promising remedies; (3) the recurrence of dry and wet periods of the climatic cycle; (4) the correlation of dry farming and stock production; (5) the economic and social consequences of the present haphazard method of land settlement. Clearly Clements' investigations and his science aimed to achieve practical results from which the society could profit. Clements believed that such studies were critical because "settlement of the arid West during the past twenty years has resulted in little but slow tragedy for the settler's family, and social and economic waste for the state and nation."[14]

Both Clements and J. E. Weaver, his student and successor at the University of Nebraska, began to receive recognition for the applied values of their work in various publications. Clements' *Plant Indicators* received praise for its philosophical point of view "and the relation of the subject to other branches of science, as well as to practical affairs." Clements himself was pleased that the line of investigations he initiated appeared to be moving ecology beyond the materials of the "literary ecologist," a figure he regarded as an "amoosin critter" who could not be taken seriously. He hoped that his and Weaver's work in "the experimental attack upon vegetation" would move the science beyond such descriptions and popular folk knowledge of vegetation. He wrote: "Ecologists, like the rest of the world, mostly think with their prejudices, and nothing but overwhelming experimental evidence can help them. However, for those who turn to the voyageurs and men who `grew up' in the prairie, nothing short of trepanning can help."[15]

It was, however, Arthur W. Sampson, one of Clements' early students at the University of Nebraska, who brought the full implications of Clements' "dynamic ecology" and the succession of plant associations to the study and development of range science. He helped organize an experiment station for the Forest Service in 1911 (initially called the Utah Experiment Station and later the Great Basin Experiment Station) devoted mainly to range research on the Manti Forest near Ephraim, Utah. Sampson remained there as director of the experiment station until 1922. His numerous publications eventually led to his appointment as professor of range science at the University of California, Berkeley, in 1923. His work, *Range and Pasture Management* (1923), was the comprehensive textbook in the field and won him the reputation as the "father of range science."

Sampson's writings, often in USDA bulletins, publicized in a practical, understandable manner how the concepts of Clements' dynamic ecology could help range managers achieve greater use of their resource. His grasp of the material and ability to put it quickly into educational bulletins usually ran ahead of the efforts of Clements and Weaver to publish findings that usually found outlets in academic publications. Weaver complained to Clements in 1918 that "brother Sampson rather 'beat us to it' on the phytometer proposition" (a piece that Weaver and Clements had been working on in relation to plant growth and climate). In articles written in 1917 and 1919, Sampson related the idea of plant succession to effective range management. The title of his article "Plant Succession in Relation to Range Management" summed up the entire thrust of his career and his work to translate theoretical concepts about range and pasturage plant cover into practice. His instructional writings described a progressive development of a simple, sparse vegetation into more complex, abundant forms. All of this was part of a process in which plant communities matured from their first colonization efforts into more complex systems of vegetation undergoing various seral stages of growth. With this understanding of the vegetational community well in mind, he explained that progressive development (re-vegetation in many cases) "may be greatly expedited by cropping the herbage in such a manner as to interfere as little as possible with the life history and growth requirements peculiar to the different successional plant states." He also explained the deterioration of the range in terms of this same paradigm. If factors adverse to progressive development intervened—i.e. overgrazing—the vegetation would revert eventually to a "first weed stage." Erosion accompanied this process until the soil was practically carried away, leaving the "pioneer stage" of soil formation—bare rock, algae, and crustaceous lichens. A change of plant cover down the scale from the more complex to the more simple and primitive was termed "retrogression, retrogressive succession, or degeneration."[16]

Those in the field struggled to understand the application of terms related to the successional process. This revealed a need for the type of educational efforts Sampson's writings offered. Forest Service employee Leland S. Smith declared to Weaver in 1918 that a knowledge of ecology appears to be a prerequisite in dealing with range questions. He admitted that he had only "skimmed thru the *Plant Succession*" book by Clements several times, "to get an idea of what part I wanted to dig on." He said that he had made use of the term "Backward Succession" in his reports and memoranda to the District Office in San Francisco in an attempt to describe what is taking place on a number of the ranges. But he knew that Clements would say that this is the wrong use of the word "succession." He was puzzled how to describe a general continued change from the climax (bunch grass) stage backward through the several stages to a ruderal stage. He said this could be traced on several ranges with old photographs of certain areas that were in climax ten years ago and were now in the early weed stage. Here was a distinct "retrogression," and he concluded that, "Perhaps this same word would be sufficient, leaving off succession altogether." Sampson's publications confirmed that this field manager's inclinations were correct.

Managers learned a new technical language and a workable application of the terms from the dynamic ecological studies being promoted by Clements and his students. Sampson's writings bridged the gap between research theory and those working with the user community in the application of science to remedy the deterioration of the ranges. His career at the experiment station and then at the university became totally involved with using the resource to take maximum advantage of the processes of nature in vegetational growth. Often his user orientation obscured the details and intricacies of the subject matter itself—plants and their healthy growth in the environment. Noting the tone and content of Sampson's writings, research scientist Weaver remarked to Clements that "one gets the feeling that the plants are rather incidental!"[17]

Sampson offered a chart that depicted stages of vegetational development. The first was the initial or pioneer state, the second a transitional stage, the third a first weed stage, the fourth, a second weed stage, and lastly a final or climax stage. Working knowledge of these stages of plant succession could assist in judging whether land was being overgrazed or undergrazed according to the health and density of the herbage. Grazing animals could be applied to the land in accordance with its ability to support animals, and a carrying capacity could be gauged and "the herbage cropped on the basis of a sustained yield." This cropping should be done in a manner that would not interfere with the maintenance of a desired successional stage. With this knowledge, "progressive succession" could be promoted and protected. Managers could determine, on the basis of the life history of the various plant species, the season of grazing with reference to the time of seed maturity and other factors required for good range management. With this knowledge, a range could be managed to keep it in a stage of subclimax over long periods of time. When less desirable vegetation began to appear, this would be an indicator that grazing pressures were too heavy and must be eased. Overgrazing often moved the range into a brush condition as stock destroyed the ability of grass to compete.

In like manner the absence of grazing might permit the ecological succession to proceed unobstructed. Plant communities mature and react to their surroundings, "and in many cases drive themselves out." When this occurred the resource was essentially wasted. This result underlined the importance of maintaining the subclimax condition for the ongoing use and productivity of the range by the intelligent intervention of man into the successional process. Theories of plant succession laid the foundation for a general theory of ecological succession eventually adopted by wildlife managers, because it appeared to hold true for wildlife and its relation to habitat as well.[18]

A later critic noted that this view of the land saw it in terms of products it could produce—livestock or desirable species of wild game. And of course, range science aimed to maximize those products with an emphasis upon domestic stock. The same critic asserted that, "Successional theory transformed range ecology since it offered researchers a way of thinking about history." This is also to say that it gave researchers a way of thinking about the future. Few boast to have a key to the future, but Clementsian plant science could clearly make bold claims to it. A successional process once set in motion would predictably lead to an envisioned result.[19]

To determine the stage or direction of a process under way on a range or in a plant community, Clements and his students initiated counting and measurement. This involved counting of species, frequency of their occurrence, and extent of foliage and even root growth. Clements developed the quadrat method to measure and count range vegetation. It called for the marking off of plots in which plant measurement and counting occurred; records were kept over a period of months or years. Sometimes fences protected the plot from grazing, and at other times grazing was permitted. Ultimately the quadrat method attempted to confirm through statistical analysis of plant composition the condition of the range and the direction of the succession. Observers determined whether more valuable perennial grasses were on the increase or in decline as compared with the less valuable annual grasses and forbs or even the increase of brush. The quadrat method was the logical outcome and manifestation in the field of the quantitative methodology of the Nebraska school of ecology expressed in Clements' *Research Methods in Ecology* in 1905.[20]

The attempt to classify range conditions occurred early. They emerged in the studies of J. G. Smith presented in the *Yearbook of Agriculture of 1895,* in David Griffiths' study in 1903 of Washington, Oregon, and Nevada range, and in E. O. Wooten's study of the New Mexico range in 1908 for the New Mexico Agriculture Experiment Station. In the early attempts by the Forest Service to draw range-survey maps, based upon range survey and evaluations, starting in 1911, range condition was indicated to be either satisfactory or unsatisfactory. "Unsatisfactory" referred to those ranges producing less than their capability. The ability of ranges to sustain stock production appeared to be the chief objective, and that became the definition of range health. Sampson classified conditions into four broad stages of plant succession in his 1919 article: (1) the wheat-grass consociation (subclimax stage); (2) the porcupine-grass-yellow-brush consociation (mixed grass-and-weed stage); (3) the foxglove-sweet-sage-yarrow consociation (second or late weed stage); and (4) the ruderal-early-weed consociation (first or early weed stage). These stages for the range under study corresponded roughly to the more modern condition classification of excellent, good, fair, and poor.[21]

As Clements undertook his new duties with the Carnegie Foundation in ecological studies immediately after World War I, he doubted the ability of professionals to communicate effectively with laymen on the value of ecological research:

> The task of making ecological work show a direct connection with practical everyday problems to the extent that the layman can get the application, is an extremely difficult one. I confess that I do not see my way clearly in this matter yet and that I do not know any one who does. In fact, it is pretty hard for the layman to get the application of a lot of the supposedly practical work that is done at the Stations already, and to put him in understanding touch with the ecological basis beneath it is much more difficult still.[22]

As if to answer this quandary that Clements expressed in 1919, Sampson's bulletin "Succession in Relation to Range Management" and James T. Jardine's and Mark Anderson's equally important "Range Management on the National Forests" appeared in the same year. In 1950 longtime range expert with the Forest Service W. R. Chapline observed that Sampson's ideas about plant succession in relation to range management "brought the close interrelation of ecological processes to range deterioration and restoration to the fore."[23]

In 1910, to coordinate and promote range research and the collection and use of technical range management information, the Forest Service established the Office of Grazing Studies, naming Jardine as its director. The office received official approval from the secretary of agriculture in late 1911. Jardine launched a grazing "reconnaissance" of specific forest ranges. In later years, this job was called "range survey" or "range-resource inventory." The range reconnaissance did more than observe the conditions of the ranges. Ultimately the survey hoped to find a simple, reliable method for empirically (visually) rating the condition of the range and also for determining whether the range was improving or deteriorating. The goal was to provide a stronger factual basis for improved range management on the areas covered. The range reconnaissance became a systematic survey that collected field data for final compilation in the office. The field work classified grazing lands into vegetation types. The resulting maps showed location, acreage, topography, amount and character of vegetation, condition of the range, available watering places, and cultural features. On the basis of these data, rangers developed grazing-management plans and periodically adjusted them for discrepancies. In the early years when no university or college taught range management, it was necessary to use capable students, preferably with botany background, to conduct range reconnaissance under careful supervision. According to Chapline, who was for many years chief of the Division of Range Research in the Forest Service, reconnaissance work was a well-planned undertaking, requiring trained personnel in charge of each party and effective training of assistants at the start of each year's work. During the evening, each field observer checked his type lines, density estimates of adjoining types on his maps, and the major plant species listed on each type against those of the adjoining observer. If there were major differences, they were usually checked in the field by the chief of the party. On rainy days and often on Sundays, the field records were assembled for range allotments with calculated forage-acre values for types. In addition, a collection was made of every plant found on the forest being covered. For each plant a form was filled out relating to the characteristics of the location where it was collected, the soil, the topography, and evidence of use by animals. Such data were not ordinarily analyzed during the field work, but the information proved helpful later in developing published material on range plants.

Range reconnaissance differed from range inspection. The latter concentrated on analysis of a specific allotment. Its purpose was to discover weaknesses in the current management plan and recommend solutions. The inspection proved most useful in fixing adjustments that must be made either upward or downward in the number of stock grazed. Although the Forest Service admitted in 1910 that many years would be required to reconnoiter the ranges of the national forests, range inspections served to produce field maps showing the relationship of the allotment boundaries to the topography. Much of this information appeared on overlays showing allotment boundaries, number of stock within each, acreage with acres per head of stock, and permittee's name. By learning the carrying capacity for certain range types through continued observation, the capacities of similar ranges could be estimated. And the basic question was how to administer the range to maintain optimal carrying capacity to sustain the yield from season to season.[24]

The central problem was to determine when a range was overstocked. In those early years, it was believed, the condition of the stock reflected an overstocked or overgrazed range. Depleted ranges seldom produced fat cattle. Later it was realized that instances occurred when such ranges temporarily continued to produce fat stock. This slow realization reflected the early range manager's greater attention to the condition of stock rather than the resource that supported it. In 1911, to meet the practical challenges of range management on its extensive domain, the Forest Service in 1911 expanded the Office of Grazing Studies into regional offices. The four branch offices included the Pacific Northwest, Intermountain, Rocky Mountain, and Southwestern regions, and all emphasized a reconnaissance of the ranges. These offices of regional studies promoted the implementation of uniform grazing practices by adjusting them to special regional needs and study of the peculiar resources of the regions.[25]

By 1915 the Forest Service achieved recognition as the national leader in range research. In that year the USDA transferred responsibility for range research from the Bureau of Plant Industry to the Forest Service. The wider scope of the assignment yielded studies relating to climate and plant growth and range management in semi-desert ranges, notably the Jornada and Santa Rita experimental ranges in southern New Mexico and Arizona and studies in the Blue Mountains of Oregon. From the latter came Jardine and Anderson's study "Range Management on the National Forests" in 1919 that meshed principles of range management with the advances in range science theory arising from the studies of Clements, his successors, and students at the University of Nebraska. The study summarized what would later be known as the "Prairie School of Range Science." It became the "range bible," modifying management procedures as originally outlined in the grazing section of the early Forest Service's *Use Book*.

After Jardine's resignation from the Office of Grazing Studies in 1920 to head the Oregon Experiment Station, Chapline took the post. In 1925 the research duties came under the Branch of Research in the Forest Service. The administrative operation of the Office of Grazing Studies, relating to range reconnaissance and development of range-management plans, remained in the Branch of Grazing. Funds for these projects represented about two-thirds of the monies allotted to the Office of Grazing Studies in 1925, when Chapline transferred to the Branch of Research to expand this work.

In 1937 the Division of Forest Influences was established, linking forest and range watershed studies.[26]

Applying the discoveries of the range scientists to everyday problems of range management often involved frustrations. In 1915 Assistant Chief for Grazing Will C. Barnes, who viewed himself as a practical cowman, complained of the temperamental natures of the scientists who worked on range problems when he first headed the Branch of Grazing: "Confound these scientific chaps anyhow with their temperamental weakness. I'd rather be a plain cow puncher individual without temperament or brains." On the other hand, Barnes knew the value of research workers in developing better methods of handling stock and for assessing range productivity and condition. He pushed for more appropriations to extend range-survey work. He was quick to point out that the Grazing Branch turned into the U.S. Treasury almost as much money as did timber sales ($1,130,500 compared to $1,175,00 in fiscal year 1915), but got only one sixth of the money necessary to carry on the work. An increase in grazing fees and appropriations to his office, he pointed out, justified the work's expense because such surveys usually resulted in the more efficient use of the range by greater numbers of stock.[27]

By World War I, the Forest Service's Office of Grazing Studies had attempted to provided some factual and empirical answers in several general areas where knowledge was needed, if the ranges were to be properly managed for optimum stock production:

1. Opening and closing dates for grazing to harmonize range readiness with nutritional requirements of livestock;
2. Determining grazing capacities of western range types;
3. Determining whether forage cover and soil were improving or deteriorating;
4. Deferred and rotational grazing practices to permit seed maturity and root system growth for the survival of perennials;
5. Improved methods of grazing sheep and goats in open and quiet herding; also new procedures for bedding herds in different locations each night;
6. Better management of cattle through well-placed watering and salting sources;
7. Elimination of damage to timber reproduction and other forest resources from grazing.

Research into outdoor public recreation had not yet begun, but concern for game animals and the impact of domestic grazing on their food resources was already becoming a point of contention between recreationists and the grazing interests.[28]

The data gathering, the quadrat studies, plant experiments, and statistical tables did not impress many old-time range men. They still insisted that experience and intelligent "rule of thumb" guessing offered better guides to proper range management. Reliance upon the "college boys" seemed a needless expense involving time consuming procedures and doubtful experiments. Still, in order for its officers in the field to administer with confidence, the professionals of the Forest Service needed to make decisions rooted in information derived from objective, scientific study. Theories had to be supported by practical observations and results. And with the growing acceptance of Clementsian vegetational science, a frame work emerged in which research could proceed and from which practical solutions and advice to range problems could be derived. As John H. Hatton, assistant regional forester for Region 2, emphasized in the mid-twenties: "Such expressions as 'the range was overgrazed but not permanently injured' must be supported by that permanent sample plot whether inferior and unpalatable species are displacing it." Administrators understood that professional range management needed a body of knowledge from which they could derive authority other than the obvious authority that their uniforms asserted over the public forest lands. Range science was the necessary source of authority and, most importantly, power for aggressive range-management policies.[29]

Still the scarcity of the research and the slowness of its application left the "rule of the thumb" principles dominant in many of the forest ranges in the 1920s. Until 1927 the Forest Service employed fewer than forty full time forest range research and technical workers in any given year. Range research doubled in the next two years with the passage of the McSweeney-McNary Forest Research Act in 1928 and expanded thereafter to meet needs in the stock industry and public land management. Scientists believed they were the best qualified to guide management's decisions in stocking the ranges in accordance with their assessment of "ecological health." The Forest Service hoped these findings would protect it from political pressures, but science itself, in retrospect, often found itself shaped by political considerations as definitions of "carrying capacity increasingly permitted fullest possible use."[30]

When the Great Depression brought economic hardship to millions of

Americans including those making their living from the western ranges, government and science responded. Drought coupled with depression made the crisis particularly acute and gave it an ecological backdrop. An ecological picture had already been sketched by the 1920s showing a relationship amongst plants, soils, water, weather, and terrain. As scientists became increasingly interested in soil conservation and erosion, the great dust storms struck the plains seemingly confirming the improvident use of soil resources. The responses of the New Deal included the creation of the Soil Conservation Service in 1935. Soils had always been an important consideration for range scientists. From the earliest studies in the Wasatch Mountains in Utah the importance of ground cover for erosion and destructive flooding appeared. Grazing, of course, played a role in causing the destruction of soils, if unregulated. The Forest Service's Copeland Report in 1933 entitled *A National Plan for American Forestry* emphasized the concept of "forest influences." It projected an interrelationship of forest resources and called for rehabilitating the range and reversing soil erosion.[31]

The decade offered enormous opportunities for science to knit together the interdependence of forest, water, range, and even wildlife resources in western grazing lands. The construct of Clementsian ecological processes stood ready to be utilized. But the economic crisis combined with the drought in the decade drove scientists and policy makers alike into defensive positions. Policy makers sought ways to ease economic hardship. This meant permitting stock numbers to remain high in the face of drought. Congressional legislation in the form of the Taylor Grazing Act of 1934, while a response to the drought and overgrazed public domain outside of the Forest Service lands, mainly secured the presence of those already using the public domain and excluded newcomers. Scientists, however, trained in the Clementsian paradigm experienced their own crisis. They stood in disbelief that the vegetational processes in the great prairie grassland communities had been disrupted by an enduring drought. Not only had the process of vegetational succession been disrupted, it seemed in many areas that it had been obliterated and replaced by a "Dust Bowl."

By the 1930s, however, from Lincoln, Nebraska to Berkeley, California range scientists trained in the Clements' school dominated the range studies departments in the schools of agriculture. As the economy revived under wartime demands in the 1940s and as the rains came again, their faith in the Clementsian ecological community was renewed, but significantly the amount of published research outwardly employing the paradigm sharply declined. Interest shifted primarily to "technological concerns" that found expression in the new *Journal of Range Management* that appeared in 1948 under the sponsorship of the newly formed Society of Range Management. Science historian Ronald C. Tobey asserts that the formation of the Society represented a sundering of the "long-term Clementsian synthesis of theory and practice . . . with practice split off, and range managers took most of the manpower out of grassland ecology."[32]

In 1950 the geographer Carl O. Sauer noted in the *Journal of Range Management* that the concept of a stable vegetational community was "misleading." He wrote, "Ecology has instructed us that plant societies may strike such happy balance with their environment and between their members as to form a stable, indefinitely reproducing order, called a climax vegetation." He questioned this concept and the idea that under a given climate there will form a stable self-perpetuating plant complex. His main argument held that the cultural introduction of fire by man often shaped the course of grassland vegetation. "Plant associations are contemporary expressions of historical events and processes," he said and concluded that, "A real science of plant ecology must rest not only on physiology and genetics, but on historical plant and physical geography."[33]

In 1952 Sampson's book *Range Management: Principles and Practices* appeared. Great Plains and Kansas historian James C. Malin expressed skepticism in reference to Sampson's acknowledgment of the importance of Clements in the founding of American ecology and noted Sampson's commitment to the "Clementsian-USDA school of ecology." Not a believer, Malin said, "The day of reckoning and re-examination of principles at the hands of the younger generation are long over due." He made a further perceptive observation that the book dealt with both principles and practices. In the realm of practice the range manager was primarily interested in the description of accepted practices. Fortunately, Malin observed, the practitioners often paid little attention to theoretical correctness. This, in a sense, is what management proceeded to do in the post-war period. With the Clementsian paradigm coming under attack as more of a belief system than a demonstrable scientific fact or occurrence, descriptions and instructions in sound range practices moved to the forefront of the literature with little attention to theoretical underpinnings. All of this was readily apparent in the articles of the *Journal of Range Management*.[34]

Yet the idea of ecological principles underpinning range knowledge and especially playing a role in determining range condition for grazing purposes persisted. E. J. Dyksterhuis in an important article in 1949 admitted there were many ways of determining range condition and that, "Some no longer have the original ecologic basis." But unmistakably Dyksterhuis believed "quantitative ecology" the way to determine range condition. He held to the concept of climax vegetation and defined it as "the highest point, or culmination, of plant succession." Furthermore, he defined "applied ecology" as "a simple medium for classifying ranges and for interpreting common range phenomena." He believed that range conditions classes could be based on percentages of "decreasers, increasers, and invaders, [plants] as measured from relative amounts in the climax for the site." Dyksterhuis's innovative standards for range condition suggested that estimates of plant species within the three classes should occur in the field and be compared to standard cover for a location. And different sites should have different standards for comparison. Forest Service range scientist Linda A. Joyce asserted in a 1993 article that Dyksterhuis's range condition concept differed from other ideas during the post-war period because "1) site classification was not dependent upon current vegetation 2) forage production was considered a basis for range condition classification 3) empirical groups of plants were replaced with an ecological classification, and 4) the widely used concept of forage density was replaced with the 'the quantitative relation of vegetation based on total annual forage production'." All of this suggested that grazing capacity, condition, and trend could be appraised and evaluated better after grazing occurred.[35]

Despite his new conception of range condition, Dyksterhuis continued to operate within the Clements-Sampson-USDA paradigm of successive stages of vegetation communities that either gravitated toward a climax stage or retrogressed backwards toward more primitive beginnings. Along with trial and error it was still the means for classifying ranges in terms of conditions —poor, fair, good, excellent. Quantitative ecology proved the best method for taking account of an increase or invasion by annuals and woody plants that was one of the most common results of range depletion. And Dyksterhuis approvingly quoted Clements in his reverence to the concept of the plant community: "There can be no doubt that the community is a more reliable indicator than any single species of it. . . . The significant species are the dominants and subdominants which give character to definite communities." This raised to the fore the importance of evaluating range condition and made this the central question in range management.[36]

As the Clementsian paradigm of the "mono-climax" with climate as the primary factor in its attainment declined, other scientists adjusted to the discrediting of this basic underpinning of range science by saying that it was important "to maintain a sharp distinction between facts and their interpretation." By 1968 ecologist Rexford Daubenmire asserted in *Plant Communities: A Textbook of Plant Synecology* the necessity of evaluating environmental variation in bringing about vegetation differentiation. By this he meant evaluation of soils and the contour of the land in explaining variation. He acknowledged that "Clements' vigorous exposition of his philosophy quickly captivated North American workers," but "research since the heyday of the monoclimax hypothesis" has pointed out its problems. He believed the "present viewpoint represents a return to the pre-Clementsian concept" and cited a study by R. Hult in 1885 that in a part of Finland where there were seven habitat types, each had a distinctive climax. Yet he observed that many "still passively accept" the Clementsian terminology, "with indeed some still accepting the major premise."[37]

From the standpoint of most range scientists, range condition was about production, measured "directly in terms of forage production and indirectly in pounds of meat or wool produced."[38] Range condition, along with the concept of trend, became a point of contention between stock operators and government agencies like the Forest Service that sought to make judgments on the number of stock to allow on ranges according to their condition and trend. Like the old concept of carrying capacity, these newer concepts were crucial in terms of numbers of stock permitted on government grazing lands. In 1948, the research arm of the Forest Service quietly initiated a study to obtain a method of demonstrating the condition and trend of ranges to ranchers. Range conservationist Kenneth W. Parker developed the so-called Parker three-step method. The goal was to develop a standard method of measurement. Stock operators generally maintained that they were the best judges of range condition because their livelihood depended upon it. They resented federal officials making what they regarded as snap judgments on the condition of the range and sometimes reducing the numbers of stock grazing on federal lands. The Forest Service continued to call upon science and its research arm to provide an objective methodology that would gain the confidence of the stock industry.

In 1951, Chief Forester Lyle F. Watts announced the completion of the three-year study for measuring trend in range condition. In Denver on August 11, 1951 Parker presented the results of his study, entitled, "A Method for Measuring Trend in Range Condition on National Forest Ranges." Recommended for perennial grasslands, mountain meadow, open-timber lands, and sagebrush-grass types found in national forest ranges in the West, the methods were the following:

1. Measurement and observation of the essential features of vegetation and soil as recorded on permanently established transect lines and plots located on the usable parts of the range. Measurement is made by means of a small ring or loop at one hundred points along each transect line.
2. Classification of these data as to condition of vegetation and soil stability and estimation of current trend in condition.
3. General and close photographs from permanently located photo points. Application of the three-step method on an allotment basis requires mapping of the usable range areas by broad vegetative types and by condition situations within these types as a prelude to sampling. Trend is determined by comparison of records made at periodic intervals.

Comparisons of data taken over a period of several years obtained in the three steps could yield an understanding of the range condition and trend.

Forest Supervisor Joe Radel and District Ranger Emil Koledin examine the condition of cattle range in Big Whitney Meadow, along the Golden Trout Creek, Inyo National Forest in California.
(USDA FOREST SERVICE PHOTO BY DANIEL O. TODD, 1958)

Experiments showed that different individuals measuring the same transect obtained nearly identical results. This approach was particularly valuable in the view of the stock operators because it minimized chances for error and biased interpretation. If the measurement was taken in cooperation with the permittee and the statistical data shared, a climate and spirit of cooperation would be generated between the service and the stockmen as never before existed. The Forest Service made it clear that the new method could not determine carrying capacity, but it could determine trends in utilization. But representatives of stock organizations soon charged that carrying capacity was determined by the "unrealistic hoop method." They described the method as throwing down a hoop in a designated area and counting the forage contained within it. "On the basis of such inconclusive, impractical and theoretical information, drastic reductions in grazing allotment have been made," they complained. More to the point they asserted, "To base decisions on scientific facts alone has seldom proven entirely satisfactory in any field of agriculture." The organizations now asserted that range managers should emphasize instead the history of the allotment, local conditions and experience of the ranchers in utilizing the range.[39]

The stock industry testified before a Senate investigating committee that the method developed by the Forest Service had shown inaccuracies "except when correlated with actual stock-carrying performance." In support of their view, they cited the text *Range Surveying and Management Planning* by Laurence A. Stoddart and Arthur D. Smith. Stoddart had long been associated with the Department of Range Management at Texas A&M College and Smith with Utah State Agricultural College. The most significant passage seized upon by the representatives of the stock industry concerned the determination of grazing capacity: "Perhaps the best method of determining proper stocking is a study of the history of the stocking over a period of years, together with a very careful study of its effect upon the range. This requires training and knowledge of plants, but it is nothing that the observant stockman cannot master." Here was a good example of personnel in the state agricultural colleges calling into question scientific, quantitative methods of range analysis.[40]

The process of making these determinations was not the exclusive property of a professional elite, nor did any great need exist to understand quantitative ecology to make intelligent judgments about condition and trend. Such statements did little for the prestige of range science and its claims of professional and expert knowledge.[41]

The Forest Service responded by defending its experience and science. It said that the range-survey procedures had been developed from fifty years of research and practical range experience by the agency and its personnel. The service emphatically believed that its range-analysis method was sound, adequately applied, and offered a realistic evaluation of range and watershed conditions on national forest land.[42]

Clearly all of this pointed to an ongoing dispute in which the weakness of range science would be at center stage. The declining prestige of Clementsian vegetational theories and the determination of some range authorities in the state agricultural colleges to deny the validity of long standing authorities in range matters spelled a disputatious future for range studies, science, and management. This proved particularly true on the public lands when agency science called for reduced grazing pressures.

At the end of the twentieth century matters on the western range are not resolved. Range science remains adrift with no general agreement on theoretical underpinnings. Complicated and contradictory information about the interrelationship of resources in an ecosystem makes it difficult to translate chaotic ecological knowledge into management practices. Overall "range health" is a major goal. The publication of *Rangeland Health: New Methods to Classify, Inventory and Monitor Rangelands* by the National Research Council in 1994 asserted that, "Rangeland health should be defined as the degree to which the integrity of the soil and ecological processes of rangeland ecosystems are sustained."[43]

But it made no mention of an overarching theory about vegetation growth and how range management fit into it -- only the acknowledgment of the functioning of the ecosystem. The so-called working model of the "ecosystem" employs the tools of computer science to analyze the system and project proper uses or non-use. Some embrace holistic management principles citing the benefits of grazing for the renewal and vitality of range grasses. Its chief advocate, Alan Savory, suggested the range grasses developed under grazing pressures from quadrupeds and the presence of these animals was necessary for range health. His conclusions were welcomed in ranching circles.[44]

Finally, an off shoot of the space program provided one of the most sophisticated techniques or tools in the development of range analysis and monitoring. Landsat research (remote sensing) on photos taken by satellites provided the expert eye with the information that many hope is the key to answering many the controversial range questions associated with condition and trend.[45]

It is clear by century's end that range science still walks a tight rope between society and science. Value changes in American society from the beginning of the century to its end further complicates matters. The *Rangeland Health* study might glibly state that "The capacity of rangelands to produce commodities and satisfy societal values depends on the interactions of climate, plants, and animals in a given physical landscape over time," but determining either the beneficial or destructive effects of those interactions, still fuel controversy in science and society.[46]

NOTES

1 Frank N. Egerton, "The History of Ecology: Achievements and Opportunities, Part Two," *Journal of the History of Biology* 18 (Spring, 1984): 106, 103.

2 Edgar T. Ensign, *Report on the Forest Conditions of the Rocky Mountains*, Bulletin no. 2, U.S. Department of Agriculture, Division of Forestry, p. 64.

3 U.S. Congress, House, *Letter from the Secretary of the Treasury Transmitting a Report from the Chief of the Bureau of Statistics* [*Ranch and Range Cattle Traffic* by Joseph Nimmo, Jr.], 48th Cong., 2d sess., 1885, H. Doc. 267, p. 50.

4 Clarence W. Gordon, "Report on Cattle, Sheep, and Swine, Supplementary Enumeration of Livestock on Farms in 1880." *Report on the Productions of Agriculture as Returned at the Tenth Census*, June 1, 1880. Vol. III (Washington, D.C.: Government Printing Office, 1883), pp. 5-10; William H. Brewer, *Up and Down California in 1860–1864*; the journal of William H. Brewer edited by Francis P. Farquhar (New Haven: Yale University Press, 1930).

5 Thadis W. Box and Idris R. Traylor, Jr. "Grasses on the Plains: The First Range Management Experimental Stations," *Journal of the West*, 29 (October, 1990), pp. 75-81; C. H. Wasser, Early Development of Technical Range Management, CA. 1895-1945," *Agricultural History* 51 (October, 1977), pp. 64–65.

6 William D. Rowley, *U.S. Forest Service Grazing and Rangelands: A History* (College Station: Texas A&M University, 1985), p. 31; Paul W. Gates and Robert W. Swenson, *The History of Public Land Law Development* (Washington, D.C.: Public Land Law Review Commission, 1968), p. 569; Harold K. Steen, *The U.S. Forest Service: A History* (Seattle: University of Washington Press, 1976), p. 36.

7 Ashley L. Schiff, *Fire and Water: Scientific Heresy in the Forest Service* (Cambridge, Mass.: Harvard University Press, 1962); T. Swann Harding, *Two Blades of Grass: A History of Scientific Development in the U.S. Department of Agriculture* (Norman: University of Oklahoma Press, 1947).

8 Maarten Heyboer, "Grass-Counters, Stock-Feeders, and the Dual Orientation of Applied Science: The History of Range Science, 1895–1960," (Ph.D. diss., Virginia Polytechnic Institute and State University, 1992), pp. 4–8.

9 Heyboer, "Grass Counters," pp. 268–270, 278.

10 Rowley, *Grazing and Rangelands*, p. 102; Joel B. Hagen, *An Entangled Bank: The Origins of Ecosystem Ecology* (New Brunswick, New Jersey: Rutgers University Press, 1992), pp. 15–16; Richard Overfield, "Charles E. Bessey: The Impact of New Botany on American Agriculture," *Technology and Culture*, 16 (April, 1975), pp. 162–81; Overfield, *Science with Practice: Charles E. Bessey and the Maturing of American Botany* (Ames: Iowa State University Press, 1993), pp. 131–156.

11 Hagen, *An Entangled Bank*, p. 22.

12 James White, "The Population Structure of Vegetation," *Handbook of Vegetation Science*, ed. J. White (Boston: Dr W. Junk Publishers, 1985), p. 1.

13 Donald Worster, *Nature's Economy: A History of Ecological Ideas* (New York: Cambridge University Press, 1977) "Clements and the Climax Community" pp. 205-220; White, "Population Structure of Vegetation," in *Handbook*, pp. 2-5; Hagen, *An Entangled Bank*, p. 23; Clements, *Research Methods in Ecology* (Lincoln, Neb.: The University Publishing Co., 1905); Clements to A. G. Tansley, Nov. 23, 1918, Frederic E. Clements Papers, acc. #1672, Box 62, American Heritage Center, University of Wyoming, Laramie [hereafter cited as Clements Papers].

14 Clements to Franklin K. Lane, November 15, 1918, Box 62, Clements Papers.

15 C. F. Korstian, "Review of Ecological Studies and Studies in Ecotone Between Prairie and Woodland," *Journal of Forestry* (Jan. 1921): 67; G. A Pearson, "Periodical Literature," *Journal of Forestry* (October, 1919): 640; Thomas H. Kearney, "Scientific Books," *Science*, 52 (December 24, 1920): 610; Clements to J. E. Weaver (January 28, 1921) Box 63 Clements Papers.

16 Weaver to Clements (November 23, 1918) Box 62, Clements Papers; Arthur W. Sampson, "Succession as a Factor in Range Management," *Journal of Forestry* 4 (May, 1917): 593–596; Arthur W. Sampson, "Plant Succession in Relation to Range Management," USDA *Bulletin* No. 791, (August 27, 1919): 6.

17 Leland S. Smith to Weaver (c. Jan.–June, 1918) Box 62; Weaver to Clements (November 23, 1918) Box 62 Clements Papers.

18 Rowley, *Grazing and Rangelands*, pp. 102, 106; Ronald C. Tobey, *Saving the Prairies: The Life Cycle of the Founding School of American Plant Ecology, 1895–1955,* (Berkeley: University of California Press, 1981) p. 70; Charles Elton, *Animal Ecology* (New York: Macmillan, 1927) p. 21.

19 Nancy Langston, *Forest Dreams, Forest Nightmares: The Paradox of Old Growth in the Inland West* (Seattle: University of Washington Press, 1995), p. 242.

20 Rowley, *Grazing and Rangelands*, 102, 106; Overfield, *Science with Practice*, pp. 148–149.

21 E. O. Wooton, "Range Problem in New Mexico," *Bulletin* No. 66, April, 1908, pp. 4–46, New Mexico College of Agriculture and Mechanic Arts, Agricultural Experiment Station; Sampson, "Plant Succession in Relation to Range Management," p. 7; for continued application of these concepts see Paul T. Tueller, "Secondary Succession Disclimax and Range Condition Standards in Desert Shrub Vegetation." in *Arid Shrublands: Proceedings of the Third Workshop of the United*

States/Australian Rangelands Panel, Tucson, Arizona, March 26–April 5, 1973, (Denver: Society of Range Management, 1973), p. 60.

22 Clements to Weaver, March 21, 1919, Clements Papers.

23 James T. Jardine and Mark Anderson, "Range Management on the National Forests," *USDA Bulletin* No. 790 (August 6, 1919); W. R. Chapline to Marion Clawson, September 18, 1950, Sec. 63, Box 637, RG 95, National Archives and Records Administration, Washington, D.C.; Rowley, *Grazing and Rangelands*, pp. 105–106.

24 William D. Rowley, "Priviledge vs. Right: Livestock Grazing in the U.S. Government & Forests" in Harold K. Steen, ed. *History of Sustained-Yield Forestry: A Symposium* (Santa Cruz: Forest History Society, 1984): pp. 61–67.

25 Jardine and Anderson, Range Management on the National Forests, p. 82; W. R. Chapline, Robert S. Campbell, Raymond Price, and George Stewart, "The History of Western Range Research," *Agricultural History*, 18 (1944): 132; R. E. Bodley, "Grazing Reconnaissance on the Coconino National Forest," *Nebraska University Forest Club Annual*, 5 (1913): 77.

26 Rowley, *Grazing and Rangelands*, p. 108.

27 Ibid., fn. 18, p. 109.

28 Chapline, et al., "The History of Western Range Research," pp. 127–143; Chapline, "Range Management History and Philosophy," *Journal of Forestry*, 49 (September, 1951): 636; R. S. Campbell, "Milestones in Range Management," *Journal of Range Management*, 1 (October, 1948): 6.

29 Heyboer, "Grass Counters," addresses the question of power over nature: "People who want power over nature become dependent on the experts who develop the knowledge that provides the means to acquire power over nature," p. 26.

30 John H. Hatton, "Economic Results of Improved Methods of Grazing" [ca. 1924], Box 5, Fd. 15, Work Project Administration, History of Grazing Papers, Utah State University Special Collections, Logan, Utah; Rowley, *Grazing and Rangelands*, p. 111; Chapline, "Range Research of the U.S. Forest Service," *American Society of Agronomy*, 21 (June,1929): 648–49; Thomas G. Alexander, "From Rule of Thumb to Scientific Range Management: The Case of the Intermountain Region of the Forest Service" *Western Historical Quarterly*, 18 (October, 1987): 409–428; Langston, *Forest Dreams*, pp. 207, 211.

31 Charles Peterson, "Small Holding Land Patterns in Utah," *Journal of Forest History*, 17 (1973): 9-10; *A National Plan for American Forestry*, 73d Cong. 1st Sess. Sen. Doc. No. 12 (Washington, D.C.: Government Printing Office, 1933); Reed W. Bailey, "Floods and Accelerated Erosion in Northern Utah," *USDA Misc. Publication, No. 196* (August, 1934): 1–21.

32 Tobey, *Saving the Prairies*, pp. 152–154.

33 Carl O. Sauer, "Grassland Climax, Fire, and Man," *Journal of Range Management*, 3 (January, 1950): 16–17.

34 James A. Malin, "Grassland Conservation," *Scientific Monthly* 74 (May, 1952):

308–09; Frank E. Egler, "A Commentary on American Plant Ecology, Based on the Textbooks of 1947–48," *Ecology*, 32 (no. 4, 1951): 673–95.

35 E. J. Dyksterhuis, "Condition and Management of Range Land Based on Quantitative Ecology," *Journal of Range Management*, 2 (April, 1949): 104–115; Linda A. Joyce, "The Life Cycle of the Range Condition Concept," *Journal of Range Management* 46 (March, 1993), "More than just a methodological technique to determine overgrazing, range condition was used as a stratification of the ecosystem in order to test whether hydrological, botanical, or physiological aspects differed across condition class," 134–35; Wasser, "Early Development of Technical Range Management," 75.

36 Dyksterhuis, "Condition and Management of Range Land,"; Jerry L. Holechek, "A Brief History of Range Management in the United States," *Rangelands*, 3 (February, 1981): 17.

37 Rexford Daubenmire, *Plant Communities: A Textbook of Plant Synecology* (New York: Harper & Row, Publishers, 1968), pp. 240–42.

38 R. R. Humphrey, "Field Comments on the Range Condition Method of Forage Survey," *Journal of Range Management*, 2 (January, 1949): 1.

39 Forest Service, *Annual Report*, 1951, and 1952, p. 63 and pp. 24–25, respectively; J. M. Jones, "Measuring Range Trends," *National Wool Grower*, 41 (September, 1951): 41; U. S. Congress, Senate, Subcommittee on Public Lands of the Committee on Interior and Insular Affairs, *Hearings on Grazing on Public Domain Lands*, 86th Cong. 2d sess., 1960, part 1, vol. 3, pp. 1–7.

40 Heyboer, "Grass-Counters," notes the differences between the science of the state agricultural colleges and Forest Service researchers.

41 Laurence A. Stoddart and Arthur D. Smith, *Range Surveying and Management Planning*, p. 223, as quoted in p. 9, note 32, of Stoddart and Smith, *Range Management* (New York: McGraw-Hill, 1955), which identified two methods of determining grazing capacity, the ocular reconnaissance and the plot method, and concluded: "It is evident that little reliance can be placed upon grazing capacities arrived at by either survey method" (p. 175).

42 Rowley, *Grazing and Rangelands*, p. 220.

43 See *Rangeland Health: New Methods to Classify, Inventory, and Monitor Rangelands* by Committee on Rangeland Classification, Board on Agriculture, and National Research Council (Washington, D.C.: National Academy Press, 1994), p. 34.

44 Wasser, "Early Development of Technical Range Management," 76–77; Allan Savory, *Holistic Resource Management* (Washington, D.C.: Island Press, 1988).

45 Paul T. Tueller, "Application of Remote Sensing Techniques for Analysis of Desert Biome Validation Studies," (Reno: Renewable Resources Center, University of Nevada, 1972), also issued as Research Memorandum, U.S. International Biological Program, Desert Biome, RM 72-7, March, 1972; Patricia Moore, ed., *Computer Mapping of Natural Resources and the Environment: Including Applications of Satellite-derived Data*, (Cambridge, Mass.: Harvard University, Laboratory for

259

Computer Graphics and Spatial Analysis, 1980); James A. Brass, *Wildland Inventory and Resource Modeling for Douglas and Carson City Counties, Nevada, Using Landsat and Digital Terrain Data* (Washington, D.C.: National Aeronautics and Space Administration, 1983); A. D. Wilson and G. J. Tupper, "Concepts and Factors Applicable to the Measurement of Range Condition," *Journal of Range Management*, 35 (1982) 684–689; Paul T. Tueller, "Remote Sensing Technology for Rangeland Management Applications," *Journal of Range Management* 42 (November 1989): 442–452.

46 *Rangeland Health*, p. 34.

GERMAINE REED

CHARLES HERTY AND THE NAVAL STORES INDUSTRY

The history of the naval stores industry in North America is practically as old as the history of permanent settlement on this continent. But long before that time, the use of turpentine, rosin, tar, and pitch was common, according to records that antedate the Christian era. Turpentine, a light volatile oil, is distilled from oleoresin (or crude gum), or it may be obtained from resinous wood by distillation or extraction. Rosin is what is left after turpentine is distilled from the oleoresin. Tar is the crude resinous liquid distilled from wood during slow combustion, and pitch is a partially carbonized and condensed product obtained by boiling or burning tar. Essential for the construction and maintenance of wooden ships, tar and pitch were the products most sought after until well into the nineteenth century when industrial change made turpentine and rosin the most important naval stores commodities.

The term "naval stores" probably arose in the era of exploration when the needs of expanding navies and merchant fleets greatly increased the demand for tar and pitch. At the time, Sweden and the Baltic countries provided the bulk of Europe's requirements, and timber-poor countries like England searched eagerly for ways to free themselves from dependence on outside suppliers. Therefore, when explorers reported the existence of vast pine forests along the eastern coasts of North America, entrepreneurial Britons saw opportunity. Jamestown was established in 1607, and the following year the London adventurers who financed it sent eight Dutchmen and Poles to the colony with orders to teach the settlers how to manufacture "glass, pitch, tar, soap, etc." The meager output and its subsequent shipment to the mother country marked the beginning of the naval stores industry in English America. Despite repeated urgings from London, however, tar and pitch production never became a major enterprise in Virginia because the pines suitable for making naval stores were too scattered, and tobacco culture soon proved immensely more profitable.[1]

Meanwhile the English had established more colonies to the north. Given the somewhat hostile agricultural environment and the need to become self-sufficient, settlers in New England quickly exploited their forest resources. But most of what they turned out went to satisfy their own needs as they engaged in more lucrative activities such as lumbering, trading, and shipbuilding. Furthermore, what they did export could not compete with Swedish or Baltic naval stores in the English market because of higher transportation costs, higher production costs due to the scarcity of labor, and inferior quality—the last largely the result of ignorance, indifference, and adulteration. Not until the end of the seventeenth century did naval stores become important in the colonial export trade; by that time production in North Carolina, permanently settled by 1665, was well under way.[2]

It took a while for authorities in England to appreciate the potential of North Carolina. Efforts on both sides of the Atlantic to develop a cheap and more dependable source of naval stores from Virginia northward had been disappointing. But late in the seventeenth century England's impending involvement in the War of the Spanish Succession (1701–14), the ambitions of the Swedish king, Charles XII, and French support for the Stuart pretender to the English throne made a steady and secure source of naval stores imperative. In 1697 and again in 1699 the Board of Trade sent investigators to the colonies, and one of them reported after visiting New England, New York and the Carolinas:

> [S]ince my arrival here [Charleston] I find I am come into the only place for such commodities upon the Continent of America; some persons have offered to deliver in Charleston Bay upon their own account 1000 Barrels of Pitch and as much tar, others greater quantities provided they were paid for it in Charles Town in Lyon Dollars. . . , Tar at 8s. pr. Barrel, and very good Pitch at 12s. pr. Barrel, and much cheaper if it once became a Trade. The season for making those Commodities in this Province being 6 mos. longer than in Virginia and more Northern Plantations; a planter can make more tar in any one

year here with 50 slaves than they can do with double the number in those places, their slaves here living at very easy rates and with few clothes. [3]

The adoption of the Naval Stores Bounty Act of 1705 stimulated naval stores production in the Carolinas, which enjoyed the advantages of a longer operating season and cheaper labor. The act offset the higher costs of transportation from America to England, compensating shippers and English importers rather than colonial producers. Because it made the American product competitive with the European, however, it seems to have increased production and export considerably, especially from North Carolina. The British navy alone received more than 35,000 barrels of Carolina tar, pitch and turpentine in 1719, and by 1768 that figure had climbed to 135,000 barrels. That these commodities were legally recognized as payment for debt and taxes provides perhaps the best evidence of the importance that naval stores products had achieved in the economies of both North and South Carolina by 1720. [4]

North Carolina was the largest naval stores producer in America by the middle of the eighteenth century. But northern cities like Philadelphia, New York, and, especially, Boston continued to be thought of as the centers of the industry until about 1800 because most of the tar, pitch, and oleoresin or crude gum from North Carolina and other southern locations was sent there in exchange for manufactured goods, for domestic consumption or for re-export to Europe. [5]

Originally, most of the crude gum from North Carolina was shipped to England for processing. Later some of it was distilled into spirits of turpentine and rosin or pitch in northern port cities and, still later (1819), the first still appeared in coastal North Carolina. Carried on in cast iron retorts, the process was very inefficient, but after the introduction of the copper still in 1834, the yield of volatile oil greatly increased. Fortunately, so did demand for the product. Spirits of turpentine was being used in the manufacture of India rubber goods and as an "illuminator." Mixed with alcohol and called "camphene," it furnished a cheap light source until petroleum products replaced it. Unfortunately, there was no corresponding demand for rosin, and its price dropped so low that still operators often disposed of it in ponds, streams and ditches. [6]

The Revolutionary War disrupted naval stores production for years; so did the War of 1812 and the Civil War. But after each conflict North Carolina regained its premier position, reaching its peak between 1870 and 1880. South Carolina ranked first in 1880 but only briefly. Georgia took the lead until 1905, and Florida replaced her until 1923 when Georgia resumed first place. Meanwhile production began in the Gulf States around the turn of the century, extending finally to Texas. [7]

The methods used to produce naval stores in America remained remarkably static for almost three centuries, a situation which, while appropriate to frontier conditions, threatened the very existence of the industry by 1900. From the beginning naval stores were obtained from both living and dead coniferous trees, the pitch pine or *Pinus rigida* in the northern colonies, the longleaf pine or *Pinus palustris* from southeastern Virginia through the Carolinas. Tar was extracted from the resinous wood of dead trees, called lightwood, by slow combustion in crude kilns. The lightwood was piled several feet high over a saucer-like mound of earth containing a funnel at the center which fed into a buried drain pipe. The wood was then covered with clay and ignited at the top. In time tar seeped through the funnel into the drain pipe which conveyed it to a clay-lined ditch or hole in the ground. There it was scooped into wooden barrels or made into pitch by setting the tar on fire and allowing it to burn long enough for conversion to occur. Both products usually contained enough sand, rainwater, and trash to set off vociferous complaints from the buyer.[8]

Crude gum or oleoresin was obtained from the living tree by blazing or scarifying the trunk and later scraping off the congealed gum adhering to the scarified surface or face. The colonists soon found that the flow of gum could be increased considerably by renewing the wounds regularly, placing each a little higher on the tree. Collecting the augmented flow of gum, however, required a receptacle. The first settlers simply made a hole in the ground below the scarified face, but this was quickly replaced with the box, a notch or recess cut into the base of the tree. These developments greatly increased the output of oleoresin but they also limited the productive life of the trees if they did not kill them outright. The box weakened trees at their most vulnerable point, making them susceptible to forest fires, windstorms, and insects. Even when they survived, the trees were usually not chipped more than four or five feet above the ground because operators found it impractical or unprofitable. As long as a seemingly endless supply of virgin timber beckoned, few seemed interested in more conservative methods. As the original longleaf pine of North Carolina began to disappear, however, new tools like the hack and the longhandled puller were introduced in order to extend chipping to a height of twelve or fifteen feet. Unfortunately, the destructive box system remained, and nothing managed to replace it until the beginning of the twentieth century.[9]

It was not for lack of effort. More than a dozen devices to replace the box were registered with the United States Patent Office by 1900. Most of them were inspired by the Hughes system used to collect oleoresin in the Landes (southwestern) district of France. Once a barren and swampy wasteland, the area had been reclaimed in the post-Napoleonic era, forested with maritime pine (*Pinus pinaster*) and developed as a naval stores producing region. Collected in a hole at the base of the tree, French oleoresin could not compete with the American product on the world market. But the Civil War shut down the American industry, and France moved to fill the gap. M. Hughes developed a less wasteful collection method in the form of a clay pot, supported on its bottom side with a nail driven into the tree and at the top by a 2 x 4-inch gutter, slightly curved and driven into the tree at an upward angle. Shallow and narrow scarification or chipping (about four inches wide) took place just above the gutter which conducted the gum into the clay pot. At the end of each chipping season the pot was raised, thereby insuring a product whose quality was superior to its American rival because it was not adversely affected by exposure to the atmosphere or discolored by traveling down the full length of the tree to a box below. [10]

None of the American patents for box substitutes proved commercially successful, but interest in the Hughes system lived on. In 1893 the United States Department of Agriculture published the report of B. E. Fernow, chief of its Division of Forestry. Fernow's section on the naval stores industry provided an exhaustive account, detailing naval stores practice in several countries, and a severe indictment of the American system. "The present practice . . . in the United States," he declared, "is not only wasteful, but highly prejudicial to present and future forestry interests." He doubted that anything could be done unless the turpentine "orchardists" themselves realized what they were doing to their livelihoods. Nevertheless, he urged the abandonment of the boxing system and the substitution of "the movable pot." Finally, he thought more experimentation was needed "before we can determine the most profitable and least injurious practice for the turpentine industry." [11]

In North Carolina someone was listening and indeed reading Fernow's report. W. W. Ashe, in charge of Forest Investigation for the 1894 North Carolina Geological Survey, decided to test the relative merits of the French and American systems of turpentining. Using longleaf pine he conducted limited experiments which, while hardly conclusive, convinced him that the quality and quantity of product would increase significantly if cups replaced boxes. Encouraged, he planned to continue his experiments on a larger scale in 1895, but other commitments interfered.[12]

The task fell to a college professor in Georgia. Charles Holmes Herty was born in Milledgeville, Georgia, and educated at the University of Georgia and Johns Hopkins, from which he received the Ph.D. in 1890. Trained as an inorganic chemist he had hoped to continue working in that field. But his first academic position at the University of Georgia provided little incentive in salary and even less in laboratory and library facilities. After a decade of frustration he was casting about for a new position and a new direction in chemistry when a chance came to spend a postdoctoral year in Europe. The experience changed his life.

Herty spent half of his sabbatical at the *Technische Hochschule* at Charlottenburg in Berlin and the other half in Zurich at the Swiss Federal Polytechnic and the university. There he immersed himself in technical or applied chemistry and met some of the outstanding chemists of the era. One of them, Professor Otto N. Witt, was a particular favorite. In casual conversation one day, Witt dismissed the American naval stores industry as a "butchery." Unless drastic changes occurred soon, he declared, the longleaf pine upon which it was based would be wiped out. Herty knew nothing about the naval stores industry although it was centered in Georgia and Florida in 1900. Within a year, however, naval stores and trying to prevent Witt's prediction from coming true became the center of his existence. [13]

When Herty returned to America he steeped himself in everything he could discover about naval stores. He searched the literature, visited south Georgia to see what impact, if any, recent government studies like the Fernow report had made on naval stores practice, and he contacted men in every branch of the industry from New Orleans to North Carolina. The results were disheartening. Longleaf pine was disappearing at an alarming rate due to the ruinous box system. Georgia and Florida produced the bulk of American oleoresin in 1900, but rapid depletion in those states was driving the industry ever westward. Together with inefficient distillation and overproduction, the box practically guaranteed what Professor Witt predicted: the end of the industry in ten or fifteen years. "We are not only killing the goose that laid the golden eggs," Herty noted, "we are actually failing to pick up all of the wealth during the dying process."[14]

The normal routine on all turpentine farms, as Herty observed it in 1900, involved box cutting as a first step. Chopped into the base of a tree in order to capture the resin or crude gum flowing from the scarified surface

Charles Herty, a trained chemist, made major technical and scientific contributions to the naval stores and newsprint industries of the South.
(FOREST HISTORY SOCIETY PHOTO, HERTY PAPERS)

above, boxes were about fourteen inches wide, seven inches deep, and three-and-a-half inches from front to back. The box was cut in the winter, usually by black laborers. The second step, cornering, required the removal of two triangular chips immediately above the box in order to provide smooth surfaces to direct the resin flow into the box. Chipping came next. Begun in March and continued weekly through November, it was done with a sharp tool called a hack and served to open fresh resin ducts. With laterally inclined strokes the chipper removed strips of bark and sapwood just above the cornered surfaces. The exposed surfaces of sapwood or "streaks" met just above the center of the box, forming an angular point known as a "peak." As chipping proceeded the distance between the fresh gum-producing streak and the box increased, ultimately requiring the chipper to replace the hack with a long-handled tool called a puller. Every three or four weeks the gum or "dip" accumulating in the boxes had to be removed to buckets and then transferred to barrels. Later the barrels were hauled to a nearby still. At the close of the chipping season, workers removed hardened resin called "scrape" from the face of the tree and raked the area around its base as a precaution against fire. [15]

Herty's study of the literature, his observations in the woods, and his scientific training convinced him that the "great evil of this industry," the box, had to be replaced. Furthermore, whatever replaced it would have to be something that would not wound the tree appreciably when installed; could be moved upward easily as chipping progressed; was simple to use, and above all, cheap. Neither the French system nor the several devices registered with the U. S. Patent Office met his criteria adequately, especially when taking into account the type of chipping practiced in the American South. Consequently he devised his own system which he described as "consisting of a simple cup suspended, through a hole near its rim, on a common nail." The gum was directed into the cup "by two shallow galvanized iron troughs or gutters, to be inserted about one-quarter inch deep by one of their long edges in shallow inclined cuts across the scarified surface of the tree." [16]

Early in 1901, through contacts with a group of Savannah naval stores factors, Herty secured the use of some turpentine land at Statesboro, Georgia, for preliminary experiments to be conducted that summer. The factors put up $150 to defray field expenses; he worked for nothing. Meanwhile, on a trip to Washington in the spring of 1901, Herty made contact with Gifford Pinchot, chief forester in the Bureau of Forestry, United States Department of Agriculture. Intrigued with Herty's plans for the Statesboro experiments, Pinchot asked him to become a "collaborator" with the bureau, a position that paid a nominal amount ($300 a year) but that would provide Herty with travel vouchers, stationery, the use of scientific instruments, and publication of his results. Like Herty, Pinchot was interested in a more efficient system of turpentining. Above all, both men wanted to save the forests. [17]

At Statesboro, Herty made comparative studies of first, second, third and fourth year boxes in sets of 100 each, 50 cups and 50 boxes, to measure the quantity and quality of the gum collected with each system. He also kept a record of the impact of temperature and rainfall on the flow of resin. Besides demonstrating to his satisfaction that his cup and gutter could be employed successfully on a commercial scale, the 1901 experiments gave Herty a better grasp of the problems to be solved, a better understanding of the men engaged in the naval stores business, and more knowledge about the nature of pine trees. For example, he satisfied himself that the darker colored rosin produced by the box system after the first season of turpentining was not due to physiological changes in the tree but solely to the way the oleoresin had been collected. This was important because darker colored rosin brought lower prices. Pinchot was so interested in Herty's results and his future plans that he asked him to join the Bureau of Forestry full time, and by February 1902, Herty was in Ocilla, Georgia, ready to conduct commercial scale experiments on land belonging to Powell, Bullard and Company. [18]

The apparatus employed at Ocilla, essentially the same as that used at Statesboro, consisted of two galvanized iron gutters and an earthenware cup with a hole in its rim. The gutters, each two inches by six to twelve inches long, were crimped lengthwise through the center to form v-shaped troughs and served to conduct crude turpentine to the cup. Hung on a nail just below the gutters, the cup was designed to hold about as much gum as the standard box.

The system worked on "round" (never boxed) trees as well as on those previously boxed. A pair of laborers, one right-handed and one left-handed, installed it on round timber by using cornering axes to produce two flat faces. The first strokes made were the same as those used in cornering a box. Next, the workers removed enough bark and sapwood with upward strokes of the cornering ax to make a flat face one-half the width of the full face, the whole corresponding in width to that of the box which would have been cut in the tree. Switching to single-edged broadaxes, each man made a downward-inclined cut on his side of the tree. The gash on one side was placed about an inch below that made on the other, and both were positioned about three inches below the chipping surface so that enough space remained to use the hack when chipping began. Workers then inserted gutters into the incisions made by the broadaxes by slipping each endwise into the upper end of the cut and pushing it down its full length. The upper gutter was forced slightly beyond the center, thus forming a spout to direct the resin into the cup below. When used on trees previously boxed the gutters were placed on the surface left by chipping done the previous year. From this point, the usual procedures were followed. But in time for the next season cups and gutters were raised, thus avoiding the loss by evaporation and discoloration that routinely occurred with the box system in the second, third, and fourth years. [19]

The Ocilla tests were designed to discover the relative yield and resulting market value of turpentine collected by the box and cup-and-gutter systems. The land and timber belonged to Powell, Bullard, and Company; the Bureau of Forestry furnished the cups and gutters and installed them on the trees. Powell, Bullard's regular work force did the chipping, dipping, scraping and stilling. The company kept careful records, retaining any profit, and the Bureau published all results. Powell, Bullard's manager chose four "crops" (or tracts of 10,500 boxes each). The first had never been turpentined, and the second, third and fourth had been worked by the box system for one, two and three years, respectively. But in 1902 one-half of each crop was worked by the box system, and the other half was cupped. For trees already boxed, the intent was to discover, as between cups and boxes, the practicability of the cup system at varying heights on the tree; the difference in quality, as to color, of the rosin produced by the two systems, and the relative loss suffered from evaporation. But primary interest centered on virgin, or never turpentined timber. There the box system and the cup-and-gutter system operated under fully equal conditions. To make the test as fair as possible, one chipper worked both cups and boxes in the virgin crop. Crude turpentine was collected separately from cups and boxes, distilled separately and carefully labeled. The rosin was then sent to the Southern Naval Stores Company in Savannah, where each lot was graded separately and the report returned to Ocilla. [20]

By mid-summer the results at Ocilla began to come in, proving so favorable that Gifford Pinchot wanted a preliminary summary so that he could submit it with his annual report. Seasoned turpentine men were equally impressed. The man whose firm had received and graded the rosin from Ocilla visited the site to see for himself. Another wanted to know where he could buy a machine to manufacture cups for the coming season. Meanwhile, Herty mounted a campaign through newspapers and trade journals, announcing his intention to address the second annual Turpentine Operators Association when it met in Jacksonville on September 10, 1902. There he reported that the cup system out-produced boxes on "virgin" or first year trees by 23 per cent and that the increase in profits more than offset the cost of cups and gutters. Herty's tabulated results for the 1902 season at Ocilla were published by the Bureau of Forestry in 1903. Table 1 shows the results.

The big problem facing Herty and the Bureau of Forestry in late 1902 was keeping interest alive in the cup-and-gutter system until a manufacturing difficulty could be overcome. Finding a firm willing and able to produce earthenware cups for clamoring operators who wanted "to get in on" what promised to increase their income while it saved their trees proved almost impossible. Here Herty's special abilities as a communicator and a facilitator became especially important. Just as he had managed to persuade a few

tools used to scarify the trees and of the very small wounds produced by them; both differed markedly from what was common in America.

Meant to mature for timber, trees in France were not turpentined before thirty years of age unless scheduled for removal because of irregular shape or overcrowding. Scarification of a single face about four inches wide took place for four years; the tree rested in the fifth year, and the following year another four-year scarification cycle began about a third of the circumference away from the first. The process was repeated until the tree was fifty-five to sixty years old; then it was intensively turpentined (with many cups) and finally cut for timber. "Bled" timber, Herty advised Pinchot, was prized for construction projects because the slower growth caused by scarification resulted in smaller annual rings and timber that was stronger and heavier.

Herty also studied French turpentine stills, the most common of which was essentially the same as that used in the United States: distillation in copper alembics with naked fire aided by injection of hot water. A second, relatively rare, method employed steam distillation, both direct and indirect. Finally, a third, intermediate type used naked fire under the copper alembic with distillation aided and the temperature regulated by a mixed injection of hot water and steam. Steam distilleries, too costly to be common even in the stable French industry, were out of the question in America where the industry shifted constantly from point to point. The mixed injection system, however, offered a real advance. Only slightly more expensive than what Americans were using, it eliminated the "personality" of the stiller from the process because a thermometer immersed in the molten resin allowed the temperature to be kept constant by varying the injection of water or steam.[26]

Herty investigated the French system primarily to determine what, if anything, could be borrowed advantageously by American naval stores operators. Under existing economic conditions, he concluded, the system would not be practical here because: French timberland averaged $300 an acre compared to the Florida rate of $4; French labor costs ranged from 30 to 50 cents a day whereas Americans earned $1 to $1.50; and finally, annual yield per crop in the United States averaged 350 barrels per season compared to 72 in France. Eventually French trees would greatly out-produce those in America, but unless timber and land values skyrocketed and labor costs fell, the French system was not practical in the United States in 1903. Applied to American timber, it might prove more productive because of the warmer climate and the nature of the trees, Herty speculated, but only experimentation could establish that. He thought it would be easy to get French *resiniers* to run such tests in the United States if the bureau thought it would be worthwhile.[27]

While in Europe, Herty also went to Bern to see Tschirch whose work to determine how resins were formed in conifers had attracted considerable attention. Tschirch's investigations "demonstrated beyond a doubt," Herty reported, the existence in untapped trees of primary resin ducts; scattered throughout the tree, the primary ducts contained a resin formed as a physiological or natural process of the living tree. In trees which had been wounded, however, large numbers of secondary resin ducts formed immediately after the wounding in the new wood. "It is from these latter that the main portion of the commercial resin is obtained," Herty explained, "such resin being a true pathological product." Tschirch's explanation for resin formation in the cambium layer of the tree provided Herty and the Bureau of Forestry with the theoretical bases for experiments they planned to begin in the 1904 season and which they hoped would ultimately prove valuable from both a "scientific and commercial standpoint." Briefly, they intended to measure the relative yield of resin from trees scarified to various depths in order to determine whether less destructive shallow chipping could be practiced without unduly sacrificing yield.[28]

In the fall of 1903 Herty moved to Jacksonville, Florida, planning to continue supervision of the work going on at Ocilla as well as his instructional activities among new converts to the cup system. He would also oversee the shallow chipping experiments and conduct laboratory analyses on resin samples collected in the field from longleaf and "slash" (*Pinus heterophylla*) pine, the two most important turpentine pines. On a much larger scale than the Ocilla experiments, the work in Florida was expected to last several years. But a bureaucratic delay in Washington which forced the postponement of the experiments, a manufacturing problem at the cup factory, and Herty's desire to realize some income from his patent on the cup-and-gutter system (something he could not do as a government employee) led him to resign from the Bureau of Forestry in April 1904. He took a full time position with the cup factory and within months a chance came for him to reenter academe as professor of applied chemistry at the University of North Carolina.[29]

Meanwhile, George B. Sudworth, Herty's closest friend at the Bureau of Forestry, informed him that the government intended to go forward with the shallow chipping experiments in the 1905 season. Herty was delighted on two counts: his former assistant at Ocilla would be in charge of the field work, and Herty and Sudworth, who had planned the experiments, would continue to collaborate. Soon Herty suggested lines of research that could be pursued in North Carolina's new laboratory by graduate students under his direction. Washington liked the idea, and by 1906 a master's candidate was doing a comparative analysis of gum samples taken from various species of pine, comparing samples taken from trees of the same species but of different ages and, finally, analyzing samples taken from specific trees at different times in the same chipping season. These were designed to compare the percentage of spirits of turpentine in the gum obtained from ordinary and shallow chipping. Herty's former assistant shipped the gum specimens, carefully collected and labeled, to Chapel Hill, and Herty visited Washington from time to time to discuss results with Sudworth. He also visited Florida "to go over the work in the woods." The plan called for Herty to present a full account of both field and laboratory work at conventions of the Turpentine Operators Association and ultimately, to prepare a formal bulletin for the erstwhile Bureau of Forestry, renamed the United States Forest Service in 1905.

Relation of Light Chipping to the Commercial Yield of Naval Stores was not published until 1911. It began with a summary of the Ocilla experiments, explained the chipping procedure in turpentining and why it seemed reasonable to expect greater yield and longer life for the tree if shallower chipping were employed, how the French naval stores industry operated, and why that method was impractical in the United States. Next, it outlined the shallow chipping research plan. A tract of timber had been divided into four equal "crops" of 8,000 faces each, and the crops had been designated A, B, C, and D, respectively. Crop A was the standard crop in which every method normally employed by the participating private firm (the Walkill Turpentine Company) would continue to be used for the next four years. Crop B would test the effect on yield by reducing the average depth of chipping from the normal (Crop A) 7/10 of an inch to 4/10 of an inch. All else would remain the same, meaning that any resulting variation in yield could be attributed only to the chipping factor. In Crop C the aim was to determine the relative yield from chipping which proceeded more slowly up the tree's trunk than was the case in Crop A. If, after four years of the same number of annual chippings, the yield from Crop C approximated that from Crop A, turpentine operators and timber owners would realize a gain of one-third. Finally, the aim in Crop D was to determine the practicability of a second working of the same crop. The chipping in Crop D was the same as that in Crop A. But the minimum diameter of trees to be turpen-

Table 2. Summary of chippings and yield, seasons 1905, 1906, 1907, and 1908. [30]

Year	A			B			C			D		
	Chipping	Dip	Scrape	Chipping	Dip	Scrape	Chipping	Dip	Scrape	Chipping	Dip	Scrape
		Lbs.	Lbs.		Lbs.	Lbs.		Lbs.	Lbs.		Lbs.	Lbs.
1	31	63,615	9,570	31	61,162	7,650	31	62,587	7,245	31	73,704	8,888
2	35	64,583	12,795	35	62,309	12,409	35	63,172	10,411	34	81,672	13,765
3	29	43,675	10,209	30	50,534	9,742	30	50,875	8,207	26	62,119	11,006
4	30	34,362	15,168	27	35,633	13,772	28	37,861	13,240	28	48,716	17,312
Total	125	206,235	47,742	123	200,638	43,633	124	214,405	38,103	119	266,211	51,001

tined (six inches in Crop A) was raised to ten inches, and the minimum diameter of trees carrying two chipping faces (thirteen inches in Crop A) was raised to sixteen inches in Crop D. No tree in Crop D carried more than two faces. Table 2 depicts the summarized results of the shallow chipping experiments.

Because the experiments had proved so successful it was decided to conduct a further experiment for one year comparing the yield from a standard crop, like Crop A, and that from two others, designated G and H. In the latter, both modifications, shallow chipping and chipping of reduced height, were combined. These results are portrayed in Table 3.

Herty thought the outcome of the Florida experiments was "sensational" and far more significant than anything he had done at Ocilla. He regretted that the Forest Service closed down the work in 1909, allegedly for lack of funds. At Sudworth's request he submitted cogent arguments against the decision but to no avail. Later, Sudworth apologized for the paltry sum ($450) Herty was paid for writing up the Florida experiments, but Herty

Table 3. One-year yield from a standard crop (A) compared with yields from two other crops (G and H). [31]

Crop	Number of cups	Number of chippings	Yield of dip (lbs.)	Increase (%)
Walkill Turpentine Company	9,880	35	90,094	..
G....................................	9,880	35	124,292	38
H....................................	9,880	35	121,474	35

answered that he had not taken on the job to make money. The only reason he agreed to do it at all was out of friendship for Sudworth and his own (Herty's) interest in what good the entire undertaking might do the southern people.[32]

In summarizing the results of the shallow chipping experiments, Herty expressed cautious optimism. "If operators will adopt the methods which have now been proved to be practicable," he predicted, "they will (1) substantially increase the yield per crop of crude turpentine obtainable in a four-year period; (2) make possible the indefinite prolongation of the turpentining period; and (3) cause far less sacrifice of merchantable timber." Together with the substitution of the cup for the box, the modified chipping studies pointed toward a "revolutionary change in the turpentine industry," which, if fully realized, meant "nothing less than a prospect of being able to turpentine a given tract permanently. . . ." By wounding the trees as little as possible, by limiting operations to trees at least twelve inches in diameter and by restricting carefully the number of cups applied to each tree, Herty thought that turpentining could go on with "trivial loss" of merchantable timber; that it could be continued on the same tract indefinitely, extending to new trees as they matured enough to be cupped, and that when timber was cut, reproduction having taken place meanwhile, the new trees could mature while operations went on elsewhere. That is what the Forest Service expected to occur as it prepared to turpentine trees in the Choctawhatchee National Forest.[33]

Herty's active cooperation in the field work of the Forest Service was over by 1909. But his research activity at the University of North Carolina continued to focus on the pine tree. With few exceptions, all of his publications and scholarly presentations underscored his determination "that the name of this laboratory should be chiefly associated with all matters pertaining to turpentine." They included such titles as "The Volatile Oil of *Pinus seratina*," "The Optical Rotation of Spirits of Turpentine," "The Chemistry of Scrape Formation," "Isometric Abietic Acid," and "Further Studies upon the Resene of *Pinus heterophylla*." Most were read at chemical meetings and later published in the journals of the American Chemical Society.[34]

The work of Tschirch also continued to interest Herty as long as he stayed at North Carolina. Tschirch regarded the formation of secondary resin ducts and the gum they produced as a pathological process, a process which took four or five weeks to complete, and Herty's observations of woods practice seemed to bear out Tschirch's theory. Operators noted that the first dippings from cupped timber never matched that from boxed trees. If Tschirch was right, it was because box cutting was done in the winter and the chipping season, the first significant wounding of cupped trees, did not begin until spring. Boxed timber, therefore, had several weeks' head start over cupped trees in forming secondary resin ducts and hence produced a full flow of gum on the first chipping. Consequently, Herty advised several operators using cups to "put one good wide streak" on their trees at least four or five weeks before regular chipping began. It seemed to work. But Herty had no laboratory evidence to explain it, so in the winter of 1906-1907 he resumed contact with Tschirch and asked the Forest Service to send him sections of unturpentined longleaf and slash pine trunks, as well as microscopic sections, both cross and longitudinal, of the scarified "faces" of trees of the same species, worked for one, two, and three years respectively. The plan was to study and compare the development of "resin tubes." Unfortunately, inadequate staff on Herty's campus kept the work from being done. [35]

Herty continued to subscribe to the Tschirch theory and his own practical experience regarding the formation of secondary ducts and resin flow until May 1916. Then he met Eloise Gerry, a young microscopist from the Forest Products Laboratory at Madison, Wisconsin. She brought her slides and microscope along and the next day Herty wrote Sudworth that they would have to find a new theory. Gerry's slides showed no secondary ducts formed above the preliminary streaks applied before the first chipping. "Evidently," he said, "there is some other reason for the beneficial effect of the preliminary streak." Herty and Gerry remained friends for the rest of his life. He admired her skill and did everything he could to see that her superiors supported her investigations of resin formation which he considered "the real scientific basis for all further work on the production of naval stores." [36]

A few years after her first meeting with Herty, Gerry published an article in Thomas Gamble's *Naval Stores: History, Production, Distribution and Consumption* which disputed Tschirch's explanation for resin formation, questioned the techniques he employed in preparing specimens, and concluded that "not only his anatomical observations but also his deductions are open to question, especially on physiological grounds." Tschirch stated that he had found no resin formed in the parenchyma cells, attributing its manufacture instead to a special membrane lining the resin duct. But Gerry's photographs revealed that gum exuded freely "from clusters of parenchyma cells which showed no separation or formation of any . . . duct." Wounding did not cause the formation of ducts; instead it "stimulated the activities of the living parenchyma cells already present." It was this response, Gerry maintained, that seemed to account for the results secured by opera-

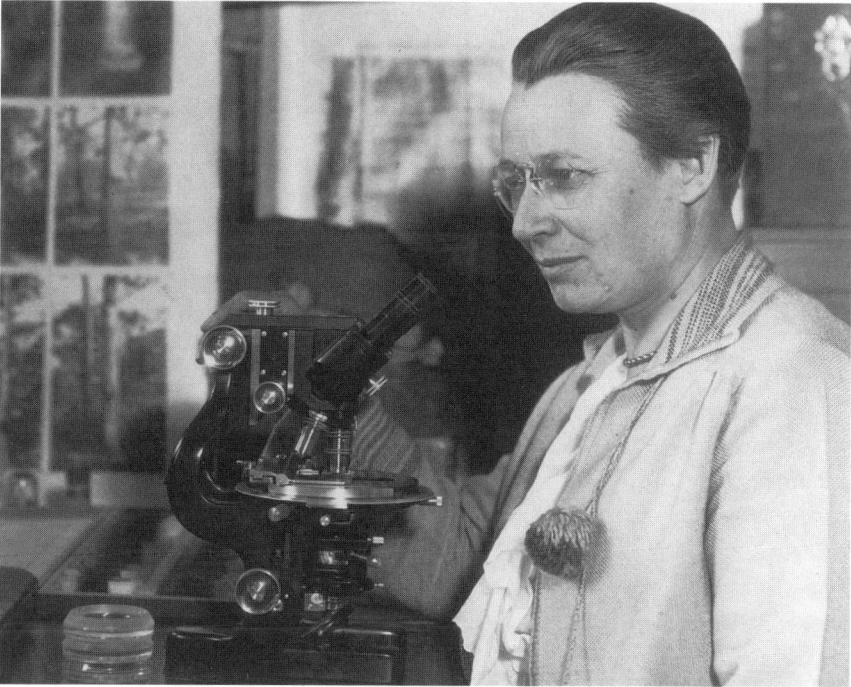

*An advocate for "more turpentine, less scar, better pine," Dr. Eloise Gerry
pioneered microscopal studies of resin-yielding pines. She began her
44-year research career at the Forest Products Laboratory in 1910.*
(USDA FOREST SERVICE, FOREST PRODUCTS LABORATORY PHOTO, CIRCA 1935)

tors who took Herty's advice to "put one good wide streak" on their trees
several weeks before they began regular chipping.[37]

Late in 1916, having just completed two years as president of the American Chemical Society, Herty left the University of North Carolina for New
York to become editor of the Society's *Journal of Industrial and Engineering
Chemistry*. For the next ten years, regardless of the posts he held, his attention centered on the establishment of a synthetic organic chemical industry in America, the achievement of tariff legislation to protect it, and the
education of Americans about the importance of chemistry in making their
country healthy, wealthy, and secure. He became so involved on the national
scene that issues like the South's economy and the condition of the naval
stores industry claimed little of his attention. But in 1926 he reestablished
contact with the region and his old friends in the turpentine industry when
he traveled south to address the American Ceramic Society in Atlanta. From

there he visited Jacksonville and Savannah, speaking informally to the naval stores section of the Board of Trade in the latter city.

The trip was an eye-opener for Herty. He came away impressed with the "progressive spirit" encountered among the naval stores men and soon began trying to catch up with developments in the industry. He asked the United States Forest Service for naval stores bulletins, he resumed correspondence with Thomas Gamble, editor of the *Naval Stores Review*, and he began lobbying Washington and the Forest Products Laboratory to increase funding for Gerry's research work. He also contacted Carlton Speh, secretary-manager of the Turpentine and Rosin Products Association, and an officer of the Pine Institute of America (PIA), which Herty learned was sponsoring research at the Mellon Institute in an effort to develop new products from naval stores. The "new spirit" that seemed to pervade what he remembered as a profoundly conservative industry delighted him. "I would certainly like to get back in touch with the industry again," he wrote Gerry in February 1927. And he did. When the PIA president and several others invited him to participate in a Naval Stores Get-Together at Jacksonville, he accepted with enthusiasm.[38]

Herty was indeed "out of touch." If he had been reading Gamble's *Naval Stores Review* regularly, he would have been aware that much was changing in the industry. In 1920, for example, the secretary of agriculture reported to the Senate that at the rate of domestic and foreign demand for naval stores then current, the nation's "supply of timber now in sight would be exhausted in less than ten years." Accelerated timber cutting by lumber interests was part of the problem. But naval stores operators themselves bore primary responsibility because of the "crude, wasteful, destructive, and sadly shortsighted" methods they continued to employ. The French had been proving for eighty years that turpentining and timber culture could be carried on simultaneously and profitably without exhausting or killing the trees. More recently, the United States Forest Service and a few "progressive" private operators had proved the same thing. "Inertia, not financial obstacles," lamented the secretary, "must be regarded as the chief reason why these conservative methods have not been more generally employed."[39]

Just nine years later, *Gamble's Naval Stores Yearbook for 1929–30* published a lengthy article which struck a distinctly different note. The editor claimed that "all calamitous prophecies heretofore made [about the future of the naval stores industry] have become obsolete by undreamed of developments." Everything being done, he noted, rested upon two things: reforestation and improved processes of production. Progress was uneven but the "drift," he thought, was unmistakably toward stability of the industry. In 1920 "intelligent reforestation," light chipping and other "progressive steps" had been smiled at, and the steam distillation system used in France was written off as impractical and inapplicable in the United States. Ten years before that the box was still common and research was "literally an unknown word in . . . [the industry's] vocabulary." Yet in 1929, thanks to "forward looking" operators aided by "devoted students of the pine," there was a new spirit of progress which promised a bright future for the industry. Specifically, Gamble cited "artificial" reforestation projects involving longleaf and slash pine underway in Louisiana and Mississippi, and the recent establishment of "full fledged" forestry departments from Texas to North Carolina, all staffed with trained foresters, funded by the states, and supported enthusiastically by the public. Ten years earlier the entire turpentine belt of the South spent $15,000 for state forestry work, counted the number of foresters on one hand, and did nothing about protecting forest land from fire. In 1929 there was $725,000 available for state forestry work, more than 140 men were engaged in some aspect of it, and in excess of seven million acres were under fire protection with state cooperation.[40]

Behind the "new spirit of progress" which Gamble celebrated in 1929 lay several contributory factors which developed in the previous decade. The naval stores industry was generally unprofitable during the first half of the 1920s. Necessity forced operators to become more efficient or get out of business. Improved tools and equipment, less destructive practices and better supervision enabled many to survive until markets improved, and their success served as an example to others. The efforts of governmental agencies, notably the Forest Service and the Bureau of Chemistry, to develop, demonstrate, and disseminate new tools and techniques also helped to educate the average naval stores operator. More of them seemed interested in what the government "experts" had to say, especially younger operators whose educational level and more frequent contact with new people and new ideas served to make them more receptive. Finally, operators who owned the timber they turpentined were more likely to adopt conservative methods than those who did not. But whether owned or leased, the increasing scarcity and higher price of turpentine forest could no longer be ignored. Not consciously driven by the preservationist's concern to save the forests for posterity, the average operator was simply doing what he had to in order to maximize his net income.[41]

The clear evidence that the naval stores producing pine tree was not going to become extinct provided one additional reason for Gamble's optimism in 1929. By the middle of the 1920s, despite bad practice, woods-burning, and the ruinous rooting of "razor-back" hogs who allegedly destroyed young seedlings, second-growth slash and longleaf pines were restocking much of the cut-over forest land. More important, operators using the most efficient tools and conservative methods had discovered that these trees, especially the slash pines, produced more resin than original-growth trees operated by the old methods. [42]

Taken together, all of these developments encouraged operators, particularly those of the younger generation, to think of their industry in permanent terms. In 1911 Charles Herty had predicted the development and profitability of a forest "continuously yielding turpentine, continuously producing lumber, and continuously renewing" itself. Austin Cary, a logging engineer with the Forest Service, found evidence in 1920 that a number of "alert, active and progressive" men in south Georgia had already embarked on such projects. As a first step they abandoned the lease system to acquire their own land, expecting to operate on the same tract perpetually. Cary thought that was a healthy sign and he saw no reason why they should not succeed, given the favorable "natural conditions" for reforestation which greatly surpassed those in France, and the degree of progress in operating methods already achieved as demonstrated by the work of Herty and the Forest Service since the beginning of the century. Finally, Gamble, preparing his 1929-30 *Year Book* early in 1929, expressed satisfaction with the progress made in the last twenty years, and confidence that in the next ten "man and nature combined, may reasonably be expected to have lifted the [naval stores] industry to a higher level than [it] has ever known in its history." [43]

It was that sense of progress and confidence that Herty felt when he "got back in touch" with his favorite industry. Early in 1927 he went to Jacksonville to attend the annual Naval Stores Get-Together, a meeting composed of turpentine producers, factors, timber growers, technical men from government bureaus, and representatives of forest product trade associations. The conventioneers listened to speeches and participated in sessions which dealt with pricing and marketing problems, research to develop new products, competition from substitutes, and the most advanced techniques in reforestation and fire control. Herty urged the membership to support Gerry's ongoing research on the causes of resin flow in the pine tree, reiterating his conviction that no industrial progress could be made without advances in basic science. He also accepted a bid from the Pine Institute of America (PIA) to serve on its technical advisory committee. The PIA was funding two fellowships at the Mellon Institute in Pittsburgh and wanted Herty's counsel about the lines of research that should be pursued. Essentially, the PIA hoped the work at Mellon on rosin and turpentine would discover new uses for those products. Herty was glad to serve because he admired the "new" naval stores industry's interest in research and its awareness that it was a part of America's chemical industry. [44]

Without doubt, the most important result of Herty's 1927 visit to Jacksonville was his introduction to Alex Sessoms. Sessoms, the nephew and namesake of a Florida turpentine operator who had been one of the first to adopt the cup-and-gutter system in 1902, invited Herty to visit his eighty-six thousand acre tract of cutover and naturally reseeded slash pine timber at Cogdell, Georgia. Organized as the Timber Products Company, Sessoms' firm was pursuing policies that Herty was sure would produce a perpetual yield of timber, turpentine, rosin, and cross ties. A professional forester oversaw operations, turpentine operators were required to use the cup system and conservative chipping techniques, and gum was processed in a modern steam distillation plant at the site. Fire-fighting teams patrolled the woods constantly, and the Hercules Powder Company of Brunswick, Georgia, leased the right to remove dead trees and resinous stumps from which it then extracted turpentine and rosin. Other lessees grazed cattle on company land, and Sessoms planned to lease hunting rights which he thought would cover his tax bills. As Herty saw it, only one problem remained: finding a use for the thinnings, "which are of size for the production of wood pulp." But that called for research. If successful, the economic impact could be tremendous, not only on the South but wherever industries depended upon wood pulp as raw material. "This I believe," Herty commented to a friend in New York, "is going to furnish the model according to which a new naval stores industry of the South will be developed." [45]

Herty's renewed contacts with the naval stores men and his inspection of Sessoms' enterprise had a profound effect on the rest of his career, if not on the future of the gum naval stores industry. Late in 1928 he became a full-time chemical consultant; two of his first clients were Sessoms' Timber Products Company and the PIA. For the PIA he agreed to continue consulting on research problems and to represent the organization before government departments in Washington. Herty had plenty of friends in the capital, and he hoped to prod the head of the Forest Service into allocating more of his budget for research, specifically for field research on the slash pine and the fundamental investigations of Gerry at the Forest Products Laboratory. For the Timber Products Company, Herty's initial efforts aimed at finding northern bankers willing to underwrite company expansion. But soon he focused on what to do with the thinnings from Sessoms' huge tract, young trees that had to be removed from the rapidly growing forest so that it could produce high-quality turpentine trees and saw timber. The thinnings were the right size for use as pulpwood in existing kraft paper mills along the Gulf Coast, but that market was becoming saturated. What Sessoms and others like him needed was a new outlet that could use great quantities of thinnings while they waited for their timber to mature. Herty was sure that the answer lay in the development of a white paper and newsprint industry in the South, something that at that time most experts considered highly improbable for economic and technical reasons. Briefly, technical men were convinced that the resinous southern pine could never replace the Canadian spruce in the standard sulfite process used to manufacture newsprint. And businessmen, pointing to falling prices and excess plant capacity in the north and Canada, turned a deaf ear to any proposals for expansion in the South. It took Herty years to acquire funding for a laboratory in Savannah so that he could disprove what the technical men maintained and years more to help convince investors that the time had come to construct the first newsprint mill in the South. By that time (1938) he was seventy years old and suffering from heart disease, but still optimistic and looking forward to what he was sure would be "wonderful, wonderful things coming up on the horizon." [46]

Herty's cup-and-gutter system of turpentining constituted what one commentator called the first forward step in American naval stores technology in a hundred years. It revitalized a dying industry and, combined with other conservative woods practice, enabled trees to survive to maturity, thus ensuring increased earnings from saw timber as well as significant reforestation through natural reseeding. In addition, his work sparked continued pure and applied research by the United States Forest Service at the Forest Products Laboratory and at experiment stations in the field. Meanwhile, enlightened private operators, influenced by government bulletins reporting that research, proved increasingly interested in more conservative methods of operation. By 1927 the cup-and-gutter system (or some variation of it) was ubiquitous; second-growth longleaf and slash pine in Georgia and Florida, again the center of the industry, supplied seventy-eight percent of America's naval stores production, and United States output earned more than fifty million dollars a year. Those circumstances would change drastically in the next five years, but by that time Herty was working hard to bring a new and infinitely more profitable industry to the South. Still interested in the material improvement of his fellow southerners, and still in love with the pine tree, he died in July 1938, happy in the knowledge that the first newsprint mill in the South was soon to be constructed near Lufkin, Texas. [47]

NOTES

1 For this discussion I have relied largely on Donald Fraser Martin, Jr., "An Historical and Analytical Approach to the Current Problems of the American Gum Naval Stores Industry," (Ph.D. diss., University of North Carolina, 1942), 36, 37, 39, 40, 42, 43; C. Dorsey Dyer, "The Physiology and Management of Naval Stores Pines and the History of the Naval Stores Industry," (M.S. thesis problem, School of Forestry, University of Georgia, 1960), 1–2; Thomas Gamble, "Early History of the Naval Stores Industry in North America," in *Naval Stores: History, Production, Distribution and Consumption*, Thomas Gamble, comp., (Savannah, Georgia: Review Publishing and Printing Company, 1921), 17–18; and A. Stuart Campbell, Robert C. Unkrich and Albert C. Blanchard, *The Naval Stores Industry*, Studies in Forestry Resources in Florida, ed. A. Stuart Campbell (University of Florida Publication, 1933), vol. I, no. 5, pp. 7, 8. This chapter is drawn from the author's book, *Crusading for Chemistry: The Public Career of Charles Holmes Herty* (Athens: University of Georgia Press, 1995).

2 Martin, "American Gum Naval Stores Industry," 42–45.

3 Quoted in Ibid., 48, n. 30; see also Ibid., 46–49, and Williams Haynes, *American Chemical Industry: A History*, 6 vols. (New York: D. Van Nostrand Company, Ind.), I, 20–22.

4 Martin, "American Gum Naval Stores Industry," 51–52; Haynes, *American Chemical Industry*, I, 23, 24; Gamble, "Early History of the Naval Stores Industry in North America," 19.

5 Martin, "American Gum Naval Stores Industry," 59–60.

6 Ibid., 67–68; B. E. Fernow, *Report of the Chief of the Division of Forestry* (United States Department of Agriculture, Washington: Government Printing Office, 1893), 334; Ostrom, "History of the Gum Naval Stores Industry," *The Chemurgic Digest* IV (July 15, 1945): 222.

7 Martin, "American Gum Naval Stores Industry," 58–60; Haynes, *American Chemical Industry*, I, 155, 156, 157; Campbell et al., *Naval Stores Industry*, 9–11.

8 Martin, "American Gum Naval Stores Industry," 61–64; Ostrom, "History of the Gum Naval Stores Industry," 217, 219; Dyer, "The Physiology and Management of Naval Stores Pines and the History of the Naval Stores Industry," 18

9 Ostrom, "History of the Gum Naval Stores Industry," 219–20; Martin, "American Gum Naval Stores Industry," 65–67.

10 I. F. Eldredge, "How the French Turpentine System Looked to an American," in *Naval Stores: History, Production, Distribution and Consumption*, Thomas Gamble, comp., (Savannah, Georgia: Review Publishing and Printing Company, 1921), 167; Ad. Genvrain & Co., "The Naval Stores Industry in France," Ibid., 159–60; Charles Herty, "The Past, Present and Future of the Naval Stores Industry," *Journal of Industrial and Engineering Chemistry* 5 (January 1913): 65–68.

11 Fernow, *Report of the Chief of the Division of Forestry*, 346–51, 352, 353, 354, 356.

12 W. W. Ashe, *The Forests, Forest Lands, and Forest Products of Eastern North Carolina*, Bulletin 5 (Raleigh: North Carolina Geological Survey, 1894), 100–105. Ashe thought anyone who wanted to know more about the Hughes system should contact Washington for a free copy of Fernow's "Clear and concise account."

13 For biographical details on Charles Herty, see Reed, *Crusading for Chemistry*, especially chapters 1 and 2.

14 Quoted in Germaine Reed, "Saving the Naval Stores Industry: Charles Holmes Herty's Cup-and-Gutter Experiments, 1900–1905," *Journal of Forest History* 26 (October 1983): 170. See also Herty's correspondence with J. F. Lewis, J. P. Williams, and John M. Egan from October 1900 through January 1901 in Charles Holmes Herty Papers, Box 39, Folders 1 and 2, Special Collections, Robert W. Woodruff Library, Emory University.

15 Charles Herty, *A New Method of Turpentine Orcharding*, Bureau of Forestry, Bulletin No. 40 (Washington, 1903), 10–11.

16 Charles H. Herty, "The Turpentine Industry in the Southern States," *Journal of the Franklin Institute*, 181 (March 1916): 344–46.

17 Herty, *A New Method of Turpentine Orcharding*, 16. See also Herty's correspondence with Gifford Pinchot, George B. Sudworth, and others in the Bureau of Forestry late in 1900 and through August 1901, in Herty Papers, Box 39, Folder 1, 2.

18 Herty, *A New Method of Turpentine Orcharding*, 16–17; Herty, "Turpentine Industry in the Southern States," 346–48.

19 Reed, "Saving the Naval Stores Industry," 172.

20 Ibid., 173–74. See also Herty, *A New Method of Turpentine Orcharding*, 18-19, and Herty, "The Turpentine Industry in the Southern States," 350.

21 Reed, "Saving the Naval Stores Industry," 174. Table 1 reproduces Table 3 in Herty, *A New Method of Turpentine Orcharding*, 20.

22 Reed, "Saving the Naval Stores Industry," 174: Herty, *A New Method of Turpentine Orcharding*, 22.

23 "A New Method of Turpentine Orcharding," Bureau of Forestry, Circular No. 24, February 2, 1903 (Washington: United States Department of Agriculture), 1–8. The circular was a necessary stopgap to give operators time to order cups and gutters for the 1903 season. Herty's full scale bulletin for the Bureau was finished early in the year, but government red tape held up publication until May. See Herty correspondence with George Sudworth, February-March, 1903, Herty Papers, Box 39, Folder 4.

24 Herty, "Turpentine Industry in the Southern States," 356–57.

25 Herty, *A New Method of Turpentine Orcharding*, 42.

26 This account is taken almost verbatim from Reed, *Crusading for Chemistry*, 30–31.

27 Ibid., 32.

28 Ibid.

29 Germaine M. Reed, "Chemistry and the Southern Pine Industry: The Contributions of Charles Holmes Herty," paper delivered in November 1983 at the Southern Historical Convention, Charleston, South Carolina.

30 Charles H. Herty, *The Relation of Light Chipping to the Commercial Yield of Naval Stores*, Bulletin 90 (Washington, D.C.: U.S. Forest Service, 1911), 13. Table 2 reproduces Table 4 in Bulletin 90.

31 Table 3 reproduces Table 8 in Herty, "Turpentine Industry in the Southern States," 363.

32 Reed, "Chemistry and the Southern Pine Industry."

33 Herty, *Relation of Light Chipping to the Commercial Yield of Naval Stores*, 30, 32. Herty did note that such management would require "some readjustment" in the thinking of timber owners. Instead of trying to get as many crops as possible out of a given acreage they would have to extend operations over larger tracts, i.e., lease larger tracts to operators. But they would benefit because the increased annual output per crop would justify higher rents for a longer period.

34 Herty to Geo. B. Sudworth, April 3, 1907, Herty Papers. See also Frank Cameron, "Charles Holmes Herty," *Journal of the American Chemical Society* 61 (July 6, 1939): 1624.

35 Reed, "Chemistry and the Southern Pine Industry;" Herty, *Relation of Light Chipping to the Commercial Yield of Naval Stores*, 25–28.

36 Reed, "Chemistry and the Southern Pine Industry."

37 Eloise Gerry, "The Production of Crude "Gum" by the Pine Tree," in *Naval Stores: History, Production, Distribution and Consumption*, Thomas Gamble, comp., (Savannah, Georgia: Review Publishing and Printing Company, 1921), 147–53. For a lucid discussion entitled "The Physiology of the Turpentine Pines" see Dyer, "The Physiology and Management of Naval Stores," Chapter 3.

38 Reed, *Crusading for Chemistry*, 259.

39 Edwin T. Meredith, "The Life of the Naval Stores Industry as at Present Carried on in the South," in *Naval Stores: History, Production, Distribution and Consumption*, Thomas Gamble, comp., (Savannah, Georgia: Review Publishing and Printing Company, 1921), 89–90. For a fuller description of what the U. S. Forest Service was doing, see U. S. Department of Agriculture, "U. S. Government's Turpentine Experience in the Florida National Forest," in *Naval Stores: History, Produc-*

tion, Distribution and Consumption, 227. For a more optimistic appraisal of the future of the naval stores industry in 1920, see Austin Cary, "A Look Ahead: A Study of Naval Stores Problems" in the same publication, 119–22.

40 Thomas Gamble, ed., *Gamble's Naval Stores Year Book for 1929-30*, (Savannah: Thomas Gamble, 1930), 4–6. More than half of the $725,000 came from federal appropriations under the Clarke-McNary Act and private sources for prevention of forest fires, growing and distributing seedlings to farmers, and assistance to farmers in managing their farm woods. See also Harold K. Steen, *The U. S. Forest Service: A History* (Seattle: University of Washington Press, 1976), 185–89, on the Clarke-McNary Act.

41 Martin, "American Gum Naval Stores Industry," 130–37. For discussions of the activities of the Forest Service and the Bureau of Chemistry in behalf of the naval stores industry, see F. P. Veitch and V. E. Grotlisch, "What Uncle Sam does for the Naval Stores Industry," and L. F. Hawley, "Forest Service Investigations of Interest to the Naval Stores Industry," both in *Naval Stores: History, Production, Distribution and Consumption*, Thomas Gamble, comp., (Savannah, Georgia: Review Publishing and Printing Company, 1921), 135–38 and 139–41, respectively. For work between 1920 and 1930, see Thomas Gamble, "Naval Stores Industry Has Made More Progress in the Last Twenty Years than in Centuries of its Previous History," and Eloise Gerry, "The United States Forest Products Laboratory," in Gamble, ed., *Gamble's International Naval Stores Year Book for 1929–30 and 1930–31*, pp. 4–10 and 149–54, respectively.

42 Martin, "American Gum Naval Stores Industry," 138–39.

43 Herty, *Relation of Light Chipping to the Commercial Yield of Naval Stores*, 30; Cary, "A Look Ahead," 119–20; Gamble, Naval Stores Industry Has Made More Progress in the Last Twenty Years than in Centuries of its Previous History," 4, 5.

44 Reed, *Crusading for Chemistry*, 259–60; Gamble, "Naval Stores Industry Has Made More Progress in the Last Twenty Years than in Centuries of its Earlier History," 9.

45 Reed, *Crusading for Chemistry*, 260; Charles H. Herty, "The Realization of a Vision," reprinted from *Manufacturers' Record*, September 15, 1927, Box 145, Folder 3, Herty Papers.

46 Reed, *Crusading for Chemistry* 259, 267, 269, 366–67. For Herty's efforts to secure a laboratory and to convince others to support a southern newsprint industry, see chapters 10 and 11.

47 Reed, "Saving the Naval Stores Industry," 175, and *Crusading for Chemistry*, 367, 368.

ROBERT L. YOUNGS

DEVELOPMENT OF THE SCIENCE AND TECHNOLOGY OF WOOD AND FIBER AND THEIR PRODUCTS

INTRODUCTION

Research in wood science and technology is based on the fundamental belief that the forest is an essential natural resource and that effective use of forest products proves vital to the economic, social, environmental, and material well being of the nation.

"Forestry is the Preservation of Forests by Wise Use" was the proclamation emblazoned over the entrance to the Forest Products Laboratory (FPL) of the Forest Service, US Department of Agriculture (USDA) when its new facility in Madison, Wisconsin, opened in 1932. The proclamation reflected well the urgent need to restore the nation's deteriorating forests and use them wisely to meet requirements for wood in the economy and social structure of the country. Under this policy, foresters, scientists, and engineers had been laboring together since the laboratory was established in 1910 and independently for decades before that to both preserve the nation's forests and use their timber productively and efficiently. Research at that facility and in university and industry laboratories was often conducted cooperatively to solve problems of mutual concern.

Research in forest products has traditionally been designed to meet rather specific pressing needs—needs that change depending on the economy, the state of the resource, the political situation, and the nation's international role, as well as the general state of science and technology. Cross ties for railroads, oak for the carriage builders, paper for records and communications, structural materials for aircraft, ships, commercial and public structures, and homes, hundreds of wood products and the packages to ship them in, and more have been at the focal point of forest products research since its initiation in the United States toward the close of the nineteenth century.

Research in forestry and forest products, especially in public institutions, has traditionally been based in agriculture. This stems from forestry's concern with sustaining the supply of timber as a crop by using it wisely and efficiently. The designation of the land-grant universities by the Morrill Act of 1862 and the Hatch Act of 1887, which established the state agricultural experiment stations, exerted major impacts on university research in forest products.

THE ROOTS OF WOOD SCIENCE AND TECHNOLOGY

Although uniquely American developments have occurred in the growth and application of wood science and wood technology, the roots lie deep in human history. In an informal sense, when humans in the valleys of the Tigris and Euphrates Rivers and in China began to use wood, they began to learn ways to use it better and extend its life as a material of construction, daily use, and international trade. The most direct ancestry of American wood science and technology, however, comes through Europe, enhanced by connections of schooling, tradition, literature, politics, colonial expansion, and trade. The most powerful driver of forest products research has been the need to maintain the timber resource and use it wisely in the face of increasing economic pressure and shortages of the timber supply. Europeans went through this stage much earlier than did Americans and in the process planted the seeds of wood science and technology that have provided the basis for much of this field in the United States. The need to protect wood exteriors against fungi and insects led to the development of creosote treatment in Europe in the seventeenth century and to use of the Bethell full-cell process by the British Royal Navy in treating ship timbers in 1838. That process was used commercially in the United States in 1875 to treat crossties for the railroads. The groundwood process of pulping wood for paper making, and later the sulfite and sulfate pulping processes, were developed in Europe in the eighteenth and nineteenth centuries. Mechanical testing of wood and its application in the design of structures also began in Europe during that time. It is not surprising, then, that as our forestry traditions came from Europe, so too, did our traditions in the use of wood and research on wood and fiber.

THE BEGINNINGS OF WOOD SCIENCE AND TECHNOLOGY IN THE UNITED STATES

Forest products research, and the science and technology derived from it, proves an integral part of forestry. It underpins the activity that gives forests their most direct economic value and that provides the economic incentive for sound, creative, and sustainable forest management. Much of the wood science and technology research and progress in this country, especially in its early days, began as adaptation of European developments to America's unique timber resource and utilization needs. As with forestry in this country generally, the roots in programs of research can be traced to the work of Franklin B. Hough. His efforts began in a formal sense in New York State in 1872, extended to the American Association for the Advancement of Science in 1873, were strengthened by the 1877 Appropriation Act, and led to his *Reports Upon Forestry* in 1878-82. Information in those reports dealt with the use of wood by the nation's railroads, in paper making, in the charcoal industry, and the imports and exports of forest products. His reports helped to convince Congress of the need for forestry investigations and led to the establishment of the Division of Forestry in the Department of Agriculture in 1881. Efforts of that Division under Hough and Nathaniel Egleston provided the first government actions to promote a rational approach to improving the rapidly declining condition of the nation's forests. The Army Ordnance Department began tests of timber at the Watertown, Massachusetts, Arsenal in 1881. These included not only strength tests but also investigations of bolts, nails, and other means of joining lumber. C. S. Sargent, of Harvard, described many American woods in his 1884 *Report on Forests*.

When Bernhard E. Fernow took charge of the Division of Forestry in 1886, he brought to it scientific knowledge and experience in the field that his predecessors had lacked. Fernow soon learned and applied a basic principle in advancing wood research, or any research to be applied in the private sector—the incentive to apply the results can only come from a basic desire to make it pay. Among other initiatives, he began a study of the technological characteristics of American timbers. In 1892 he outlined a new program to study "timber physics," by which he meant studies dealing with the physical, mechanical, and chemical properties of wood; the relationship of those properties to wood structure; and the relationship between properties, structure, and the conditions of growth. Filibert Roth, of the University of Michigan, worked out the details of the study and began the research. This resulted in the publication in 1895 of Bulletin No. 10 of the Division of Forestry, *Timber: An Elementary Discussion of the Characteristics and Properties of Wood.* This may rightly be called the first American textbook in the field of wood science and technology. Roth's accomplishments included the determination that the shrinkage of wood was due to loss of water from the cell wall, not from the cell lumen, a basic fact of the physical behavior of wood.

Mechanical testing of wood began under the leadership of Professor John Johnson, of Washington University in St. Louis. This became an important part of the "timber physics" program, because it began to systematize research on the strength and related properties of timber. The Carriage Builders Association of America raised a critical question regarding the relative value of northern- and southern-grown oak for use in carriage construction. In one of its early accomplishments, the timber testing program established that no real difference existed, thus expanding the resource for this application. The program also determined that longleaf pine was twenty-five percent undervalued as far as strength was concerned; that proper seasoning could increase strength of timber by fifty to seventy-five percent; that the proportion of latewood in softwoods provides a good criterion of strength; and that there is a good relationship between specific gravity and strength— all basic concepts in structural use today. In the years that followed, Fernow went to Cornell, Roth joined him, and Johnson went to Wisconsin, all to concentrate on teaching future generations the science and technology uncovered during the program.

Based on these first values for strength, engineers well into the first decade of this century used very conservative working stresses for timber, ranging from 900 to 1200 pounds per square inch (psi) for fiber stress in bending with no variation for grade. This contrasts with stresses of 900 to 3300 psi available later in the century using machine stress rating.

1901–1908: FOREST PRODUCTS RESEARCH EXPANDS

A new period of expansion saw the division of forestry converted to a bureau and a section of Special Investigations formed. Beginning with biological studies of trees, the section diversified in 1901 and began again to look into forest products. In cooperation with the federal bureau of chemistry, research was undertaken on important resin and tannin bearing trees.

Graduates of university programs led by those who had left the division of forestry and its cooperating laboratories returned to carry forward new programs.

Government research in wood preservatives began in 1901, in cooperation with universities, industry, especially the railroad companies, and the federal Bureau of Plant Industry. The idea of preservative treatment, practiced in Europe for many years, had been introduced into this country in the mid-nineteenth century. However, as long as the railroads could buy at reasonable prices the white oak they needed for crossties, preservative treatment was not considered economically feasible. But, by the turn of the century, white oak supplies were running short, and prices increased rapidly. Studies examined the preservative treatment of many wood species, including some that had been considered to be "inferior." The Rueping and Lowry empty-call processes were developed and commercial application, concentrating on oilborne preservative systems, began early in the twentieth century.

A more specific focus on the needs of wood users strengthened the timber testing program in 1902. The same awareness of the need for conservation and better use of the timber supply drove the new program. Timber testing was approached as a practical problem in industrial engineering. Harold Betts, a graduate of Stevens Institute of Technology, organized a timber testing laboratory in Washington, D.C.. The laboratory undertook tests on southern pine beams in cooperation with the Bureau of Public Roads, drawing on that Bureau's interest in timber bridges. Betts then organized another timber testing laboratory in New Haven at the Yale Forest School.

Harry Tiemann, also a graduate of Stevens, took over leadership of the Yale laboratory in 1903. Additional timber testing laboratories were established in 1903 at Purdue University, Lafayette, Indiana, and the University of California at Berkeley. Assistant Forester of the Bureau of Forestry Frederick E. Olmsted took the lead in outlining a comprehensive timber testing program that emphasized mechanical properties of structural timbers and the effect of testing procedures and conditions as well as effect of treatments such as preservation and drying. The first recorded research to improve shipping containers occurred at Purdue in 1905. Additional timber testing laboratories were established at the University of Washington in 1905 and at the University of Oregon in 1906. These laboratories developed basic procedures for standardized testing, significant new information on timber properties, and basic improvements in wood preservation techniques.

Tiemann and his colleagues at Yale developed the concept of "fiber saturation point," and worked out the basics of the internal fan lumber dry kiln.

Timber tests during the eighteenth and nineteenth centuries in England and France led to the observation that strength of wood is related to the rate and duration of loading. The faster load is applied and the shorter its duration the higher is the apparent strength. Timber testing programs at Yale and Purdue early in this century included formalized determination of wood deformation and failure under long term loading, impact loading, different rates of loading, and vibration. This led to reduction of design stresses for structural timber to take into account long term loading.

Research efforts also began in pulp and paper and wood chemistry. These built on the developed art and technology of paper making, which had begun in China about 105 AD and included techniques of groundwood, soda, sulfite, and sulfate pulping developed in Europe in the nineteenth century. By the turn of the twentieth century, wood, rather than cloth, had become generally accepted as a material for paper making, and the demand for wood to make paper added to the adverse impact on the resource. So, the first organized research in pulping and paper making aimed at broadening the resource pool. A small experimental sulfite plant was established in Boston in 1906 to determine the paper making possibilities of several American woods. The U.S. Forest Service operated it in cooperation with Arthur D. Little. After two years it moved to Washington, D.C. The Bureau's wood chemistry research begun at Yale, dealing with wood distillation and chemical problems of wood utilization, also moved to Washington, D.C., in 1908. L. F. Hawley and Ernest Batemen, both to continue as leaders in wood chemistry research, served on the staff of this laboratory. Technology for making rayon from cellulose of wood and cotton, first developed in France in the 1880s, was applied in this country in 1910.

Veneer and plywood have an even longer history, dating back to at least 3000 BC in Egypt and somewhat later in Greece. Then a means of conserving valuable decorative woods, the modern concept of plywood originated by gluing together knife-cut thin sheets of wood to make a product that would replace lumber for some applications. In the United States, Douglas-fir plywood was introduced as a novelty at the Lewis and Clark Exposition in Portland, Oregon, in 1905. Plywood began to be made in the East from hardwoods, but the lack of moisture resistant adhesives limited its use to a few interior applications.

By mid-decade, concern arose about the uncoordinated results of research and testing programs in several universities and at laboratories in Washington, D.C. The universities provided essential help by maintaining the equipment and by providing personnel necessary to conduct the work. Discussions with the industry and with universities pointed more and more strongly toward the need for a centralized forest products laboratory where the necessary equipment could be concentrated and research coordinated. The Forest Service outlined the needs of such an institution and five universities responded—Purdue, Cornell, Minnesota, Michigan, and Wisconsin. After months of technical and political discussion and negotiation, Secretary of Agriculture Wilson announced in March 1909 that Madison and the University of Wisconsin would be the site of the new laboratory. The university's excellent reputation for practical scientific research proved a critical factor in the selection of Wisconsin. Plans then moved ahead, and the new Forest Products Laboratory (FPL) opened on June 4, 1910, beginning a new era in federal forest products research.

1910–1917: PRE-WAR YEARS

Use of timber at the turn of the century remained wasteful and profligate. The 1908 forest inventory indicated that when using trees for lumber, twenty-five percent of the tree remained in the forest, twenty-two percent was lost as slabs, trim, and edging at the mill, fourteen percent became sawdust, and thirteen percent was bark. With the reduction of the timber resource, such waste in utilization of the forest clearly needed correction. Many in the wood industry recognized and lamented this situation, but argued that the economics of resource utilization limited them. The twin goals of efficient and wise use of the resource and improved economics of conversion dominated forest products research during those early years (a situation that in many respects continued into the 1990s).

There was no cadre of specialists in forest products science and technology to undertake the needed research. Instead, engineers, chemists, foresters, botanists and others with scientific and technical training took up the cause. Many who had led university programs moved to Madison; others remained on several campuses to continue research. Cooperation with universities continued with a strong focus on work at Madison. Facilities for fundamental studies were now improved, and research policy favored that approach as a basis for subsequent practical applications. Research developed systematic information on the basic mechanical properties of various wood species. Fundamental studies of "timber physics" carried forward the work begun at Yale on wood-moisture relations and basics of wood drying. Yet the practical application never strayed far from mind, and the border between "fundamental" and "applied" became indistinct and easily crossed. The research in drying fundamentals led to new approaches to air drying and kiln drying. The research in mechanical properties led to improvements in structural applications of wood. The bulletin *Tests of Structural Timbers* was issued in 1912, reporting on strength tests of several wood species that had been conducted over the previous nine years. From 1910 into 1913 the emphasis on basics continued, adding to the meager store of information about American woods and their properties.

In 1913 increased focus on industrial cooperation entered the picture, calling on experience and interest developed to deal with the practical problems of utilizing the resource. This took the form of either field tests with industries, universities, or units of government; or contributions of materials for expanding laboratory research into practical areas. In the field of wood preservation, which had been sadly lacking in basic information, studies of the chemical, physical, and toxic characteristics of the preservatives of the day were subsequently applied to developing improved treating procedures and to field testing treated crossties, poles, and other ground-contact products. Timber testing that had focused on the basic mechanical properties and behavior of wood turned more toward testing manufactured products and developing standards and grading rules for uniformity in engineering applications. Basic studies of wood chemistry, wood distillation, and resin production in the tree produced practical improvements in the production of alcohol and naval stores. Research at the Forest Products Lab in pulp and paper became even more closely attuned to the needs of industry, due to the strong research needs and research consciousness of that part of the field. The pulp and paper industry had begun its own research and also began establishing programs of research and education at universities. Improvements in the groundwood process, used for newsprint, received first attention because the spruce supply was dwindling, and imports from Canada made up increasing proportions of the supply needed to meet escalating demands for newsprint. The Forest Products Lab established and operated a pilot laboratory in Wausau, Wisconsin, in cooperation with the American Pulp and Paper Association on a semi-commercial basis. This led to the successful use of several additional species for groundwood. Studies of chemical pulping also moved ahead, especially in use of the sulfate (kraft) process, and established that a market kraft pulp could be made from southern pine.

1917–1920: WOOD GOES TO WAR

America's entrance into World War I in April 1917 resulted in a major change in the research program at the Forest Products Lab and, indirectly, in the programs of research in universities and industry. Fundamental information gathered from 1910 to 1917 was applied to solve many of the new problems. Others required new research directions. Military construction, aircraft, shipping, gunstocks, wagon and artillery wheels all required wood, and wood by-products were needed for munitions, gas masks, and medicines.

Wartime research to improve adhesives for plywood from the animal and starch based products with poor moisture resistance led to the development of blood albumin and casein adhesives. This greatly expanded the applicability of plywood in critical use conditions, especially for aircraft. Mechanical properties studies revealed the basic fact that the strength of plywood in tension is due primarily to the layers of veneer having their grain orientation parallel to the force. All in all, advances in adhesives, coatings, veneer, plywood and laminated construction not only contributed substantially to the war effort but led to subsequent industrial developments which applied the research results.

Research in kiln drying of wood and related research in wood-moisture relations, water and heat transfer through wood, intensified with special emphasis put on drying of aircraft wood and walnut gunstocks. Both of these were critical to the war effort and required more knowledge about how to dry wood as quickly and economically as possible while retaining its structural integrity. Also critical were satisfactory supplies of oak for artillery wheels and vehicles. Tiemann and his colleagues had made a good start on this and, through refining earlier research results and assisting in their application to practical drying problems, made dramatic decreases in drying time while achieving increases in the quality of the finished product.

The request in June 1917 for twenty-two thousand combat airplanes within a year called for a marshaling of research information and capability unheard of in the early days of forest products research. The planes were made of a wood framework, usually spruce, with steel connectors and cables. Although only half that number were built during 1917, by the end of that year the industry was producing at the desired rate as a result of intensive application of the limited basic information already derived from research on wood properties and engineering . In the rush of moving from a militarily unprepared nation to one determined to have a major impact on the outcome of the war in Europe, research in glues and gluing, plywood manufacture and properties, and the adaptability of previously unused species for plywood use, moved forward intensely and was applied quickly,

Research also intensified in packaging and its application to meeting urgent and very practical needs. The United States, not yet accustomed to its role as an international power, lacked an infrastructure for moving goods across the Atlantic Ocean. Military and other government agencies turned especially to the Forest Products Lab and its connections to the packaging industry to help overcome serious shortages in essential supplies. Research results showed that many species other than white pine could be used for boxes. Results of engineering research were quickly applied to redesign of containers for safer handling and transport and for conservation of shipping space.

The fact that intensive application of research results to meet wartime needs could move ahead so rapidly attested to the value of the fundamental research in many areas of wood and wood products during previous years. The intensified wartime efforts had other benefits as well. Accomplishments during the war provided a strong base for post-war research and developments in waterproof glues, waterproof coatings, veneer, plywood, and laminated construction. New product possibilities were both visualized and realized, and with them new research needs provided a strengthened framework for forest products research in the coming years.

THE 1920S: NEW INDUSTRIAL NEEDS

The vigorous research effort during the war had laid the groundwork for new technical advances in the post-war use of forest products. The major increases in research funding during the war declined, and in peacetime research staffs in forest products that had been greatly augmented (e.g. more than doubled at Forest Products Lab) were reduced again to near prewar levels. The emphasis shifted from a heavy concentration on fundamental research to research application and close federal-state-forest industry cooperation. This new cooperation proved clearly evident when Forest Products Lab celebrated its tenth anniversary in 1920, as leaders in all major parts of the industry and academia spoke to the need for, and possibilities in, such cooperative efforts. The awareness of the degree to which wartime demands had depleted the forests, resulting in wood shortages and sharp increases in prices, provided further incentive.

Following the statement of President Calvin Coolidge, "A tree saved is a tree grown" at the 1924 National Conference on the Utilization of Forest

Products, focus shifted to identifying waste in the conversion of wood to products and either eliminating it or finding new uses for the residual material. A study conducted with the furniture industry showed that small dimension stock of the appropriate sizes could be used more economically and efficiently than standard lumber sizes in furniture manufacture. Wood researchers and silviculturists joined in efforts to determine how growth conditions affected wood properties to advance the notion of "growing wood to order." Researchers in timber mechanics and engineering expanded the knowledge of the basic mechanical properties of wood by evaluating most of the commercial American wood species using standardized, replicable procedures. This led to the ability to compare species for various applications and to develop working stresses that could be used as guides for safe engineered structures. Also, research on large wood structural members led to advances in the ability to design them for efficient and safe structural use. In 1923 the Forest Service published *Basic Grading Rules and Working Stresses for Structural Timbers* as a basis for safe and efficient structural use in the wood industry.

Research to improve the use and performance of the more water resistant glues that had been developed during the war and testing of glued products of many kinds provided a basis for industrial advancement in plywood and a variety of laminated products. The new concept of building up products from small pieces, which reduced drying time, expanded the source of wood supply, and enabled use of the best quality of available wood, received much research attention and subsequent application.

A new field of research begun in the 1920s was wood finishing for exterior use. This led to greatly improved understanding of what finishes could be expected to do in protecting wood, of how to manage the composition of the finish and its application, and of how substrate characteristics affect finish performance.

A process for manufacturing a hard-pressed fiberboard (hardboard) was developed in 1926. This high-density board (density >50 pounds per cubic foot) is being used now as prefinished paneling and siding and as molded products. Other forms of hardboard (fiberboard) have been developed over the years to meet specific needs. Medium density board (31 to 50 pcf), for example, primarily finds use as house siding. Originally derived from pulpwood, the fiber for this board can be obtained from a variety of sources, especially waste from primary forest products processes.

Increased emphasis on basic research in wood chemistry during the 1920s led to significant new information on the location of the basic components

—cellulose, hemicellulose, and lignin—in the cell wall, separation of lignin and cellulose, and the colloidal nature of wood. This work established that lignin is concentrated most strongly in the secondary wall of wood cells. The name "holocellulose" was developed to include all of the carbohydrate content of wood substance.

As the railroads expanded and demand for crossties and other timbers in severe service grew, researchers broadened their view of possibilities for wood preservative chemicals. Acid copper chromate as a preservative was introduced in the late 1920s. The development in the early 1930s of pentachlorophenol treatment exploited the fungal and insect toxicity of the chlorinated phenols. Chromated copper arsenate (CCA) treatment followed a few years later and ammoniacal copper arsenate (ACA) in the late 1930s. Research driven by the need to improve wood protection led to more complete understanding of the physics, chemistry, and biology of wood deterioration.

More efficient and economical wood utilization also provided a theme in pulp and paper investigations. New understanding of the role of temperature led to improvements in both the yield and the quality of sulfite pulp. A process for making a strong white pulp from southern pine by modified sulfate pulping and two-stage bleaching was developed. This opened the door for southern pine to be used for book and magazine paper in addition to the brown kraft paper commonly used for bags and wrapping. The change from bulk handling of goods to individual packaging created a demand for strong, inexpensive paper that unbleached kraft could supply. Two important related discoveries substantially improved the economics of kraft pulping: (1) the modern chemical-recovery furnace and (2) the ability of the unbleached kraft process to be used with almost any wood.

A major accomplishment of the 1920s was the development of the neutral sulfite semichemical process, combining chemical and mechanical pulping to enable hardwoods to be used for paper. In the course of finding a way to pulp chestnut chips from which the tannin had been extracted, FPL researchers developed this process, which yields pulp that gives superior stiffness to corrugating board. Researchers found that neutralization with an alkali, reduced corrosion problems in sodium sulfite pulping of hardwoods and also reduced the amount of cooking needed in preparation for the grinding phase of pulping. Use of this process for pulping other hardwoods (following the demise of chestnut) as a source of corrugating medium has continued and grown to the present.

The passage of the McSweeney-McNary Forest Research Act of 1928 marked an important legislative development to recognize and enhance research capability. Federal appropriations for such research gradually increased to one million dollars per year. The increased interest of the forest products industries, however, began to be reflected in cooperative efforts, which added both funds and in-kind contributions to the research program. The end of the decade, with the depression, brought plans for the new building for the Forest Products Lab in Madison, as part of a public works program initiated by President Hoover.

The practical, productive, and science-based development of effective wood use during the early years soon stood in clear need of more basic information on wood. Samuel Record at Yale exerted notable leadership in organizing world wide efforts in wood anatomy. This led to the formation of the International Association of Wood Anatomists in 1930. The Yale wood collection, which he established and which is now at the Forest Products Lab, is one of the world's major collections. That and the Harvard collection, begun by I. W. Bailey, and the Field Museum collection (now at the Forest Products Lab), together with the Forest Products Lab collection itself, provide an outstanding resource for basic wood studies. H. P. Brown, who had also studied tropical woods intensively, moved to the New York State College of Forestry at Syracuse to continue this area of work in wood technology. During this period, instruction in wood technology and wood utilization became prevalent in forestry school curricula. These were based on significant compilations of literature, such as the *Wood Handbook*, first published in 1936 and periodically revised since that time, as well as literature in the related sciences and engineering. Faculty members wrote textbooks in wood technology to facilitate their instruction in the field. Students of the subject became familiar with the names and the publications of leaders in the early phases of curriculum development, such as Harry P. Brown, Nelson C. Brown, Ralph Bryant, Arthur Koehler, Emanuel Fritz, Carl Forsaith, George Garratt, Samuel J. Record, and Louis E. Wise. Soon graduates of such programs entered the industrial workforce, taking up further research in the field.

THE DEPRESSION AND NEW MARKETS FOR WOOD

The stock market crash of 1929 had a major impact on the wood products industries. The nation's third largest employer, with 3 million employees, it suffered thirty-five to forty-five percent employment loss by 1932. Thus, research programs—governmental; academic; and industrial—in wood properties, timber engineering, wood chemistry, wood processing, and wood protection were designed to assist wood producers, forest products manufacturers, and consumers by reducing the cost of forest products, and by improving properties and quality for increased satisfaction and extended service life.

Particular problems existed in the area of building and construction. Further research on strength of structural lumber led to the publication by the Forest Service in 1934 of a *Guide to the Grading of Structural Timbers and the Determination of Working Stresses*, which embodied principles still used in such applications as structural lumber. The need for low cost housing also prompted the adaptation of aircraft technology developed during and after the war. The "stressed skin" approach to construction used plywood faces and a light structural core. This reduced the need for materials and expanded home-owning opportunities for people with little money to buy housing. Research on reducing moisture condensation in walls led to the development of vapor barriers, now a standard feature of house construction.

Laminated construction, accomplished by gluing together boards to make timbers of any size needed for construction, was developed in America in the 1930s, based on earlier work in Europe, which had produced laminated wood as early as the turn of the century. T. R. C. Wilson, of the Forest Products Lab, went to Europe, especially Germany, in the early 1930s to gather information on this technology and to provide a basis for the necessary engineering of processing and standards. Working with the newly forming laminating industry, research in this country quickly built on the European work with modern adhesives and fabrication techniques. A notable product was the glued laminated wooden arch, now a standard feature in the design of churches and other large open-space buildings. Many forms of straight and curved laminated beams could now be made to meet the needs of particular construction projects. This approach enabled the use of a wide array of lumber species and sizes, though Douglas-fir and southern pine were most commonly used. Development and use of such heavy timber construction led in turn to the need for improved timber connectors, since the connections between such members was usually the weakest point of the construction.

Preservative treatment of poles became common during the 1930s to preserve substantial investments in electric power and telephone lines, as well as the poles themselves. Bell Telephone Laboratories conducted tests of poles treated with greensalt (chromated copper arsenate), pentachlorophenol, and creosote. These led to the observation that the preservative did not adversely affect strength of the poles, but steaming used in the treatment process did.

Engineering design and construction principles for wood containers that had been worked out during the war became, during the decade of the 1930s, increasingly applied to fiber containers. Most of the container market then called for wood fiber-based products rather than natural wood. Those opened new opportunities for the pulp and paper industry in producing the fiber materials used for such containers.

Research in wood drying continued to refine the kiln drying procedures that had been under development since early in the century. A new development in drying was chemical seasoning. In this procedure, pieces of wood to be dried were soaked in solutions of sodium chloride or urea before drying. Such treatment greatly facilitated the drying of large timbers and refractory species, such as some of the oaks.

Phenolic resin adhesives had been developed to a commercial level in the late 1920s, and during the 1930s became the subject of intensive research to learn how to use them most effectively in glued wood products. Phenolic resin adhesives permitted use of glued wood products in moist atmospheres without serious loss of strength in the adhesive joints.

Development of adhesives and techniques for laminating heavy timber beams made possible engineered structures from lumber to meet many design needs.
(USDA FOREST SERVICE, FOREST PRODUCTS LABORATORY PHOTO)

Wood chemistry took basic steps toward utilizing the forest more effectively through expanding the usable resource and using forest and mill residues during this decade. Effort concentrated on lignin, since progress had been made in solving the mysteries of cellulose. Lignin remained a challenge to wood scientists. It could not be broken down into "building blocks" as could cellulose. Tons of lignin in spent sulfite pulping liquor were going to waste, not to mention the tons of lignin in the form of logging and manufacturing residues. It was discovered that lignin could be hydrogenated to a clear liquid by high pressure and temperature in the presence of a catalyst and that this contained methanol and other useful compounds. Further research showed how to make a useful lignin plastic. Greater knowledge of lignin, added to the existing knowledge of cellulose, provided a basis for chemical conversion of wood to a variety of useful products.

In the late 1920s it had been observed that second-growth southern pine timber reached pulpwood size before an appreciable amount of heartwood developed. At an industrial lab in Savannah, research indicated that the southern pines thus grown could be ground to make pulp for newsprint. During the late 1930s, a mill was built in Lufkin, Texas, to make newsprint by that method. Also, during this time the pulp and paper industry of the South expanded rapidly, largely on the basis of kraft (sulfate) pulp from southern pine. An important research accomplishment was the ability to make a strong white paper from southern pine by modifying the sulfate process. Two processes for producing newsprint from southern pine came from further research. One of these used a combination of bleached sulfate and groundwood pulp; the other used semichemical pulp from black and red gum mixed with pine groundwood. With the arrival of more favorable economic conditions during the 1940s, the newsprint industry in the South grew rapidly. The discovery that pulp strength of a softwood pulp increased with increasing proportions of earlywood led to the successful pulping of Douglas-fir by a modified sulfite process. The pulping and paper making research of this period resulted in a substantial broadening of the resource base for paper making as well as broader opportunities in management of the forest resource. Use of kraft papers and paperboards expanded rapidly in the mushrooming packaging field, assisted economically by the application of chemical recovery processes that had been developed in the late 1920s and early 1930s.

Another field of wood research which made important progress at this time involved uncovering relationships between tree growth conditions and wood properties. One example of this was the observation that wood of young second-growth softwoods did not typically have the strength of the old growth material of the same species, due to a lower proportion of latewood. Further, it was learned that this could be controlled by closer attention to stand density and control of crown size. Studies of oleoresin production in the southern pines showed the effect of damage to the tree by fire and by the method of chipping.

THE 1940S: WOOD GOES TO WAR AGAIN

America's entrance into World War II unleashed an insatiable demand for forest products. Aircraft, ships, dwelling units, containers, explosives, plastics, rayon, and an array of other products created new opportunities and new technical problems in use of the nation's wood resource.

During the preceding two decades, the aircraft industry had pretty much gotten away from wood use, primarily because adhesives did not have adequate moisture resistance. The advent of synthetic resins changed this and substantial research effort went into developing the technology for closely engineered moisture resistant plywood and understanding the mechanics of plywood construction. This was extended to refinement of "sandwich" (or "stressed skin") construction by building panels of plywood separated by a low-density core, which had been originally developed in the 1930s. This was applied in construction of aircraft such as the British "Mosquito" bomber, which saw wide use during the war.

The need for high quality kiln dried structural wood led to an intensive effort to improve drying techniques and technology. The program of technical training in wood processing and certification of dry kilns for drying aircraft wood involved not only researchers at the Forest Products Lab, but also many university wood science and technology faculty. In the process of conducting this certification much was learned about the need for clear paths of air circulation in stacks of drying lumber. New understanding of temperature and moisture effects on development of internal stresses and checks in wood during drying and of heat and moisture movement through wood led to the formation of drying schedules in which temperature was raised gradually from an initially low level, and relative humidity was lowered gradually from an initially high level.

Military support and needs led to the development of many new concepts of property improvements for critical uses that could be realized by combining wood with paper and with resins under a variety of temperature and pressure conditions. A. J. Stamm and his colleagues at the Forest Products Lab played a key role in developing a variety of products of this nature. They served as forerunners of later diversification of wood and paper products by combination with other materials.

The technology of wood and wood-based products in packaging substantially advanced during this period, due to the need to transport large quantities of a wide variety of goods to many parts of the world safely and economically. This involved engineering of boxes and crates and also interior cushioning, bracing, supporting, and cradling packed items, including prevention of corrosion of metal parts. Wet-strength fiberboard added substantially to packaging effectiveness and economy. The intense effort related to military material and other items needed for efficient and economical shipment of food, clothing, and supplies, had major benefits for transportation of goods after the war.

The shortage of large, high quality white oak and Douglas-fir timbers needed for building ships (particularly non-magnetic minesweepers) prompted new research in laminating heavy timbers from available lumber sizes. This led to improvements in selection and use of adhesives and laminating techniques, as well as new approaches to ship design and construction. By war's end this technology received intense application in the manufacture of landing craft and PT boats.

A threatened shortage of grain and of industrial alcohol pointed to the need to make alcohol from wood, based on previous knowledge of the structure of cellulose and the possibility of converting that to sugar. Chemists at the Forest Products Lab primarily developed a process known as the "Madison wood sugar process," based on the Scholler process, which had been developed in Germany before the war. It consisted of hydrolysis of wood chips with sulfuric acid under heat and pressure, removing the sugar, and fermenting it. The Madison process substantially improved the productivity of the operation. It was shown to produce large volumes of alcohol, but could not compete economically with the process of making alcohol from grain, so it was never developed commercially. Work on the chemistry and chemical engineering involved in that process led to further productive study in universities in several parts of the country of wood chemistry and wood as a chemical raw material.

After the war ended, research attention focused once again on domestic problems related to conservation and effective use of the declining timber resource. The problem of waste of wood in the forest, in the mill, and in use, recognized during the war but subverted in favor of more immediate problems, surfaced again for intensive attention. Research in sawmilling produced sawblades that would cut more smoothly with less sawdust than those in current use. The relationship of saw speed, feed rate, and power consumption was clarified and applied to improve the efficiency of sawing operations. Engineering and fabrication of housing components were improved based on wartime developments such as engineered structural plywood and improved fasteners. Those developments were applied to the production of affordable housing to replace and improve on housing deteriorated during the war period.

Many improvements in handling, grading, gluing, and laminating small pieces made it possible to produce large, high quality materials, from the available timber resource. Research related to silviculture and wood quality showed how to maintain high strength in second-growth southern pine by controlling the rate of crown growth, a follow-up to basic observations in the 1930s. New developments in pulp bleaching techniques, plus the resurgence of industry and trade, accelerated the further expansion of the paper industry, especially the production of kraft paper.

The development of the electron microscope in Germany in the 1930s provided means to look into the ultrastructure of wood and understand its basic anatomical features in ways not previously possible. This first electron microscope depended on the transmission of electrons through the specimen material, comparable to light beams used earlier, but was capable of more than a hundred times the resolution of the finest light microscope. Structural features of the layers of the cell wall and of bordered pits, involving elements of five to ten nanometers (10^{-9}m) in size, for example, could be observed.

The concept of softwood plywood had been developed before the war, although commercial applications were largely limited to interior use. During the war its military applications received much attention. After the war Douglas-fir plywood began to attract greater commercial attention in the northwestern United States as university, industry, and federal researchers joined forces to establish a standard technical basis for a structural plywood industry. This concept applied new knowledge of synthetic resin adhesives, high-speed veneer peeling on a rotary lathe, and the related processing technology, with developments in design of plywood as an engineering material.

Recognizing the need for strong communication between researchers and the industry applying such research, representatives of federal, university, and industrial interests established the Forest Products Research Society (now Forest Products Society). That non-professional technical society continues to provide effective technology transfer and awareness of research needs through a variety of media and conferences. These expanded in the 1960s and later through the journal and meetings of the professional Society of Wood Science and Technology.

THE 1950S, 1960S, AND 1970S:
RESOURCE CONSERVATION AND USE

Although the Korean War kept alive military needs, development of research and technology during this period aimed at further capitalizing on basic knowledge gained during wartime, applying this to more general uses, and conserving the timber resource by using it more efficiently to meet major consumer needs.

Effects of duration of load on structural members had been studied further since the first decade of the 1900s. Studies for wartime structures had advanced the knowledge of load duration effects. In 1951 Lyman Wood published the first mathematical model of duration of loading effect on wood strength. This has been further refined in subsequent research and remains a standard factor in establishing stresses for structural timber today. Studies of the rheology of wood in the 1960s led to refinements of that understanding by revealing some of the basic relationships between wood deformation, its conditions of loading, duration of loading, and its physical and mechanical characteristics. Wood was observed to behave essentially as a linear viscoelastic material at stress levels up to about 50 percent of its static strength. Slightly earlier research in Australia led to the concept of mechano-sorptive deformation to explain the increase in deformation of wood under stress when it is subject to moisture changes. That concept has been developed further with reference to both structural elements in use under variable humidity conditions and wood subjected to internal stresses due to shrinkage during drying. The research of J. M. McMillen developed the basic understanding of those drying stresses and related wood strength that has led to drying schedules that take into account rheological behavior as well as other aspects of physical and mechanical characteristics of wood.

Southern pine plywood marked a major development during this period. Softwood plywood began to replace lumber in many types of construction, creating a need to broaden the resource base to accommodate this demand. The industry was based on the West Coast, where large supplies of old-growth Douglas-fir, not uncommonly four or five feet in diameter, had provided an ideal resource for peeling large quantities of high quality veneer.

As that resource diminished and major markets developed in the East, attention shifted to the possibilities of using the smaller second growth pine of the South. H. O. Fleischer, of the Forest Products Lab, noted in 1951 that there was unrealized potential in peeling small logs. The research that followed during the next few years, involving also the Southern Forest Experiment Station and the plywood industry, showed that veneer could be peeled from southern pine and bonded successfully with resin adhesives. A lathe that enabled peeling of much smaller logs was introduced, based on experience in Scandinavia. By the following summer, industry representatives and the Department of Commerce had agreed on a commercial standard, which became effective in November 1963, and by December, preliminary operations had begun at a Georgia-Pacific plant in Fordyce, Arkansas. Two plants in Texas began operation the following year. From that time, the industry grew rapidly to become a major supplier of softwood construction material.

Wartime research to develop technology for strong, lightweight structural materials for aircraft resulted in improved understanding of the basic design, engineering, and construction principles of sandwich construction, which had its roots in World War I. Many more combinations of face and core materials had become possible, with faces of plywood, hardboard, aluminum, plastic laminates, or others separated by core materials such as balsa, rubber, plastic foams, or formed sheets of cloth, metal, or paper. Plywood and hardboard with a paper honeycomb core became a common application of this principle.

Cold soda chemi-mechanical pulping became a versatile process that could be used to make a variety of products ranging from newsprint to printing papers, tissue, molded paper products, and paperboard, depending on the proportion of groundwood to that of chemical pulp. It had been discovered in 1919, but was not used until further research in the late 1940s and early 1950s led to enough information about the variables of the process to make it commercially feasible. The introduction of the disk refiner and centrifugal cleaners also made it commercially practicable.

Kent Kirk at the Forest Products Lab and others carried out studies of the biotechnology of wood, leading to basic knowledge of biological reactions of the various chemical elements of wood. Of particular note are efforts to break down lignin by enzymes secreted by white rot fungi as a new approach to pulping. This has shown enough promise that it continues to be developed in cooperation with the University of Wisconsin and the paper industry.

The study of wood structure and density as indices of other properties and as aids to evaluating wood quality and variability received particular attention. The increment core technique was applied to surveying density and other aspects of wood quality of standing timber, with particular concentration on the southern pines and Douglas-fir. Moisture sorption and equilibrium moisture content, though not new concepts, became understood more completely. Related to this was new understanding of the movement of liquids and gases in wood and of the swelling and shrinkage of wood. Recent developments in the field of polymer science provided a basis for further understanding of creep and stress relaxation and of the relatively new concept of mechanosorptive creep mentioned earlier.

Research from the 1930s to 1960s introduced a number of new treatments to improve wood's durability, hardness, stability, and other properties. Some of these were complex chemicals that could not be introduced into wood in an effective form. It was found that they could be introduced in a simple form and converted to an effective form in the wood by means of nuclear radiation. Developments in nuclear physics and an improved understanding of radioactive isotopes also increased interest in radiation methods for non-destructive testing of wood. It was learned that both beta- and gamma-radiation could be used to determine wood density or moisture content. These approaches have found some application in control of paper processing, but have not been used significantly in evaluating solid wood because of the longer count times required. Equipment to measure speed of sound transmission through wood was used experimentally to measure irregularities such as wetwood—areas really wet due to bacterial infection—and internal checking. Measurement of the electrical resistance of wood is also used to detect wetwood. Studies of the vibration characteristics of wood, especially in bending, have indicated possibilities for non-destructive evaluation of wood stiffness and, by correlation, bending strength. In an approach applied more commercially, the observed relationship between stiffness and strength in bending has been used as the base for machine stress rating of structural lumber. Methods of sampling timber to evaluate its strength and related properties and inherent quality have been established on a statistically sound basis.

The need to use second growth timber in place of the large old growth on which much wood products production has traditionally been based has driven technology for moving small logs efficiently through processing operations. We have talked about the success in dealing with this and related problems in learning how to make plywood from small second growth southern pine. Likewise, technology for moving small logs through the headrig in a sawmill has undergone transformation by making use of basic information on wood growth patterns in the log and advances in electronics and computer scanning. Out of such research in the 1960s and 1970s came the Best Opening Face concept of determining the best location for the first cut in a sawlog, which has been adapted to softwood sawmills in many parts of the country.

A major step in mechanical conversion of softwood logs during the 1960s and 1970s was the chipping headrig, developed under the leadership of Peter Koch at the Southern Forest Experiment Station. This made it possible to convert logs into lumber or veneer and produce useful chips without producing slabs or sawdust.

The field of wood-based composites saw several new developments during this period. Research at Washington State University, Forest Products Lab, and many other institutions moved technology ahead to make efficient structural composites. The scientific and technical understanding of particleboard had advanced to the point where it was possible to make flakes or strands of wood and press a board called flakeboard or to orient them in processing to take advantage of their strength in making oriented strandboard (OSB). This technology, quickly adapted to major commercial application, replaced softwood plywood for many structural uses. Steam-injection pressing provided new opportunities to make structural composites economically from a variety of low quality and residual wood sources. Also developed during this period was medium density fiberboard (MDF), a platen pressed dry-process board made from wood fiber. It has been used especially as core stock in the furniture industry and as house siding.

In the early and mid 1970s, research on knife cutting of thick (½ inch) veneer to eliminate sawdust and sawkerf led to a concept called Press-Lam. This had greater uniformity of mechanical properties than the same material in lumber form. Later research integrated Press-Lam technology with plywood manufacturing and produced parallel laminated veneer (PLV) or laminated veneer lumber (LVL), now a recognized commercial structural product in many parts of the world.

Important advances were made during this period in the study of pallets and the engineering aspects of their design and fabrication. Pallets provide the major outlet for hardwood lumber in this country. They prove essential to movement of goods in trade. E. George Stern at Virginia Tech provided leadership in this area and in development of nails and timber connectors. The transmission electron microscope was refined during the 1950s and wood researchers began to take advantage of its ability to resolve details as small as a few nanometers (10^{-9} meters) to uncover details of wood cell structure. A further development in the 1960s was the scanning electron microscope, giving much greater depth of field and impressions of wood element structure than the transmission electron microscope.

Developed more recently in France and now used in this country in wood anatomy studies and wood characterization is the X-ray densitometer. This permits rapid measurement of wood density within annual rings, as well as their width, and earlywood/latewood proportion, directly from increment cores.

The passage of the McIntire-Stennis Act authorizing funding for forestry research proved of special significance to university research in forest products. Enacted in 1962, it marked the beginning of steady increases in university forest products research. Support for such programs at both the university and federal levels was further broadened by introduction of the U.S. Department of Agriculture Competitive Grants Program, which included research in wood utilization.

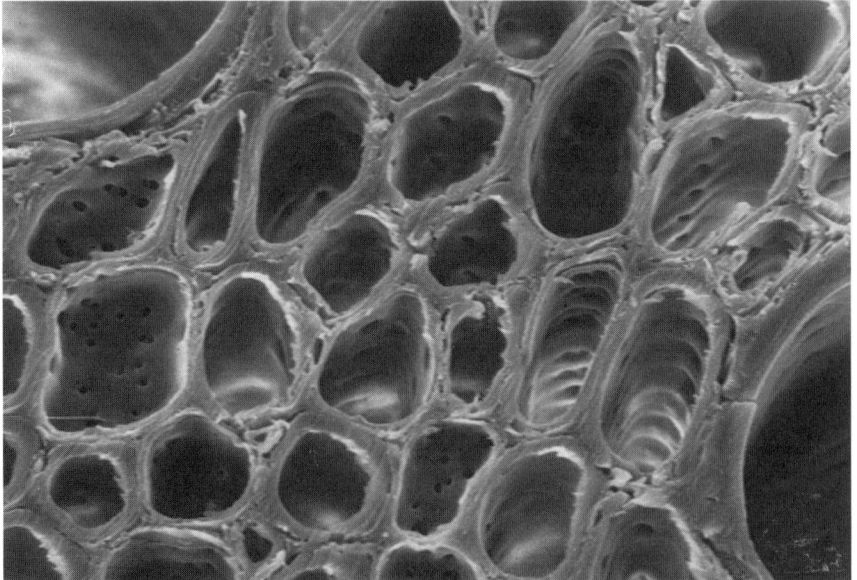

Development of the scanning electron microscope has made it possible to study the structure of wood at many levels of magnification, as in this micrograph of white oak at 1,000 magnification.
(VIRGINIA POLYTECHNIC INSTITUTE AND STATE UNIVERSITY PHOTO)

THE 1980S AND BEYOND: ENVIRONMENT, ECONOMY, SUSTAINABILITY, AND CONSUMER NEEDS

Environmental concerns, while recognized in earlier scientific and technical development, moved to center stage after 1980. They have done so, however, with a close eye on economy of operation and being aware of and meeting consumer needs. Sustainability in the face of increasing demand on the resource for all of its benefits is the keynote here as in other aspects of forestry. This has had noticeable effects in research programs and technical developments that apply their results.

The wood preservation industry, with its long history of research, is one of those most strongly affected with particular pressure on pentachlorophenol and creosote based preservatives. Development of preservatives that are non-toxic to people and animals has received special emphasis. Research with waterborne preservatives has begun to develop rapid fixation schemes for chromium- and arsenic-containing solutions to keep the active salts from leaching into the surrounding ground and atmosphere. Copper-based preservatives offer an alternative to those containing arsenic or chromium. Several new biocides have begun to show promise in oil borne preservative systems. Biological control through the use of antagonistic microorganisms has shown promise for some less severe applications. Modification of wood by introducing chemicals that reinforce the cell wall has also shown promise of providing protection while improving moisture stability and some mechanical characteristics. Treatments such as this are becoming increasingly important in the effort to sustain the wood supply while meeting the many other demands on the forest and preserving a safe, healthy environment for both producers and consumers of wood products.

The technology and economics of the 1990s dramatically reduced wood waste, the amount depending on the type of forest, the type of logging, and the market for lower grade material. With whole-tree chipping the proportion left in the woods can be essentially zero and with modern logging of southern pine it might be five percent, while virtually all of the material brought to the mill is used, either for products or for fuel to run the plant.

Recycling of wood and fiber products became the order of the day as both wood and paper industries applied the results of extensive research to make recycling both technically feasible and economically profitable. Science and technology coming from the research laboratories and industry trials provides a basis for making recycled fiber a common component of wood- and fiber-based products. These will make use of dry- and wet-formed processing, wood/inorganic and wood/plastic composites, improved de-inking processes, and new approaches to pulping, bleaching, and paper making. The increasing success of these efforts makes it possible to conserve the standing timber resource and reduce waste disposal problems while meeting the needs of a rapidly increasing world population.

Research in the late 1980s and early 1990s has indicated that thermoplastic composites made with post-consumer wastepaper or waste wood fiber as a reinforcing filler have some very useful characteristics. Wood fibers can also be combined with such materials as polyethylene, polypropylene, and others to form reinforced plastics using conventional plastics processing equipment. New interest has developed in the United States in an old concept of designing composite products that combine wood and inorganic materials such as Portland cement or gypsum. These can be made not only structurally strong but also water and fire resistant. A further new development in composites is reinforced glued laminated beams, in which high-strength fibers such as carbon, aramid, or fiberglass are oriented in a thin plastic matrix and laminated into glued laminated construction.

Pulping and paper making are seeing new science and technology to reduce pollution, increase yield, and broaden the resource base. Removal of lignin by use of oxygen rather than chlorine, ozone bleaching, elemental chlorine free bleaching, and other approaches are steps toward reducing environmental impact. New approaches to effluent treatment are being developed. Mechanical pulping, long favored for its high yield, but plagued by low strength of the resulting fiber, received assistance by chemicals and heating with such pulping process as chemithermomechanical pulping (CTMP), plus steam explosion and hydrothermal treatment to increase yield and maintain strength. Press-Dry paper making, which applies pressure during forming, and its commercial derivatives are making possible strong paper from short fibered furnish. Corrosion control in the processing equipment receives a major focus of engineering studies.

A significant trend in advancing wood science and technology came from the introduction of electronics via process control and information systems. This made possible the application of basic knowledge in many ways not previously possible. Scanning and control technologies have been brought together to improve both efficiency and economy in wood and fiber products manufacture and to use the available resource more effectively.

It has been recognized for many years that underutilized or lesser-known species that grow in many of the same regions as the preferred commercial

Development of the Press-Dry Process of papermaking made possible strong paper from short fibered hardwoods.
(USDA FOREST SERVICE, FOREST PRODUCTS LABORATORY PHOTO)

species can offer opportunities to alleviate overcutting the preferred species, extend the resource base, and provide new options for forest management, while improving sustainability of the forest's use for other benefits. Compilations of information on such species have been assembled, and marketing efforts are beginning to introduce them to the trade. Examples are cooperative studies to improve and publicize the performance of structural glued-laminated timber of southern pine, red maple, and yellow-poplar.

CONCLUSION

Through all of its history in the United States, the development of wood and fiber science and technology has fundamentally been directed toward realizing the potential of the timber resource in advancing the economic, social, environmental, and material well being of the nation through wise use of that resource. The growing awareness that the forest resource was both essential in meeting the needs of the nation and deteriorating rapidly in its ability to do so provided its primary incentive a century ago. Over the intervening hundred years the development of the science and technology of wood and fiber has provided a strong base on which to both use the resource wisely and, as emphasized by B. E. Fernow, "make it pay." That development continues, with full awareness of the changes in emphasis and breadth of the concept of "wise use" of the forest. Demand for all of the products and benefits of the forest continues to grow, presenting new challenges to those who define the policies and directives for sustaining the forest resource. Those challenges are reflected also in the growing need for "new" wood science and technology to enable the "new" forestry to meet the diverse expectations with which it is confronted.[1]

FURTHER READING

Ellefson, P. V. and A. R. Ek. "Privately Initiated Forestry and Forest Products Research and Development: Current Status and Future Challenges." *Forest Products Journal* 46 (1995): 37–43.

Fridley, K. J., M. O. Hunt, and J. F. Senft. "Historical Perspective of Duration-of-load Concepts." *Forest Products Journal* 45 (1995): 72–74.

Gorman, T. M., and W. C. Feist. *Chronicle of 65 Years of Finishing Research at the Forest Products Laboratory.* FPL GTR 60, Madison, WI: Forest Products Laboratory, 1989.

MacCleery, D. "Resiliency: the Trademark of American Forests." *Forest Products Journal* 45 (1995): 18–28.

Maloney, T. M. "The Family of Wood Composite Materials." *Forest Products Journal* 46 (1995): 19–26.

May, T. K. "Structural Lumber, Past-Present-Future." *Proceedings of the Forest Products Research Society* 2 (1952): 72–75.

Myers, G. C., and J. D. McNatt. *Fiberboard and Hardboard Research at the Forest Products Laboratory. A 50-Year Summary.* General Technical Report FPL-47, Madison, WI: Forest Products Laboratory, Forest Service, U.S. Dept. Agriculture, 1985.

Nelson, C. A. *History of the U.S, Forest Products Laboratory (1910-1963)*. Madison, WI: Forest Products Laboratory, Forest Service, U.S. Dept. Agriculture, 1971.

Perlin, J. *A Forest Journey: The Role of Wood in the Development of Civilization*. New York: Norton, 1989.

Sargent, C. S. *Report on the Forests of North America (exclusive of Mexico)*. Dept. of the Interior, 10th Census. Washington, D.C.: GPO, 1884.

Stern, W. L. "Highlights of the Early History of the International Association of Wood Anatomists," in *New Perspectives in Wood Anatomy*, P. Baas, editor. The Hague: Martinus Nijhoff/ Dr.W. Junk, 1992.

Wangaard, F. F. "Wood Science and Technology." in *Encyclopedia of American Forest and Conservation History*, R. C. Davis, editor. New York: Macmillan, 1983.

Youngquist, W. G. and H. O. Fleischer. *Wood in American Life 1776–2076*. Madison, WI: Forest Products Research Society, 1977.

Youngs, R. L. "History of Timber Use," in *Concise Encyclopedia of Wood & Wood-Based Materials*, A. P. Schniewind, editor. New York: Pergamon Press, 1989.

Youngs, R. L. "Forest Products Utilization Research: Providing Technology for Wise Use." *Unasylva*. 45 (1994): 38–45.

NOTES

1 I gratefully acknowledge the assistance of the many colleagues and friends who gave so generously of their time and memory to provide information for use in this chapter. Particularly appreciated is the effort of Frederick F. Wangaard, Alan D. Freas, John I. Zerbe, and Max Davidson, who read a draft of the manuscript and gave me valuable comments and suggestions for its improvement.

GERALD W. WILLIAMS[1]

SOCIAL SCIENCE RESEARCH IN THE FOREST SERVICE

An Increasing Emphasis

This paper reviews the history of social science research in the USDA Forest Service, the new role of social science in ecosystem management, and some current issues surrounding the use of social science in the land management agency. Alan Ewert defined social science (human dimension) research as: "The scientific investigation of the physical, biological, sociological, psychological, cultural, and economic aspects of communities and individuals in relation to the use and appreciation of natural resources."[2]

Almost every aspect of the use and management of the natural resources involves the social sciences. Simply stated, the social sciences are those disciplines that are concerned with people acting in society and how society works. Thomas Marcin wrote that, generally, the social sciences can be separated into sociology, economics, political science, history, anthropology, and human geography. Many of the disciplines have components that are tied to various laws enacted by Congress and regulations promulgated by the agency, especially those requirements that are concerned with conducting or documenting environmental impact assessment work. Traditionally the disciplines most frequently concerned with uses of the natural resources have been recreation/sociology and economics, although political science (policy) has played an important, if less evident, role.[3]

318

Personal experience has found that many managers and biological researchers tend to view the social sciences as "frill" or "fluff," which only detracts from the need to produce goods and services (although this is changing). Still others view social science research as an area which has no theory or track record and is certainly not deserving of valuable time and money to be "thrown at" problems which have no solutions. Nothing could be farther from the truth. Still others, Marcin noted, "recognize the importance of the social sciences, [but] they may not know how to apply them in their scientific research."[4]

GOVERNMENT FORESTRY IS A SOCIAL MATTER

The U.S. government first became involved in forestry matters in the late 1800s not because of a constitutional directive, but because of highly concerned and motivated individuals, organizations, congresses, and presidents. Involvement came about because of their concern about the impending loss of the nation's public domain forest land to the saw and the plow. The practice of forestry had become a social/political matter. From the beginnings of federal forests and forestry in 1891 an emphasis fell on establishing and managing national forests for people. National forests were established to conserve the diminishing forests, provide wood for human consumption, and provide clean water. Forest Service researcher Harry Gisborne wrote "as is evident, Forestry has become one form of applied sociology"[5] Henry Solon Graves, chief of the Forest Service from 1910 until 1920, noted:

> Forestry is fundamentally a public problem. The purposes of forestry are essentially public in character. Forestry aims to continue the growth and production of the forests for future needs and to secure those general public benefits arising from the mere existence of forests . . . A new principle has been introduced in our public land policy, namely, that there are certain classes of land [forests] whose management vitally affects the public interests and which cannot be mismanaged without grave danger of direct injury to the public.[6]

Since the passage of the Forest Service's Organic Act of 1897, people have not only been the major reason for establishing national forests but were the recipients of the products coming from the forests. The century old phrase "the greatest good of the greatest number in the long run" exemplifies the determination of the Forest Service to administer the national forests for the benefit of people. This utilitarian philosophy has provided the tie between the practice of government forestry, the social sciences, and the citizens of the United States for nearly one hundred years.

Gifford Pinchot and President Theodore Roosevelt were part of the Progressive Era. Samuel P. Hays (1959) noted that both men were prominent Republicans who tended to emphasize moderate social change and improvements through governmental actions as exemplified in the utilitarian philosophy.[7] Harry Gisborne wrote that:

> [In 1905] . . . when Secretary of Agriculture James Wilson transferred the old Forest Reserves . . . he too saw the social significance of our forests. And he took political action to make those forests serve those social uses. . . . Gifford Pinchot . . . stated, in his book "The Training of a Forester," that in his opinion the entire profession of forestry had in reality adopted the definitely social axiom, "For the greatest good of the greatest number in the long run.". . . this was aimed directly and specifically at PEOPLE and not just things [resources] [Emphasis in original.][8]

Forest Service research in the social sciences has, until recently, focused on recreation and fire prevention. The outdoor recreation component in section seven of the McIntire-Stennis Act of 1962, which defined the scope of federal forestry research, provided part of the reason for this emphasis. This outdoor recreation subsection (four) was generally interpreted as the *only* social science area that could be researched, thus leaving the related social science disciplines without support or funding. Fire prevention research has been studied since the 1930s when concerns about human caused fires were at a peak. For the purposes of this paper, both of these topics are subsumed under the larger sociology arena as sociological researchers contributed greatly to the literature in these important research areas.

A distinction needs to be made in regards to the notion of research. If the term implies basic strict theory construction and hypothesis testing, as Jack Gibbs suggested, then little of what is done in the Forest Service research branch qualifies. If, however, research takes on a practical or applied "bent," then there is a great abundance of research work being accomplished. The following discussion involves the latter definition of applied research methodology and results.[9]

SOCIOLOGICAL RESEARCH

Sociology as a distinct academic discipline began just after the turn of this century. In contrast to economics, the role of sociology is not as clearly defined but is implied in most of the Forest Service actions. The Forest Service has carried on Pinchot's principle of social forestry, but research on the subject has been generally lacking. The concern for the forest resources has been an ever-present management goal for the agency, but the goal has to be integrated with a concern for the people. As Benton MacKaye in 1918 phrased it:

> And if we desire to have a system of forestry in this country which is concerned only with wood supply, streamflow, and their *material* by products, then we should pay no attention to the social aspects of forest management. But if we pursue this policy we must not be surprised in future times of crisis if the labor situation, in the industry which is ultimately in our charge, becomes acute and grows worse instead of better . . . [Emphasis in original].[10]

The practice of sociology as social policy in the Forest Service became most evident during the Great Depression. Beginning in 1933, the Forest Service was heavily involved in providing work for unemployed young men through the Civilian Conservation Corps and, to some extent, through the Works Progress (later called Projects) Administration. Basic social and economic research was conducted by the regional offices during this era. At the same time, the U.S. Resettlement Administration was endeavoring to help local families and communities by "tying" the management of public forest land with one or more communities for their long-term well-being. Studies were made of the general community conditions, with the notion that baselines were needed to understand the plight of the towns and to devise ideas on how to help the families through agency actions. Here was a case where the Forest Service was at the forefront in helping millions of men and their families to get through the Depression.[11]

The Forest Service carried on the idea of social responsibility through everyday management. Lyle F. Watts, chief of the Forest Service from 1943 to 1952, first greeted the Forest Service employees in the Washington office with this social forestry message:

> You and I have a lot to do with trees in the forest. . . ; with forage on open ranges. . . ; with wildlife, and with soil erosion and the like . . .

321

The U. S. Forest Service experienced social policy firsthand, in 1933, when the George Washington National Forest in Virginia became the first site to host a Civilian Conservation Corps Camp. President Franklin D. Roosevelt visited the camp shortly after its establishment. Dining with him from left to right: Major General Paul B. Malone; Louis McHenry Howe, secretary to the president; Harold Ickes, secretary of the Interior; Robert Fechner, director of the CCC; Henry A. Wallace, secretary of Agriculture; and Rexford Tugwell, under secretary of the Interior.
(AMERICAN FORESTRY ASSOCIATION PHOTO)

I am confident that we all think and work with these things because they are tools through which PEOPLE may be served . . . I know what happens to PEOPLE in forest communities after their timber has been liquidated improperly or too fast . . . forest land resources can bring reasonable security to PEOPLE who work in a given locality and who want to own HOMES and raise FAMILIES. [Emphasis in original.][12]

Another "social experiment" effort in the mid-1940s culminated in the passage of the Sustained-Yield Forest Management Act of 1944. This act was intended to tie together national forest lands (tree harvest) with local communities in two different, but related, ways: either through establishment of cooperative sustained yield units to co-manage both national forest and private timberland within a specified area or establishment of federal units containing just national forest land, without the cooperative management aspects of private timberland. These long term agreements were intended to stabilize employment, communities, forest industry, and the local tax base, as well as to assure a continuous supply of timber at cost without competitive bidding. Despite more than sixty applications for cooperative sustained-yield units, only one such unit was established, at Shelton, Washington. Another sixteen applications were made for federal units, of which five were established in the West. What was intended to be a long term solution for the towns through an assured timber supply was quickly attacked by the timber industry and communities as contradictory to the free enterprise system because of this unique federal/private monopoly. The last federal unit was established before 1950.[13] This effort to stabilize communities still has implications today with the implementation of adaptive management areas as part of the 1994 "Northwest Forest Plan" (see the discussion on social science research in the ecosystem management era below).

Much of the recent work in sociology in the Forest Service revolves around the National Forest Management Act of 1976 (NFMA), planning at the national forest level, and compliance with the National Environmental Policy Act of 1969 (NEPA) in preparing environmental impact statements. Beginning in the late 1970s, sociologists found employment in most regional offices and a few forest supervisor's offices. Never more than a dozen sociologists worked in the Forest Service management arm at any one time, although a number of cultural anthropologists, economists, and public affairs specialists have undertaken community assessment studies. Their main functions were to lead public involvement efforts and write the social impact assessments as part of forest plan environmental impact statements (EIS).

Often agency social scientists researched and wrote socioeconomic overviews (SEOs) of particular national forests and surrounding lands, while at other times these SEOs were contracted out to professional people or firms. From the 9 Forest Service regions and 156 national forests, around 35 overviews were completed as separate reports, with another 100 added to forest plan EISs as discussions in the "affected environment" sections or as appendices. In addition, related background reports were prepared for some national forests on special social topics, for example a study of minority group relationships to the Siuslaw and Willamette National Forests.[14]

By the 1990s, manuals for projects and plans on national forest system lands that require NEPA analysis provided direction for national and regional social impact analysis.[15] The first such guidance came from a 1975 contract with Richard P. Gale from the University of Oregon. Gale produced a massive notebook/workbook entitled *Social Assessments Reference Notebook* in 1981 (an earlier pamphlet version had been printed by the Forest Service in 1977). Similar books and publications have appeared to assist researchers and the public with understanding the various types of methodologies for evaluating the impacts of natural resource management. William Burch from Yale University was employed by the agency in the 1960s to assist in developing ways to incorporate the social sciences into everyday management. Recently, Forest Service sociologists worked with social scientists from other agencies and organizations to design a methodology for doing socioeconomic assessments.[16] Although not technically "research," these assessment reports, research methodology guidance, and background materials have provided much needed information for decision makers, agency social scientists, and the public.

Recently, sociological research in the Forest Service moved away from strict socioeconomic analysis and into the "human dimension" aspects of ecosystem management (see the discussion on social science research in the ecosystem management era below). Information about how different people and groups value the forests has important applications for the everyday work of natural resource trained employees. Sociological analysis, for some employees, has resulted in a paradigm shift in their thinking and acting.

The Seattle Forestry Sciences Laboratory, part of the Research Branch, staff and natural resources research unit, emphasizes the role that values play in natural resources management. Also, David Bengston and Zhi Xu from the North Central Forest Experiment Station reported a formal content analysis study of changing environmental values. Their findings support the idea that: (1) the public values are shifting and that many people are moving away from economic definitions and toward more aesthetic definitions of forest resources and management; (2) forestry professionals and the environmentalists differ greatly, which may account for the intensity of recent conflicts; and (3) the public holds a different value set from both the working foresters and the environmental organizations.[17]

A number of attitude and opinion survey reports have published information obtained from Forest Service employees. Two of the most recent were completed under Forest Service contracts: a University of Washington study and another survey by Kaset International. Other projects completed during the last ten years have tended to focus on the NFMA/NEPA process associated with forest planning efforts, as well as the effectiveness of public participation in the NFMA planning process. Each of these reports has resulted in the agency and the public rethinking the role of the Forest Service in the complex world in which it acts.[18]

The following two subsections on recreation and fire prevention are considered for this discussion as components of the sociology area. In reality, both areas of research developed without any academic disciplinary direction or theory, but because of the needs of national forest managers. Yet the sociology field, both in methods and researchers, has heavily influenced these research topics.

RECREATION RESEARCH

Forest Service research in the social sciences began with an interest in how and why people recreate in the national forests. Outdoor recreation grew during the 1910s because of better automobiles and "good roads." Informal observations of where people were already camping and problems that they were creating resulted in the agency making recreation plans to build campgrounds to meet their needs as well as to minimize ground disturbance and sanitation problems. Arthur Magill noted that these were the first "sociological" studies undertaken by the agency:

> Some of the recreation research was directed toward environmental problems rather than strict social science. Though it is all the result of human use. Nevertheless, a large portion [of recreation research] does involve [the] social sciences even though it may not have been done by social scientists, e.g. foresters, geographers, landscape architects. . . , engineers, and possibly some other disciplines I don't know about![19]

Recreation research and informal studies have greatly expanded over the twentieth century to include basic outdoor recreation, wilderness, urban/rural interface, and visual resource management, which has its foundation in environmental psychology.

Recreation management got a boost in the mid-1910s from an effort to create an agency to manage the national parks. Forest Service Chief Henry Graves steadfastly opposed what became the National Park Service in 1916, going so far as to propose that the Forest Service take over management of the national parks from the U.S. Army, which had managed the parks since the late 1800s. The Forest Service also began a program of encouraging summer home leases on the national forests to show people the wonders of the forest land, as well as gain support from the users of the forests.

During the 1920s, many studies addressed the need for recreational sites and structures to accommodate the rapidly increasing number of people visiting the national forests by auto. Newly hired recreational specialists at the regional offices carried out many of these studies. During the Great Depression of the 1930s, the Civilian Conservation Corps men improved and expanded the use of recreation on the national forests and parks. Many of these sites are still used today by millions of visitors.[20]

This situation existed until the early 1950s when the Forest Service funded Samuel T. Dana to prepare an analysis of outdoor recreation. He completed his report, "Problem Analysis: Research in Forest Recreation," in 1957. Harry Camp wrote that the paper "provided an excellent overview of problems in forest recreation, and° was quoted often during the development of the research program."[21]

Dana's report recommended creating within the agency a division of forest recreation research, soon thereafter established.[22] Harry W. Camp

Arthur Carhart, an early Forest Service landscape architect with an interest in recreation and wilderness, built this camp for a landscape project near Beulah Valley Colorado, on the San Isabel National Forest. Carhart stands at center.
(USDA FOREST SERVICE PHOTO, 1920)

became the first director in 1959. He believed that recreation as a research subject in its own right was being ignored, and that a lack of qualified researchers existed, as well as little support by national forest administrators and staff. Assistant Chief Edward Cliff in a 1958 speech to the Federation of Western Outdoor Clubs put this issue in perspective when he noted that:

Our most pressing need at present is for more information on how to properly manage, maintain, and improve areas subject to heavy mass recreation use. Basic to this broad problem is such surveys and studies as inventory of present and potential studies on all forest lands; study of the kind of recreational facilities people desire and are necessary associated facilities; carrying capacity and possible rotation and deferment of use of camp and picnic sites; how to prevent or alleviate deterioration of vegetation and soils; how to revegetate and restore recreation sites and protect them from attacks of insects, disease, and fire; and how best to coordinate recreation with other land use.[23]

Not until 1960 did recreation research become a standard line item in the federal budget for the Forest Service. Cliff participated in the 53rd Western Forestry Conference in Seattle in December 1962, where he gave a speech entitled "Forestry and Forestry Research in the U.S. Department of Agriculture," in which he noted that the agency was only beginning to explore this new field:

a rapid expansion of the relatively new and unexplored field of research . . . will provide a better basis upon which to handle the problems of policy and management of forest recreation . . . it is long overdue.[24]

Formal research in recreation had just begun in the 1960s. The national forest managers did not fully embraced it. As agency historian Terry West noted:

Recreation research was concerned first with gaining credibility through recreation natural resource studies. . . . A lack of support for . . . this type of research by field managers was part of a legacy of opposition by "old-line" managers to the influx of specialists experienced by the agency in the 1960s and 1970s. The employment of nontraditional specialists challenged the authority of the old-line managers, accordingly it took time for new programs to be accepted. Sociological studies, although advocated by Dr. George E. Jemison, who later became Chief of Research, were postponed until the recreation research program was better established and funded . . . In this period, forest recreation research was conducted regionally by the experiment stations.[25]

By the mid-1960s, opportunities for recreation research expanded, often through affiliations with land grant universities. Five cooperative forest recreation research units were established between 1962 and 1966. The following schools of forestry established these cooperative units (and program leaders): Syracuse University (Elwood L. Shafer), Michigan State University (Hugh A. Davis), Utah State University (J. Alan Wagar), North Carolina State University (Stephen J. Maddock), and the University of Washington (J. Alan Wagar). Camp wrote that these units were designed to "strengthen curricula and stimulate interest in forest recreation, conduct recreation research, and advise forest recreation graduate students."[26] A number of these graduate students eventually came to work for the agency.

Today, at least twenty cooperative agreements exist between the Forest Service and various universities that are involved in social science research. Magill noted that the list of cooperators included Robert Lee (University of California and University of Washington), Perry Brown (Oregon State University), John Heywood (Ohio State University), Dave Simcox, Ron Hodgson and Steve Dennis (California State University–Chico), Steve Hollenhorst (West Virginia University), Jim Gramann (Texas A&M), and Robert Sommer (University of California–Davis).[27]

Beginning in the 1970s, several research and experiment stations hired social science researchers. The new researchers who joined with the Forest Service research arm greatly improved the quality of recreation research and enabled the growth of this type of research into different areas. Hundreds of new research studies ensued, with scores of articles, reports, and several books as a result. Notably during this decade were the new research areas of wilderness use, vandalism, and visual resource management.

Wilderness recreation research came into being during the late 1960s and early 1970s. Robert Lucas led many of the early wilderness studies, as explained by M. B. Dickerman:

> Bob was pretty much a one-man show with a few temporary helpers. He had helpers doing the interviewing, but he was the scientist in charge and designed his studies. He was an exceptional individual, he had very good judgment about how far you pursue [a] problem. . . . He had good judgment as to what areas to explore in order to solve the questions he was looking at.[28]

One agency-produced book had a great deal of influence on Forest Service recreation managment—*Wilderness Management,* edited by John Hendee, George Stankey, and Robert Lucas. The tome was also used as a textbook at several universities. Both wilderness managers in the Forest Service and students, many of which would eventually work for the agency, received their first knowledge about wilderness as a resource through reading this book.

Vandalism at outdoor recreation sites, as well as on other federal lands, has been noted as a problem from the very start of the Forest Service. Over the decades, many recreation specialists in the agency have studied and reported on vandalism. Land managers expressed great concern but had little information about any patterns of behavior or ways to prevent vandalism. Little in the way of formal research activities began until the 1960s and 1970s, when vandalism became a major focus of recreation research activity that culminated in two national conferences in 1976 and 1988.[29] Art Magill concluded after the second conference:

> Sam Alfano, Roger Clark, and myself independently concluded that very little new information had evolved during the 12 years between conferences . . . So far as I know, nothing new (no new FS research) has be[en] undertaken since the 1988 conference with the exception of an evaluative study by Conner & Hartig in coop. [cooperative] agreement with the PSW/PNW [Research] Stations. Research on prevention of vandalism is needed.[30]

A growing concern exists about vandalism of American Indian sites as well as historic Forest Service buildings. One recent national conference focused, in part, on this problem (see the Anthropological/Archaeological Research section below).

Visual resource management took high priority during the 1970s at the Berkeley unit of the Pacific Southwest Forest and Range Experiment Station. The concept and practice of scenic management had started as early as 1908 in the Forest Service when a 100-foot-wide strip was left along public highways during timber harvesting operations in California. The Forest Service hired its first professional to specialize in landscape management in 1916, followed by the first full-time landscape architect, Arthur Carhart, in 1919. Carhart, working in the Rocky Mountain Region, developed the first recreation plans that included scenic values as goals. After he resigned in 1922, little was completed in this area until the New Deal era of the 1930s when the massive public works programs were instituted in the Forest Service and other land management agencies. World War II saw the program essentially die, only to be revived in the mid-1950s under "Operation Outdoors".[31]

With the advent of large-scale timber harvesting during the 1950s and 1960s, the Forest Service began to see the need for scenic management around timber harvest areas. The Forest Service hired part-time researcher R. Burton Litton in 1964 to study landscape management. By the mid- to late-1960s the clearcutting dilemma was becoming apparent to the agency, as the public outcry over this method mounted. Edward H. Stone II was selected as chief landscape architect in the Washington office in 1965 to help address the clearcutting problem. In 1968, the USDA printed Litton's booklet entitled *Forest Landscape Description and Inventories.*

The following year a workshop held in St. Louis, Missouri, brought together a number of prominent researchers and managers to discuss scenic management and clearcutting. "This workshop could rightfully be called the birthplace of the Forest Service's official landscape management program."[32] This workshop led to the development of a series of national forest landscape management handbooks, which in 1974 culminated in the publication of *National Forest Landscape Management. Volume 2-Chapter 1. Visual Management System.* This handbook, adopted nationwide, has been used as official land management direction for more than twenty years.

Another landmark conference on visual management took place in April 1979 at Incline Village (Lake Tahoe), Nevada. This three-day conference resulted in more than one hundred papers presented and published in the conference proceedings.[33] The thrusts of the meeting included 1) current and future challenges to the visual resource, 2) available technologies to solve landscape problems, and 3) appropriate combinations of technology to solve problems. Magill summarized the impact of the conference and 752-page proceedings:

> The proceedings . . . had the unfortunate affect of making some FS administrators (including research administrators) believe *all* we needed to know about visual resources had been accumulated & included in the book. This couldn't be farther from the truth. In fact, there still exists considerable need for research in environmental psychology & vis. [visual] resource mgmt [management].[34]

Research studies in the area of visual or scenic resources have covered many and varied topics such as "seen areas," psychological and physiological responses to scenic views, attitudes, economics, esthetics of fire and timber harvesting, visual and scenic preferences, mapping scenic beauty, historic landscapes, and many others.[35] The visual landscape system, employed throughout the Forest Service at all levels of planning and management, has also been used extensively by the Park Service, Natural Resource Conservation Service, other federal and state agencies, and even by a number of timber companies in the management of their lands. By the mid-1990s the most current visual management system was referred to as the Scenery Management System, also referred to as "a sense of place."

Probably the most significant change in recreation research came in the mid-1980s. A new Forest Service research work unit was established at Riverside, California, to study changing recreational uses, especially by non-traditional users:

> Prior to 1986, National Forest Supervisors in southern California began noticing several changes in recreation patterns, for example, the numbers or people using the National Forests were increasing and the cultural/ethnic affiliation of the user groups was changing. The supervisors requested that a research unit be formed which could look into these and related issues. In 1986 the research unit came into existence. ... As defined in 1989, the mission of the unit is to develop effective visitor management strategies for high use wildland recreation areas with an emphasis on different cultural and user groups.[36]

The Riverside research unit has worked extensively with the wildland/urban interface to examine problems concerning how the urban culture interacts with the national forests in southern California. The research unit has actively studied the various user groups (especially the Latino/Hispanic culture) in their use of the forests for recreation. Other important studies and conferences have focused on fire and rural residents.[37]

Recreation research continues to have the highest visibility of all the social science research in the Forest Service. Recreation research also remains the most heavily funded and staffed of any of the social science research areas.

FIRE PREVENTION RESEARCH

The Forest Service has a long and rich tradition of research in fire management from the prevention end. From the very beginnings of the agency in

1905, managers expressed interest in fires and their causes, especially human caused. Bradshaw recounted that "the earliest fire prevention research I found was an economic analysis of fire prevention effectiveness published in 1927— *The Cape Cod Fire Prevention Experiment*' conducted by the FS and the State of Massachusetts."[38]

Fire research on the human causes of wildland fire got a boost in the Great Depression. In part this was due to many reports of unemployed people intentionally setting fires so that they could in turn be hired by the Forest Service or states to put out the fires. Some of the biggest names in sociology in the 1930s studied southern woods burning. At the end of the decade, the Forest Service entered into a formal contractual agreement with the American Association for the Advancement of Science for further social and behavioral research on people caused fires.[39]

Researchers conducted numerous sociological studies of fire prevention over the decades. "The attitudes and characteristics of these groups (such as arsonists, hunters, children, and rural residents), their knowledge of fire, and/or their reasons for setting fires have been documented. . . ." Land managers have used this information effectively to develop fire prevention programs aimed at groups who might be likely to start wildfires through arson, ignorance, or carelessness. Other studies have focused on how well the fire prevention messages have been transmitted and understood. Lessons from this research have been instrumental in designing effective fire prevention programs.[40]

Several sociologists have played important roles in this research area. Linda Donoghue noted that they produced a large body of information that was highly regarded and well used by managers:

> The researchers, including Shea, Kerr, Folkman, Doolittle, Bertrand, Bernardi, and others, were housed in two RWU's [Research Work Units] dedicated to fire prevention research. By examining the sociological and communication perspectives of the wildfire problem, they developed a fountain of knowledge critical to designing effective fire prevention programs.[41]

According to Bill Bradshaw, "Larry Doolittle at Starkville, MS was on the concluding end of sociology research for woods burning; Folkman received a USDA career service award for his pioneering behavioral research on fires caused by children playing with matches."[42]

POLITICAL SCIENCE RESEARCH

The role of political science cannot be clearly separated from the other social science disciplines. Part of the reason for this is that so many of the social science influences on the Forest Service have been debated in the public arena and codified into law and regulations.

Political science as an applied discipline has been involved with the Forest Service since its earliest days, but formal research into the political or public policy area has been lacking by the agency (although academics have produced scores of books and articles on the subject). Some suggest an increasing need for the agency to be proactive rather than reactive to the needs and desires of Congress and outside organizations. According to Paul Culhane, over the years, the timber industry and recently other groups have quite effectively used direct and indirect pressures on the president, Congress, and the Forest Service to get the agency to take actions that would benefit the purposes of the interest groups.[43]

After World War II, the timber industry exerted increasing pressure to enlarge the timber harvest to meet the rapidly expanding demand for new housing. Many of the private timberlands had been cutover, and the time was ripe for opening the national forests for timber production in accord with the agency's long term goals. Although the timber products industry had been basically friendly toward the Forest Service for many years, the 1950s saw an increase in political pressure applied to the agency through Congress and industry groups to increase harvest levels. At the same time, other interest groups attempted to de-emphasize timber production and equalize the roles of recreation, grazing, wildlife, fisheries, and other resources—concerns that still affect the agency today.

The most authoritative book on the political side of the agency during the earliest years was the autobiographic *Breaking New Ground* by Gifford Pinchot. This book, published a year after his death, documented Pinchot's efforts to establish forestry and forest management between 1898 and 1910. Another classic book, *Conservation and the Gospel of Efficiency: The Progressive Conservation Movement 1890–1920* by Samuel P. Hays, covered the early years of the agency and the political climate and players around the turn of the century. A more modern political or organizational analysis, *The Forest Ranger* by Herbert Kaufman, looked at several Forest Service field offices around the country in the mid-1950s to see what influenced the employees and what they were doing in communities. Since that time, several other studies have appeared as follow-ups on the Kaufman work. Harold Steen published a work on the congressional origins of the Forest Reserve Act of 1891 and the Forest Management Act of 1897. Another type of political research study recently dealt with how political cartoonists have viewed natural resource management for one hundred years.[44]

Beginning as early as 1957, a special unit, now known as policy analysis, in the Washington Office began to study the agency itself. This staff undertakes studies of emerging as well as old policy issues facing the agency. Published reports totaled 258 during 1969 through 1992, with most emphasis on timber, general management, and recreation. Much of the most recent political science writing has focused on management and policy issues relating to Forest Service long-range planning. In 1990, a special Forest Service policy analysis team studied and critiqued the recent NFMA process for national forest plans. This comprehensive study was the culmination of interviews and letters from more than thirty-five hundred people from across the country who were involved with the decade-long planning effort.[45] The recent Forest Ecosystem Management Team (FEMAT) report contained several sections discussing the sociopolitical climate for ecosystem management (see the discussion on social science research in the ecosystem management era below).

ARCHAEOLOGICAL (HERITAGE/CULTURAL RESOURCES) RESEARCH

The Forest Service began basic surveys and partial excavations on prehistoric sites during the 1970s. Although technically not research, thousands of sites and millions of acres of national forest land have been looked over (surveyed), prodded, "shovel tested," and excavated. Working cooperatively with the various universities and state historic preservation officers, thousands of reports have been generated and filed with the states as part of compliance with the National Historic Preservation Act of 1966 and often the National Environmental Policy Act of 1969 for documentation of ground disturbing projects. Forest Service employees (archaeological technicians) carry out almost all of the compliance surveys, while most of the excavation studies (research) are conducted by Forest Service archaeologists or through contracts with universities or professional contracting firms.

Cooperative efforts to document various archaeological and historical sites have been carried out with several national and local archaeological groups, as well as with members of the public through the highly successful Passports in Time (PIT) program. By the 1990s, about one hundred twenty of these projects took place every year. The PIT projects, which are scattered across the country, are designed to excavate or document hundreds of sites which will, in the long run, benefit larger scale analyses of the human use of the land. Members of the public are encouraged to assist Forest Service heritage specialists (usually archaeologists and historians) on selected projects. Although the process of gathering and documenting each site becomes slower with the use of non-professionals, the agency has received positive press coverage of the PIT projects, increased citizen involvement with their national forests, and shown a positive link between the past and present.

After completion of "test pits," "shovel tests," formal excavations, and projects, however, the reports of these investigations sit mainly in file cabinets in ranger district and state offices. Any artifacts found are stored in Forest Service offices or various university museums. Little effort has been made to review these heritage studies, compile information in larger zones or areas, or to seek a broad consensus on the meaning or interpretation of the cultural resources (heritage program) materials found on national forest land.

Yet one formal research program exists attempting to undertake this level of understanding. In August of 1986, a general management review of the Southwestern Region and the Rocky Mountain Forest and Range Experiment Station identified that, in spite of a wealth of prehistoric Indian sites, no formal research effort was being directed at the issue. The report stated, "Region 3 has a wealth of cultural resources of national significance with concentrations as high as 40 or 50 per square mile . . . [and] there is no Forest Service research program related to cultural resources." One alternative proposed in the review called for the "Region and Station [to] develop an analysis of the need for a cultural resources research program for the Southwest." By January of 1987, an action plan was approved by the regional forester to cosponsor a symposium on cultural resource management and research needs, as well as to analyze and prioritize these identified needs. The following spring, fifty-seven participants attended the Grand Canyon Symposium, with Joe Tainter as the symposium organizer. Proceedings of the symposium (*Tools to Manage the Past: Research Priorities for Cultural Resources Management in the Southwest*, GTR-RM-164) were published in October of the same year.

In 1989, Regional Forester Dave Jolly provided funding for two administrative studies in support of cultural resources research: a contract with the Museum of New Mexico for a chronological study of the Jemez Mountains and another with a private firm to study pothunting (illegal digging and collecting Indian artifacts) at Perry Mesa on the Tonto National Forest.[46]

A new cultural resources work unit was established in 1991 at Albuquerque, New Mexico. George Peterson was the acting project leader from 1991 to 1993, followed by Tainter, who became the permanent project leader.[47] The Rocky Mountain Experiment Station created a scientific advisory group to help guide the research effort. Tainter wrote that funding was provided for several studies, as well as to cosponsor a symposium at the Santa Fe Institute:

> Although this was planned initially to be a program focusing on southwestern forests, it has grown to have a national charter. This means that research can be done anywhere in the U.S. The funding that is available does not, however, allow that potential to be fully realized at the moment . . . While it may take some time for the program to realize its full national charter, much of the research has been and will be broadly applicable.[48]

Albuquerque's unique cultural heritage research work unit focuses on "sustainable development, global change in history and prehistory, cultural conflicts in land use, and heritage resource management throughout the United States."[49] In doing so, the unit encompasses archaeologists, anthropologists, and historians in an interdisciplinary effort to study and understand the characteristics of sustainable societies (how societies adapt to change), the cultural dimensions of ecosystem management (occurrence and causes of cultural conflict over land management), and the management and enhancement of heritage resources (focus on landscapes, land use, and the role of humans in ecosystems).

During the 1980s and 1990s, Forest Service archaeologists and historians completed scores of heritage studies, with numerous contracts made with private contractors (many in the academic community) to investigate, excavate, and report on various types of prehistoric and historic sites on the national forests. These reports, some of which are massive, tend to have little consistency and very little, if any, peer reviews. Research projects such as the Albuquerque unit will, it is to be hoped, bring a standard of consistency and review to this important effort at understanding the past.

Vandalism of archaeological and historical sites forms a growing area of concern. Vandalism is a problem in pre-contact American Indian sites (common in the Southwest where "pot hunters" hope to find old pottery), prehistoric rock art (both pictoglyphs and petroglyphs), and historic buildings (especially vulnerable are non-occupied buildings). Vandalism of these remote sites has similarities to problems in urban areas. The first International Symposium on Vandalism in North America took place on April 20-22, 1986, under the auspices of the Forest Service, the Institute for Environmental Studies at the University of Washington, and Vandalism Alert, Inc. This symposium was held to "encourage and stimulate the exchange of ideas, solutions to problems, and descriptions of research needs. . . ."[50]

Another research area of cultural anthropology has focused on American Indian uses of the forests. Studies of the subsistence gatherers/hunters are being carried out in the Alaska Region, originally by Robert Muth and then by Stewart Allen. Subsistence use areas in Alaska are being mapped using geographic information system technology to delineate where the historical and traditional hunting, fishing, and gathering areas are located. Continuing efforts to record American Indian uses of cultural sites on the national forests have led to varying success. A study on the Mt. Baker-Snoqualmie National Forest successfully located many traditional sites associated with vision quests, medicine plant gathering sites, vegetable food gathering, salmon fishing, and big game hunting areas. Because of the confidential nature of the cultural information, site data are not freely disclosed to anyone outside the tribe or the state historic preservation office where the reports are filed. Thus site specific information is withheld so that pothunters, or even well meaning individuals, will not vandalize sacred sites.

HISTORICAL RESEARCH

Hundreds of books, articles, papers, manuscripts, and oral histories have been produced about the Forest Service at all levels in the organization. This type of history, often called administrative or organizational history, has been a part of the agency since at least 1911 when former Chief Pinchot requested written letters from employees that he worked with before he was fired by President Taft in 1910. A second round of requests from Pinchot came in the late 1930s when he was conducting research for *Breaking New Ground*. These letters can be found in the Pinchot Collection at the Library of Congress in Washington, D.C.[51]

Former Forest Service employee Arthur Carhart, one of the first advocates of wilderness in the agency in the early 1920s, worked out an agreement in 1960 with Chief Richard McArdle to send copies of Carhart's speeches and other documents to the Conservation Library, which he had started, at the Denver Public Library. Following this agreement, a formal history program for the agency began when Robert D. King received the duty of compiling a history file in the Washington office division of information and education. An ad-hoc history committee formed about this same time. Efforts lagged until 1968 when Cliff sent a memorandum to the field offices to "compile, interpret, and preserve the essential elements of Forest Service history." Two years later, a memorandum dated August 24, 1970, established a Forest Service "history activity." The following year, the agency signed a cooperative agreement with the Forest History Society (now located in Durham, North Carolina) to do six taped oral histories with retired Forest Service officials. Shortly thereafter, another cooperative agreement with Harold (Pete) K. Steen, then assistant director of the Forest History Society, led to the publication of his book *The U.S. Forest Service: A History*.

Over the last quarter of the twentieth century, the history unit in Washington, D.C. produced or oversaw the production of a litany of agency histories. Oral histories, which have been transcribed and published, have been made with many former chiefs of the agency. Numerous oral histories have been conducted by the Forest History Society with staff and directors in the Washington office. Hundreds more have been undertaken in various regional offices, supervisor's offices, ranger districts, and research stations over the years. In addition, the Washington Office has let contracts with the academic community to conduct research and write about specific historical topics, while many other studies have been made by agency employees and retirees. Histories have been completed for several of the Forest Service research stations and regions, including Region 1 (Northern Region), Region 3 (Southwestern Region), and Region 4 (Intermountain Region). A Region 10 (Alaska) book was contracted through the regional office and was being revised and updated in 1996.[52]

Numerous studies, reports, and articles have been published about the Forest Service and related land management agencies. The centennial of the national forests in 1991 resulted in a great number of celebrations, oral histories, reports, and publications, as well as a report on the proceedings of the centennial symposium held in Missoula, Montana, that summer. Historical research has a great potential to show "how things were," as well as change over time, in the agency at any given time period. Sandra Forney, regional archeologist in Region 9, noted that the ability to show that people, agencies, and ecosystems are in constant change will greatly benefit decision makers and the public as the agency moves into ecosystem management.[53]

SOCIAL SCIENCE RESEARCH IN THE
ECOSYSTEM MANAGEMENT ERA

Social science research may finally be "rising to the top" with the advent of ecosystem management in the Forest Service in 1992. Considerable discussion by Chief Jack Ward Thomas and others has taken place about what the physical, biological, and social components of ecosystems are and how they operate. Very few people, if any, can fully understand the complexity of ecosystems, much less manage them. Yet as social science research in ecosystem management grows, it is obvious that understanding the complexity of the social/human aspects of ecosystems is just as difficult as the physical and biological components.

The social or human dimension of ecosystem management has only recently captured the imagination of a number of research scientists, as well as land managers. The literature is rapidly growing in the area, but most appears in the form of thought or discussion papers on how to integrate social science into the analysis/assessment and decision making process. Linda Donoghue has recently compiled a sixty-four page list of Forest Service human dimension (including economics) research projects from across the country. Some of these studies are ongoing, while others are new, taking advantage of new priorities in research and national forest administration in the ecosystem management era.[54]

With the advent of the Forest Conference ("timber summit") held in Portland, Oregon, on April 2, 1993, the amount of research and writing on ecosystem management has increased tremendously. The resultant massive 1993 report of the Forest Ecosystem Management Assessment Team (FEMAT) included a chapter on social assessment (chapter VII) and another on economic evaluation (chapter VI). This social chapter discussed the past and present political "climate," documented previous research and conducted new research work on hundreds of communities in western Washington and Oregon, as well as northwestern California. The FEMAT assessment, which was started and finished within three months of the forest conference, has been used to illustrate one process to incorporate social science research into decision making processes for the Forest Service and the Bureau of Land Management.[55]

Criticisms of the FEMAT process and results of its studies have led to revisions of the process for the new "Eastside" (Washington, Oregon, Idaho, and western Montana) Interior Columbia Basin Ecosystem Management Project (ICBEMP). One result has been a series of new background studies and assessment reports for the ICBEMP dealing with many aspects of the social science arena. A large-scale assessment of the southern Appalachian area was completed in 1996. It included a report which assessed the past and present social, cultural, and economic conditions of the southeastern states. A similar project was also completed during 1996 for Region 5 (Pacific Southwest) which concerns ecosystem management of the Sierra Nevada, while a slightly earlier study was conducted on the spotted owl in California. Thomas Keter has conducted extensive archaeological and historical research on the North Fork Eel River in northern California that incorporates many of these data on present day ecosystems.[56]

A human dimension team in Region 5 has recently developed a draft guidebook for ecosystem management. Judy Rose noted that "the guidebook . . . is taking an integrated look at ecosystems. . . . Our human dimension input to the framework has been generally organized around cultural (or human) ecology, but also with input from other social sciences."[57] Linda Lux reported that Region 5 is also sponsoring the development of an annotated bibliography which "will cover cultural ecology, cultural changes, paleoenvironmental reconstruction, historical research, social and cultural anthropology and ethnographic studies, and economic and social research and analysis."[58]

One area of the 1994 Northwest Forest Plan (resulting from the FEMAT report) for the spotted owl region involves "adaptive management areas." These ten areas were designated and designed to engage the public in setting management priorities and policies. These areas serve as a way to involve agency employees, research scientists, and local people, who have the greatest knowledge and experience with the area and resources for the benefit of local communities. The Pacific Northwest Forest and Range Research Station, through the Seattle Forestry Sciences Laboratory research group led by Roger Clark, provides assistance to teams of public and private members in their attempts at coming to equitable solutions to complex problems. Social "experiments" will be encouraged to vary the intensity of management and output of resources. Yet, from a sociopolitical perspective, the adaptive management areas may face the same political consequences as did the Sustained-Yield Management Act of 1944, which designated cooperative and federal sustained yield units.

Another promising area of research work involves the use of census data combined with a geographic information system. The blending of the two is available through TIGER files and census data and made available in all Forest Service regions. The Rocky Mountain Regional Office spearheads several of these efforts. The current technology can display 1980 and 1990 census data overlaid with geographic information to give a visual picture of information collected by the decadal census. Having this visual, as well as tabular, information available for Forest Service planning team members, decision makers, and the public will, it is hoped, lead to better decisions about the impacts of people on ecosystems and an estimate of the impacts of changing policies on people near the national forests.[59]

COMMUNITY RESEARCH

The wildland/urban interface, where people live in areas adjacent to national forests and rangelands, is an area where there are growing controversies over land use, water rights, ecological needs, recreational conflicts and a place where fire and flood are periodic and potentially very serious. This research area is of growing importance especially for land managers in the national forests that are near large population centers, where a large number of people recreate during the summer and winter months, or have private homes adjacent to these national lands.

A new area of social science research and funding involves urban forestry, which deals with the urban and suburban forests and parks, to make people aware of environmental issues and ways to solve them through involvement, cooperation, and partnerships. Rural development provides another arena of recent emphasis from the 1990 Farm Bill, which increased funding and assistance to rural communities. The Forest Service has taken a lead in many rural development programs in states with national forests, especially in the West.[60]

A rural development research program created in the Pacific Northwest Research Station in 1991 is anchored in the *Forest Service Strategic Plan for the 90's: Working Together for Rural America* and *Enhancing Rural America: National Research Program*. The program, according to Kent Connaughton, includes agendas and working relationships with public and private individuals, groups, and agencies who are involved in rural and community development and who see natural resource conservation as an instrument of economic and social change. Wendy McGinnis noted that members of the research team are working with the Forest Service, states, and other sources to compile relevant socioeconomic data for the Pacific Northwest. Having these data available at one location, rather than at scores of locations and agencies, should prove beneficial to social scientists as well as project planners. With the publication of the FEMAT and subsequent environmental impact statements, this rural development research program will evaluate the impacts and effects of implementation of both the Northwest Forest Plan and the efforts at community assistance provided for by Congress and the administration.[61]

A number of university studies, underwritten by the Pacific Northwest Station, have tried to understand the role of social change in resource-dependent communities. Their findings are shining considerable light on the role of internal and external factors that relate to community change. Jo Ellen Force, her associates, and Gary Machlis have studied the many types of relationships of local resource production, local historical events, and national economic and societal changes to community social change. They have concluded that social change in resource dependent communities (forestry, fishing, mining, and tourism) can be explained by knowing what national and regional societal changes are occurring and what local historical events have shaped the community. They also found that the production or non-production of natural resources are not very good explanatory variables of community stability or change.[62]

Tourism is often touted as the "savior" of resource dependent communities which are being "ravaged" by federal timber and fish policies. Studies do not exist, however, to prove such contentions. Magill noted that the Forest Service, other federal agencies, universities, states, and counties have produced a considerable number of basic recreation and tourist/tourism reports.[63] Many of the national studies include data collected for decades on the same questions so that long-term changes in recreational activities can be seen.

CURRENT SOCIAL SCIENCE RESEARCH ISSUES

A number of challenges can interfere with the development of successful social science research programs. At the national level, despite support from former Chief Jack Ward Thomas, effective leadership and coordination for social science research has not developed. Great ambiguity exists concerning what social science research is and what it can contribute to basic theory, policy-development, and decision making. Other barriers include the persistent belief that the social sciences are less important and less rigorous than the biological and physical sciences:

> Remarkably, systematic analysis and evaluation . . . appear not to be an important part of the forest research community's program. In-

stead, the research community continues to focus almost all of its resources on the management of trees; although we have yet to hear of a case where a tree has sued the forest [service]. There must be a shift to viewing the fundamental questions of the Forest Service's social context as necessary and legitimate objects of research and experimentation.[64]

When considering questions that have a social science application, a tendency exists to equate social science/human dimension research with economic viewpoints and methodologies. Consistently low funding levels for social science (non-economic) research reinforces that image. The broad range of other social sciences such as anthropology, history, social psychology, and political science are often not considered as useful in generating valuable information.[65]

National Park Service social scientists Darryll Johnson and Donald Field described similar problems during the early 1980s, yet their observations still held true in the late 1990s:

Despite interest in applied social science . . . there remain challenges, particularly in many government agencies. Among these are the following: (1) utilization of social science input is not an institutional function among agencies where it is found; (2) there are few practicing professionals who disseminate social information to managers or layman; (3) few government agencies have central locations for storage and access of social science data; (4) in many government agencies (notably land management agencies), there are few outside pressure groups to monitor agency action with regard to human issues; and (5) there remains a belief by some administrators that social science is nothing more than intuition or common sense.[66]

Many aspects of everyday Forest Service management activities deserve attention, and they all have strong social science components. Bormann suggested that concepts of the environment such as sustainability, forest health, biodiversity, and ecosystem management are essentially human constructs that serve as expressions of human values. Roger Clark and George Stankey suggested that agencies and institutions are poorly equipped to develop a thorough understanding of social issues. The National Research Council recognized that "future generations of foresters and forestry educators will need to better integrate knowledge of behavioral science and social-cultural systems into biological conceptions of forests." Machlis observed that biologists, ecologists, and other natural science professionals are now faced with a hard reality: Ultimate solutions to natural resource problems lie in social, cultural, economic, and political systems; the very systems that are the focus of the social sciences.[67] Holden has argued that:

> The social sciences have lagged far behind in assessing the interactions between physical changes and human activities. Far more is known about the processes of global warming, deforestation, resource depletion, and pollution than about the processes of the human institutions that create these effects.[68]

Terry West noted in his paper on the history of Forest Service research that:

> There is still much improvement needed in the area of employment of skilled social researchers in . . . the agency . . . I find that too often social research is assigned to experiment station staff as an additional duty or contracted out. The results are often of limited value. Social research is increasingly critical given the renewed agenda in the USDA on rural revitalization. Also, the social and economic importance of forest recreation in timber-dependent communities of the northwest is especially important given the reduced timber harvests expected there in the future. It is especially critical that agency social analysis go beyond past narrow case studies of recreation potential in timber dependent communities, we need to study the transaction costs in the transitional phase from logging to a diversified economy.[69]

What is needed is a reexamination of the role that social science research should assume in current and future natural resource management issues. How successful the natural resource community is at integrating the social sciences through a research program will determine, in large part, how effective any emerging long term solutions to these various resource issues will be.

CONCLUSION

National forest management is the practice of many applied social science disciplines and research. As stated in 1935 by Pacific Northwest Regional Forester C. J. Buck: "Forestry, therefore, is leading us into sociology [and other social sciences]. . . . Forestry is taking on a larger aspect heretofore and one which demands the best of each of us."[70]

Social science research in the Forest Service has a short but rich history. Yet much of the historical development of this type of research (and others, too) is being lost through carelessness, lack of concern, or ignorance. As recounted by Bradshaw: "I had collected an archive of all this stuff [from the 1920s and later] but, it was later pitched at the lab (actually, one week before I asked them to send it to the WO [Washington Office]). . . . five decades of work [gone!]."[71] For many researchers, it is not the process that is most important, but rather the product. How, when, and why are forgotten in favor of the printed words of a scientific publication. These books and articles are not history, but they do document the milestones of the research process.

Recreation and fire prevention research have led the way for much of the Forest Service social science research community, but other disciplines are showing much promise in the near future. It is to be hoped that the ecosystem management era will bring together many of the social science disciplines with the physical and biological disciplines in the agency.

NOTES

1 This brief history of social science research was prepared under the direction of the Research Branch of the Forest Service through the Forest History Society in Durham, North Carolina. Many thanks for comments, additions, and corrections on various drafts of this paper go to Terry West, Arnold Holden, Susan Johnson, Art Magill, Joseph Tainter, David Gillio, Geneen Granger, Chris Christensen, Linda Lux, Judy A. Rose, Linda Donoghue, Bill Bradshaw, Ellen Credle, and Joyce Casey. They are, however, not responsible for the contents.

2 Alan W. Ewart, Personal communication on December 8, 1994; This definition was derived from formal and informal discussions at the conference of Forest Service human dimension researchers held in Washington, D.C., in the spring of 1993. At the present time, this definition does not have the "official blessing" of the agency, but it does summarize the important components of any good human dimension/social science research program in the Forest Service.

3 Thomas C. Marcin "Integrating Social Sciences into Forest Ecosystem Management Research: Complex Forest Management Issues Require Increased Cooperation Across Disciplines." *Journal of Forestry*, Vol. 93, #11 (Nov 1995): 29–33.

4 Marcin, "Integrating Social Sciences into Forest Ecosystem Management Research", ibid., p. 30.

5 Harry T. Gisborne, "Sociological Shackles on Forestry." *The Ames Forester*, 31 (1943): 27–28.

6 Henry Solon Graves, "Public Aspects of Forestry." *American Forestry*, 17, #9 (September 1911): 525–526.

7 Samuel P. Hays, *Conservation and the Gospel of Efficiency: The Progressive Conservation Movement, 1890–1920*. Harvard Historical Monographs No. 40. (Cambridge, MA: Harvard University Press., 1959).

8 Gisborne, "Sociological Shackles," pp. 23–24.

9 Jack Gibbs, *Sociological Theory Construction* (Hinsdale, IL: The Druden Press, Inc., 1972); Darryll Johnson and Donald R. Field, "Applied and Basic Social Research: A Difference in Social Context." *Leisure Sciences*, 4, #3 (1981): 269–279; Donald R. Field and Darryll R. Johnson, "The Interactive Process of Applied Research: A Partnership Between Scientists and Park and Resource Managers." *Journal of Park and Recreation Administration*, 1, #4 (Oct 1983): 18–27.

In the Forest Service a perceptual split exists between "pure" research versus so-called "administrative studies," which are carried out by the national forest system and not under the direction, usually, of recognized agency researchers from the research arm. Since most of these administrative studies are not peer-reviewed or published, Bill Robbins (1993) felt that many researchers do not consider these "gray literature" reports as "scientific." Robert Buckman, retired deputy chief of research, noted that:

> Administrative studies are simultaneously a source of strength and a source of considerable tension. There's an intellectual curiosity on the parts of people who work in the National Forest System and in State and Private Forestry that should be encouraged. Administrative studies were intended to deal with site specific issues with questions of scale and operational procedures. It was a part of an RD&A [Research Development & Administration] sequence. But it hadn't quite worked out that way. In most cases administrative studies have a terrible track record. Many of them are poorly planned, poorly executed, and almost never recorded for later reference.

Because of this bias these studies are rarely acknowledged or cited in other studies. This distinction is especially problematic when it comes to social science research in the agency, as there are thousands of reports that were conducted under commonly accepted scientific standards that have never been published. The discussion on social science research does not make those fine distinctions and includes work completed by the Forest Service either under contract or by in-house social scientists. William G. Robbins, 1993. "The United States Forest Service and the Problem of History." *Public Historian*, 15 (Summer 1993): 41–48; Harold K. Steen, (ed.). 1994. *View From the Top: Forest Service Research by R. Keith Arnold, M .B. Dickerman, Robert E. Buckman*. Oral histories. (Durham, NC: Forest History Society, 1994), p. 341; Brian Stephens and Michelle Dees, 1992. "Forest Service Social Science Network Directory." Dated October 1992. (Washington, D.C.: USDA Forest Service, Environmental Coordination, 1992).

10 Benton MacKaye, "Some Social Aspects of Forest Management." *Journal of Forestry*, 16, #2 (Feb 1918): 214.

11 Alison T. Otis; William D. Honey; Thomas C. Hogg; and Kimberly K. Lakin, *The Forest Service and the Civilian Conservation Corps: 1933–42*. FS-395. (Washington, D.C.: U.S.G.P.O., 1986); John A. Salmond, *The Civilian Conservation Corps, 1933–1942: A New Deal Case Study* (Durham, NC: Duke University Press, 1965); USDA Forest Service, North Pacific [Pacific Northwest] Region, "Economic and Sociological Analysis of the Seventy Five Counties in Oregon and Washington (North Pacific Region)." Regional office report. (Portland, OR: USDA Forest Service, North Pacific Region, Section of Public Relations, 1935); Gerald Williams, "History of Social Science in the Forest Service: Reflections on the 'Greatest Good of the Greatest Number in the Long Run'." Paper presented at the USDA Forest Service Social Science Coordination Workshop, Rosslyn, VA, 14 April 1987. (Portland, OR: USDA Forest Service, Pacific Northwest Region, 1987).

12 Gisborne, "Sociological Shackles," p. 27.

13 David A. Clary, *Timber and the Forest Service* (Lawrence, KS: University Press of Kansas, 1986); Dahl J. Kirkpatrick, "New Security of Forest Communities [the Sustained-Yield Management Act of 1944]" in *Trees: Yearbook of Agriculture* (Washington, D.C.: U.S.G.P.O., 1949), pp. 334–339; Gerald W. Williams, "Community Stability and the Forest Service." Paper presented at the Oregon Planning Institute Rural Lands Workshop held at the University of Oregon, on September 30, 1987. Last revised in January of 1995. (Portland, OR: USDA Forest Service, Pacific Northwest Region).

14 Professional Analysts, *Study of Minority Groups' Relationships to the Siuslaw and Willamette National Forests*. Report to the Forest Service. (Corvallis, OR: USDA Forest Service, Siuslaw National Forest, 1980).

15 Dale J. Blahna and Lambert N. Wenner, "Social Analysis Bibliography for Forest Service Programs," Supplementary Edition. (Washington, D.C.: USDA Forest Service, Environmental Coordination, 1986); Geraldine L. Bower and Earnest R. McDonald (eds.). "Proceedings National Workshop Social Impact Analysis: NEPA Challenge for the 90's, Albuquerque, New Mexico, March 16–17, 1993" (Washington, D.C.: USDA Forest Service, Environmental Coordination, 1993); Maurice E. Voland and Wendell H. Hester (eds.), "Social Impact Analysis in the Forest Service: a Workshop Summary." (Washington, D.C.: USDA Forest Service, Environmental Coordination, 1993); Maurice E. Voland and William A.Fleishman (eds.), "Sociology and Social Impact Analysis in Federal Natural Resource Management Agencies." (Washington, D.C.: USDA Forest Service, Environmental Coordination, 1983); Lambert N. Wenner, *Minerals, People and Dollars: Social, Economic and Technological Aspects of Mineral Resource Development*. RI-92-133. (Missoula, MT: USDA Forest Service, Northern Region, 1992); Lambert N. Wenner (ed.), "Issues in Social Impact Analysis: Proceedings from an Interagency Symposium, November 1984." (Washington, D.C.: USDA Forest Service, Environmental Coordination, 1984).

16 William R. Burch, Jr. and Donald R. DeLuca, *Measuring the Social Impact of Natural Resource Policies* (Albuquerque, NM: University of New Mexico Press, 1984); Interorganizational Committee on Guidelines and Principles, Guidelines and Principles for Social Impact Assessment. NOAA Technical Memorandum and NMFS-F/SPO-16, (Washington, D.C.: USDC: National Oceanic and Atmospheric Administration and the National Marine Fisheries Service, 1994).

17 George H. Stankey and Roger Clark. *Social Aspects of New Perspectives in Forestry: A Problem Analysis* (Milford, PA: Grey Towers Press, 1992); David N. Bengston and Zhi Xu, "Changing National Forest Values: A Content Analysis." Research Paper NC-323, (St. Paul, MN: USDA Forest Service, North Central Forest Experiment Station, 1995).

18 Brian J. Boyle, Margaret A. Shannon; Robert A. Rice; Kathleen Halvorsen; and H. Stuart Elway, *Policies and Mythologies of the US Forest Service: A Conversation with Employees.* Report prepared for the USDA Forest Service and the Pacific Northwest Research Station. (Seattle, WA: University of Washington, College of Forest Resources, Institute for Resources in Society, 1994); Bruce Hammond, "Forest Service Values Poll Questions: Results and Analysis." Report prepared for the Forest Service. (Tampa, Fl: Kaset International, 1994); M. Greg Holthoff and Robert E. Howell, "Findings of an Evaluation of Public Involvement Programs Associated with the Development of a Land and Resource Management Plan for the Ouachita National Forest," in *Proceedings of Interagency Symposium: Property Rights and Public Values in Managing Natural Resources, August 7, 1993, Orlando, Florida* (Washington, D.C.: USDA Forest Service, 1994) pp. 49-75.

19 Arthur Magill, Personal e-mail note dated March 4, 1994.

20 Alison T. Otis; William D. Honey; Thomas C. Hogg; and Kimberly K. Lakin, *The Forest Service and the Civilian Conservation Corps: 1933-42.* FS-395 (Washington, D.C.: U.S.G.P.O., 1986); Salmond, *The Civilian Conservation Corps, 1933-1942*; William C. Tweed, "Recreation Site Planning and Improvement in National Forests 1891–1942," FS-354. (Washington, D.C.: U.S.G.P.O., 1980).

21 Harry W. Camp, *An Historical Sketch of Recreation Research in the USDA Forest Service.* PSW-H-1, (San Francisco, CA: USDA Forest Service–Pacific Southwest Forest and Range Experiment Station. Camp, 1983), p. 4

22 Harold K. Steen, *The U.S. Forest Service: A History*, (Seattle, WA: University of Washington Press, 1976).

23 As quoted in Camp, *An Historical Sketch*, p. 6.

24 Terry L. West, "Research in the U.S.D.A. Forest Service: A Historian's View." Paper presented on May 18, 1990, at the Third Symposium on Social Science in Resource Management at College Station, TX., p. 8.

25 West, "Research in the U.S.D.A.," ibid., pp. 8–9.

26 Camp, *An Historical Sketch*, p. 9.

27 Magill, e-mail note dated March 4, 1994.

28 Steen, *View From the Top*, p. 131.

29 Sam S. Alfano and Arthur W. Magill (technical coordinators), *Vandalism and Outdoor Recreation: Symposium Proceedings.* General Technical Report PSW-17, (Berkeley, CA: USDA Forest Service, Pacific Southwest Forest and Range Experiment Station, 1976); Arthur Magill, Personal e-mail note dated November 16, 1994.

30 Magill, e-mail note dated November 16, 1994.

31 USDA Forest Service, (draft), *Landscape Aesthetics: A Handbook for Scenery Management.* Agriculture Handbook Number 701. (Washington, D.C.: USDA Forest Service, 1993).

32 USDA Forest Service, *Landscape Aesthetics,* p. 9.

33 Gary H. Elsner and Richard C. Smardon (technical coordinators), *Proceedings of Our National Landscape: A Conference on Applied Techniques for Analysis and Management of the Visual Resource,* General Technical Report PSW-35, (Berkeley, CA: USDA Forest Service, Pacific Southwest Forest and Range Experiment Station, 1979).

34 Magill, e-mail note dated November 16, 1994.

35 Robert G. Lee, *Assessing Public Concern for Visual Quality--Landscape Sensitivity Research and Administrative Studies* PSW-19 (Berkeley, CA: USDA Forest Service, Pacific Southwest Forest and Range Experiment Station, 1976); USDA Forest Service, *Landscape Aesthetics*; Richard S. Smardon; Michael Hunter; John Resue; Mary Zoeling; and Richard B. Standiford (ed.), *Our National Landscape: Annotated Bibliography and Expertise Index.* Special Publication 3279 (Berkeley, CA: Division of Agricultural Sciences, 1981).

36 USDA Forest Service, "Brief History and Mission Statement," *Recreation Research Update,* Issue #14 (October 1993): 1.

37 Alan W. Ewert, Deborah J. Chavez, and Arthur W. Magill (eds.), *Culture, Conflict, and Communication in the Wildland-Urban Interface* (Boulder, CO: Westview Press, 1993); Robert D. Gale and Hanna J. Cortner (eds), *People and Fire at the Wildland/Urban Interface [a Sourcebook]: From the Wildland/Urban Fire Interface Workshop for Social Scientists, Asheville, North Carolina, April 6–8, 1987* (Washington, D.C.: USDA Forest Service, 1987).

38 Albert J. Simard and Linda R. Donoghue, "Wildland Fire Prevention: Today, Intuition–Tomorrow, Management," in James B. Davis and Robert E. Martin (technical coordinators) *Proceedings of the Symposium on Wildland Fire 2000, April 27-30, 1987--South Lake Tahoe, California.* Berkeley, CA: USDA Forest Service, Pacific Southwest Forest and Range Experiment Station, 1987), pp. 187–198; Bill Bradshaw, e-mail note dated November 28, 1994).

39 Bradshaw, e-mail note dated November 28, 1994.

40 Simard and Donoghue, "Wildland Fire Prevention," p. 188.

41 Linda R. Donoghue, e-mail note dated November 25, 1994.

42 Bradshaw, e-mail note dated November 28, 1994.

43 Paul J. Culhane, *Public Lands Politics: Interest Group Influence on the Forest Service and the Bureau of Land Management.* (Baltimore, MD: The John Hopkins University Press for Resources for the Future, 1981).

44 Hays, *Conservation and the Gospel of Efficiency*; Herbert Kaufman, *The Forest Ranger: A Study in Administrative Behavior* (Baltimore, MD: The Johns Hopkins University Press, 1960); Harold K. Steen "The Beginning of the National Forest System." FS-488. (Washington, D.C.: USDA Forest Service, History Unit, 1991); Gerald W. Williams "100 Years of Political Cartoons: Political Cartoonists' View of Natural Resources Management 1897–1994." Paper presented at the 4th North American Symposium on Society and Resource Management held at Madison, WI, on May 17–20, 1992. Revised in 1994. (Portland, OR: USDA Forest Service, Pacific Northwest Region).

45 Alan Carter, "Forest Service Policy Analysis: An Evaluation." Senior Evaluation Service candidate development project report. (Washington, D.C.: USDA Forest Service, Policy Analysis Staff, 1993); Gary Larsen, Arnold Holden, Dave Kapaldo, John Leasure, Jerry Mason, Hal Salwasser, Susan Yonts-Shepard, and William E. Shands, "Critique of Land Management Planning." FS-453. Eleven volumes/pamphlets. (Washington, D.C.: USDA Forest Service, Policy Analysis Staff, 1990).

46 Richard V. N. Ahlstrom, Malcolm Adair, R. Thomas Euler, and Robert C. Euler. *Pothunting in Central Arizona: The Perry Mesa Archeological Site Vandalism Study*. Cultural Resources Management Report No. 13. (Albuquerque, NM: USDA Forest Service, Southwestern Region, 1992).

47 Judy Propper, "Background on Proposal for Cultural Resources Research Work Unit." (Albuquerque, NM: USDA Forest Service, Southwestern Region (R-3), 1993); Joseph A. Tainter, "Cultural Heritage Research Program." *The Forest Service Heritage Program Times*, 4 (Oct 1993): 1-4.

48 Tainter, "Cultural Heritage Research Program," p. 3.

49 USDA Forest Service, "RM-4853: Cultural Heritage Research [Unit]" (Washington, D.C.: USDA Forest Service, 1994).

50 Harriet H. Christensen, Darryll R. Johnson; and Martha H. Brookes (technical coordinators), *Vandalism: Research, Prevention, and Social Policy*. PNW-GTR-293 (Portland, OR: USDA Forest Service, Pacific Northwest Research Station, 1992), p. v.

51 Gerald W. Williams, "Gifford Pinchot Papers, Library of Congress - Madison Library, Washington, D.C." Revised on May 1, 1987. Roseburg and Eugene, OR: Umpqua and Willamette National Forests.

52 Terry L. West and Dana E. Supernowicz (compilers), "Forest Service Centennial History Bibliography, 1891–1991." (Washington, D.C.: USDA Forest Service, History Unit, 1993); Lawrence Rakestraw, *A History of the United States Forest Service in Alaska* (Anchorage, AK: Alaska Historical Society and the USDA Forest Service, Alaska Region, 1981).

53 Ronald J. Fahl, *North American Forest and Conservation History: A Bibliography* (Santa Barbara, CA: Forest History Society and ABC-Clio Press, 1977); Edward N. Munns, *A Selected Bibliography of North American Forestry*. Two volumes. USDA Miscellaneous Publication No. 364 (Washington, D.C.: GPO, 1940); Gerald R. Ogden, *The United States Forest Service: A Historical Bibliography, 1876–1972* (Davis,

CA: University of California, Agricultural History Center, 1976); Judith A. Steen, *A Guide to Unpublished Sources for a History of the United States Forest Service* (Santa Cruz, CA: Forest History Society, Inc., 1973); West and Supernowicz "Forest Service Centennial History Bibliography;" Gerald W. Williams, *Selected References Concerning the usdaForest Service: Social, Political, and Historical Sources of Information.* (Portland, OR: USDA Forest Service, Pacific Northwest Region. Third edition, 1995); USDA Forest Service, *Remembering the [1991] Centennial [of the National Forests],* FS-535 (Washington, D.C.: USDA Forest Service 1993); Terry West, "Centennial Mini-Histories of the Forest Service." FS-518. (Washington, D.C.: USDA Forest Service, History Unit, 1992); Harold K. Steen, editor, *The Origins of the National Forests* (Durham: Forest History Society, 1992); Sandra Jo Forney, "Heritage Resources: Tools for Ecosystems Management." in Geraldine L. Bower (ed.) *Proceedings of Interagency Symposium: Property Rights and Public Values in Managing Natural Resources, August 7, 1993, Orlando, Florida.* (Washington, D.C.: USDA Forest Service, 1994), pp. 137–141.

54 David N. Bengston, "Changing Forest Values and Ecosystem Management." *Society and Natural Resources* 7 (1994): 515–533; Fred Cubbage, "Integrating People's Values Into Management Decisions Conflict of Problem Statement Suggestions," (Research Triangle Park, NC: USDA Forest Service Southeastern Research Station, 1993); B. L. Driver, "A Planning and Analysis Framework for Integrating Social and Biophysical Data Into Sustainable Ecosystems Management Recommended by R-3 Human Dimensions Study Group" Manuscript dated October 3, 1993. Ft. Collins, CO: USDA Forest Service, Rocky Mountain Forest and Range Experiment Station; Jo Ellen Force, Gary E. Machlis, L. Zhang, and A. Kearney, "The Relationship Between Timber Production, Local Historical Events, and Community Social Change: A Quantitative Case Study" *Forest Science,* 39 (1993): 722–742; Gary Machlis, "The Contribution of Sociology to Biodiversity Research and Management" *Biological Conservation,* 62 (1992): 161–170; Gerald W. Williams, "Ecosystem Management: Putting the People Back In." Paper presented April 7, 1993, at the USDI Bureau of Land Management's "Arid Land Management: An Ecosystem Approach Workshop" in Hood River, OR. Revised in 1994. (Portland, OR: USDA Forest Service, Pacific Northwest Region); Gerald W. Williams, "Ecosystem Management: How Did We Get Here?" Paper presented February 22, 1993, at the Society of American Foresters (Indiana) winter meeting in Indianapolis, IN. (Portland, OR: USDA Forest Service, Pacific Northwest Region); Gerald W. Williams, "Ecosystem Management and People: How to Incorporate the Human Dimension." (Portland, OR: USDA Forest Service, Pacific Northwest Region, 1994); Linda R. Donoghue, "Socioeconomic Research in the USDA Forest Service: A Status Report." Paper presented at the Fifth International Symposium on Society and Resource Management held in Ft. Collins, CO, June 7–10, 1994; Linda R. Donoghue, *Directory of Social Science Research in the USDA Forest Service.* Report dated June 1995. (Washington, D.C.: USDA Forest Service, 1995).

55 Roger Clark and George H. Stankey, "FEMAT's Social Assessment: Framework, Key Concepts and Lessons Learned" *Journal of Forestry* 92 (1994): 32–35; Richard P. Gale, "Not Scientifically Sound [FEMAT's Social Assessment]," *Journal of Forestry* 92 (1994): 33; Gerald W. Williams, "FEMAT to SEIS: Transforming the Social Assessment." Paper presented at the Integrating Social Science in Ecosystem Management: A National Challenge Conference held December 12–14, 1994 in Helen, GA. To be printed in the proceedings volume.

56 William G. Robbins and Donald W. Wolf, "Landscape and the Intermontane Northwest: An Environmental History" PNW-GTR-319 (Portland, OR: USDA Forest Service, Pacific Northwest Research Station, 1994); Gary Machlis, Jo Ellen Force, and Jean E. McKendry, *An Atlas of Social Indicators for the Upper Columbia River Basin [Idaho and Western Montana]*. Contribution Number 759 (Moscow, ID: University of Idaho, Idaho Forest, Wildlife, and Range Experiment Station, 1995); Kevin S. McKelvey and James D. Johnson, "Historical Perspectives on Forests of the Sierra Nevada and the Transverse Ranges of Southern California: Forest Conditions at the Turn of the Century," Chapter 11 in Jared Verner, Kevin S. McKelvey, and Barry R. Noon (technical coordinators) *The California Spotted Owl: A Technical Assessment of Its Current Status* (Albany, CA: USDA Forest Service, Pacific Southwest Research Station, 1992); Thomas S. Keter, "An Interdisciplinary Approach to Historical Environmental Modeling," Paper presented at the Society for American Archaeology meeting in St. Louis, MO, on April 18, 1993. (Eureka, CA: USDA Forest Service, Six Rivers National Forest).

57 Judy A. Rose, Personal e-mail note dated November 10, 1994.

58 Linda M. Lux, "Ecosystems Projects in Region 5 Incorporating the Human Dimension," Personal e-mail note dated November 10, 1994, p. 1.

59 Steven Shultz and John Regan, "The 1990 Census, GIS Technology, and Rural Data Needs." *The Rural Sociologist*, (Winter 1991): 23–29; According to Wendy J. McGinnis by personal e-mail to author dated April 28, 1997, TIGER "stands for Topologically Integrated Geographic Encoding and Referencing (system). In cooperation with the US Geological Survey, the Census Bureau created a computer-readable, seamless, geographic database covering the United States, Puerto Rico, and the outlying areas. This made it possible for the Census Bureau to automate geographic support for the 1990 and subsequent censuses and surveys, including producing maps for data analysis and display."

60 Ewert, Chavez, and Magill, *Culture, Conflict, and Communication*; Dwyer, John F. and Herbert W. Schroeder, "The Human Dimensions of Urban Forestry," *Journal of Forestry*, 92 (1994): 12–15.

61 USDA Forest Service, *Forest Service Strategic Plan for the 90's: Working Together for Rural America* (Washington, D.C.: USDA Forest Service, 1990); USDA Forest Service, *Enhancing Rural America: National Research Program* (Washington, D.C.: USDA Forest Service, 1991); Kent Connaughton, "Rural Development Research," On file at the Pacific Northwest Research Station, Portland Forestry Sciences Lab, Social and Economic Values Research Program. (n.d.); Wendy J. McGinnis, "Rural Economics Research Database," Personal e-mail note dated May 13, 1994.

62 Jo Ellen Force and Gary Machlis, "Understanding Social Change in Resource-Dependent Communities," Paper presented at the Forestry and the Environment: Economic Perspectives II conference in Banff, Alberta, on October 13, 1994; Force et al., "The Relationship Between Timber Production."

63 Arthur W. Magill, "Natural Resource Managers: Their Role in International Tourism and Rural Development," paper presented at The Chief's Interagency Conference on Tourism at Park City, Utah, September 21–24, 1992.

64 Donald E. Voth, Kim Fendley, and Frank L. Farmer, "A Diagnosis of the Forest Service's 'Social Context'," *Journal of Forestry* 92 (1994): 20.

65 Donoghue, "Socioeconomic Research in the USDA Forest Service"; Alan W. Ewart and Gerald W. Williams, "Getting Alice Through the Door: Social Science and Natural Resource Management," manuscript in review.

66 Johnson and Field, "Applied and Basic Social Research," pp. 269–270.

67 B. Bormann, "Is There a Social Basis for Biological Measures of Ecosystem Sustainability?" *Natural Resource News* 3 (1993): 1–2; National Research Council, *Forestry Research: A Mandate for Change* (Washington, D.C.: National Academy Press 1990), p. 38; Machlis, "Contribution of Sociology;" Clark and Stankey, "FEMAT's Social Assessment."

68 C. Holden, "The Ecosystem and Human Behavior," *Science* 242 (1988): 663.

69 West, "Research in the U.S.D.A. Forest Service," p. 9.

70 Clarence John Buck, "New Year's Greeting," *Six Twenty-Six*, 19 (1935): 3.

71 Bradshaw, e-mail note, 1994.

MARION CLAWSON

HISTORY OF RESEARCH IN FOREST ECONOMICS AND FOREST POLICY

Research in the history of forest economics and on the history of forest policy are separate but closely related subjects. Some research seeks to provide an explanation of past events or an understanding of present situations and practices, without explicit policy connotations or suggestions. Some policy formation rests on accumulated experience of the policy-makers, with no direct research undertaken to support it. More commonly, the research concludes with suggestions for new policy or for modifications in existing policy; and typically policy-makers collect data and make analyses to develop their policy conclusions.

This chapter focuses on the history of forest economics research but gives limited consideration to the history of research on forest policy. Forest economics, like other specialized branches of economics—e.g., agricultural economics, labor economics, and welfare economics—both contributes to and draws upon general economic theory and practice. Forest economics has contributed only modestly to general economics but has drawn extensively on such general economics. Both forest economics and general economics have evolved and developed over the past century, but forest economics has not been strongly influenced by some of the newer general economics—e.g., macro economics management (often associated with Keynes).

Economics is marked by the interchanges of personnel among its specialized branches. Over the past few decades forest economics has gained many workers (like the present author) who were trained and did earlier research

Marion Clawson began his career as an economist with the USDA's Agricultural Economics Bureau, moved on to serve as the director of the Bureau of Land Management; spent time in Israel as economic advisor, and finished his career as a senior fellow with Resources for the Future.
(FOREST HISTORY SOCIETY PHOTO)

in other fields. They have typically acquired sufficient knowledge of forestry to be able to bring to their research the expertise developed elsewhere.

GERMAN INFLUENCE ON FOREST ECONOMICS

American forestry and American foresters were much influenced, especially during the later nineteenth and early twentieth centuries, by the German experience. As this influence has been well described elsewhere, there is no need to repeat it here. Forest economics has been influenced less directly by the German experience than have some other branches of forestry, but some German influence is often implicit, often unconsciously, in the work of American forest economists.

The German forest economist Martin Faustmann developed a theory of the optimum timing of timber harvest in 1849.[1] He gave no consideration to demand for forest products, nor to organization of the timber industry, nor to environmental considerations. As Zivnuska says: "This traditional emphasis in forest management on regulation through technical determination and control of the annual cut either ignores the crucial problem of markets or involves an implicit assumption as to the existence of current demand for the amount the forester desires to cut. Such an assumption may have been valid under the particular conditions of nineteenth-century Germany, but it is not applicable in the case of a nation subject to pronounced cycles in lumber production."[2]

During that period, Germany was a large net importer of lumber, and variations in demand could be offset by variations in imports, especially from Poland. Technology of wood conversion was nearly unchanging and the industry simply organized. German forests had been managed and harvested for decades, so there existed an on going process of converting old growth forests into managed rotation forests, as was the case in the United States. Faustmann produced basically a silvicultural rather than an economic solution to the single problem of optimum age for timber harvest, but some of his concepts have been highly influential in American forestry. A later section of this chapter will explore the re-discovery and modernization of Faustmann for American forestry.

BEGINNINGS

In 1953 Duerr and Vaux edited and published their classic book, *Research in the Economics of Forestry*.[3] They had cast their net widely to include all persons and projects which could reasonably be described as forest economics. The result is a highly inclusive account. It is authoritative, and it provides a full account of the history and status of forest economics up to that time.

These editors list 3 "associates," 67 "contributors," and 19 "advisors," and acknowledge the help of 27 others. This total of 116 names includes few duplicates; it includes essentially everyone who by the most generous definition were working in forest economics. From personal knowledge I identify nine persons trained in both economics and forestry: E. M. Gould, Jr., G. R. Gregory, Sam Guttenberg, Lee M. James, H. R. Josephson, Alf Z. Nelson, Charles H. Stoddard, Albert C. Worrell, and John Zivnuska. Some of the others were primarily agricultural economists, with varying degrees of involvement in forestry; some were Forest Service employees, with varying degrees of training in economics; some were with universities; and some with forest industry firms. A highly varied lot!

The editors identify 861 "studies" that they regard as being forest economics, and as being all such studies up to that time. Of these, 223 were concerned with forest management, 207 with supply and demand, 192 with the "forest economy at large," 154 with timber harvesting and processing, and 85 with the agents of production. They further subdivide these general fields into some 25 specialized categories. One may reasonably conclude that this division of studies into categories reflected the interest of forest economists at that time.

This study shows further that seventy-five percent of the studies were a single study by the researcher(s) concerned. "Obviously, research in the economics of forestry is being led, not in the main by specialists, but by workers—mainly technical foresters—who enter the field in passing. Few research workers devote a major share of their labor to forest-economics research". They further show that half of the researchers had only a bachelor's degree and that only seventeen percent had a Ph.D. Further, most studies were relatively short, with fifty-one percent of one year or less duration. The general picture which emerges is that, at mid-century, forest economics in the United States was a relatively undeveloped professional field, with only a few well-trained workers doing continued research, and with a larger number of other workers, less well-trained in economics, incidentally studying particular problems, often briefly. To this writer, the experience of forest economics up to 1950 has many similarities with agricultural economics of fifty years earlier.[4] In each case, there was a growing and developing field, struggling to define itself and its terminology, and trying to develop useful theory. The leaders in each case had been trained in something other than the field they were trying to develop—by definition, the pioneers could not have been trained in a field which did not exist.

GROWTH OF A PROFESSION

The quarter century after the Duerr/Vaux book saw a substantial growth of forest economics as a professional field. It is impossible in a short chapter to list all the projects and name all the persons involved; I have chosen those which I think were most important. It is also impossible to say how far the developments in these years were influenced by the Duerr/Vaux book. Certainly, many of the subsequent studies made followed suggestions made in that book, but they might have had the same orientation had there been no book.

Most influential in these years was the appearance of three high class textbooks on the economics of forestry: Duerr, 1960; Worrell, 1970; and Gregory, 1972.[5] New and original research findings are ordinarily not reported in a college text, and from this point of view textbooks are not research. But a text does require the summarization and synthesis of past research, and to this extent is itself research. At the least, a text presents the author's concept of the field. Above all, and in this case, these good texts provided instructors, students, and practitioners with theory and method superior to what had previously been available. A considerable number of foresters better trained in economics surely resulted. Of the three, I will here examine only Duerr.

Duerr emphasizes the dynamics of forestry. He considers supply, demand, markets, and institutions. He traces the role of the productive factors of land, capital, and labor. He gives particular attention to capital: "one characteristic of forest capital is that the timber product is also the timber-growing machine, and ultimately the machine is the product." In deciding whether to harvest a tree or a stand of trees, or to let it or them grow a few years longer, the forester should consider the income that could be earned in alternative investment of the funds that could be obtained by harvest. Until his book, this viewpoint was not common, even among economists.

Duerr's approach to forestry was similar to that of many agricultural economists toward agriculture, especially in the consideration of the factors of production and of the dynamics of the activity. In this, Duerr was influenced by his major professor at Harvard, John D. Black, who had pioneered in some of these ideas for agriculture twenty-five years earlier.

Duerr and associates edited an eclectic text in 1979.[6] The contributors were nearly all new persons that were not involved in the 1953 book. In part, of course, the passage of time had removed from active work many of the men involved in the earlier book; but in large part it also reflected the emergence of a new and generally better trained lot of professional workers. In 1957 Gaffney resurrected and modernized Faustmann.[7] In particular, he develops the influence of site rent (the value of the site in other uses) and of interest as affecting the economically optimum timing of timber harvest.

Mead applied modern monopolistic/competition theory and analysis to the timber sales from national forests.[8] He shows clearly that the allegedly competitive timber sales were not truly competitive in the economic sense. There was but a single seller (the Forest Service) instead of the larger number of competing suppliers required by theory of competition. Often, there was but a single effective buyer, again instead of the number of potential buyers required by theory of competition. There had been, and still is, much loose talk about the possibilities of competitive sale of products from federal land. Truly competitive sales from federal land are rare indeed or impossible for timber, grass, and minerals.

Over the years the Forest Service has published a number of appraisals of the forest situation.[9] These appraisals have included a large amount of data on the forest situation at the respective dates, and as such they have been highly useful to scholars and businessmen. Indeed, to a large extent these have been the only comprehensive and continuing source of data. But they have also included analyses of the prospective demand and supply of forest products, and as such have clearly been economic research. Pinchot in the early years of the century had dramatized "timber famine"—the United States would run out of timber in twenty years if it did not mend its ways. His extreme rhetoric was muted by the Forest Service, but the conviction of a coming timber scarcity dominated Forest Service thinking until 1952. These various appraisals have persistently underestimated future timber supply.[10] These appraisals have grown in sophistication of analysis over the years, the later ones being much more defensible to an economist than the earlier ones.

Many forest product companies collected data, especially on their own holdings but sometimes in areas of potential supply; made economic analyses of scale and methods of operation; and put into effect their findings. This was not research, in the specialized use of that term, and the results were rarely published. But it was application of forest economics to real situations. The distinguishing character of such private inquiry, as contrasted with publicly-sponsored studies, was that the conclusions of the firms' studies were more likely to be put into operation.

The forest industry made great advances during this quarter century in professional competence.[11] As early as 1952, the Paley Commission recommended: "That the Federal Government raise the level of silvicultural work on its commercial timber land at least to the level maintained on intensively managed private forest land of comparable value."[12] This viewpoint was a far cry from the attitude 50 or 100 years earlier, that public forests were the sure source of future timber supply and that private forestry was exploitive with no consideration for the future. Perhaps not everyone would have agreed with this recommendation, but it is significant that it was made by an official body. It was also a far cry from the new concepts of ecosystem management. In 1956 Dana produced his classic book, *Forest and Range Policy*.[13] It is as much political science as it is economics, but is has been very useful to scholars and others who want to know the history of policy in these fields. Its citations of legislation have been particularly helpful. It did not suggest how policy might best be changed, but was concerned with how present policies have evolved.

ECONOMICS OF OUTDOOR RECREATION

The economics of outdoor recreation provides a good case study of the development of a subfield of economics. Outdoor recreation takes place in many natural settings but a forested area, especially a relatively open forest, is one setting preferred by many people. Increasingly, they have used such sites and have insisted that publicly owned forests be readily available for their use. At the same time, outdoor recreation is but one of several kinds of outputs from forests. Thus, outdoor recreation and forests are not always associated but there is a substantial overlap between these two concepts.

In 1949 the National Park Service engaged Roy A. Prewitt to devise a method for valuing national parks. By his own statement, he had not the foggiest idea how to proceed. He wrote to a dozen of the leading economists of the day, asking them how they would make such valuations. With one exception, these economists replied that, although they were sure the national parks did indeed have value, they had no idea how such value could be measured. This was the state of the economics of outdoor recreation at mid-century: nothing.

The one exception was Harold Hotelling, who thought that the basis for a demand curve for outdoor recreation might lie in the data on the travel of visitors to an area. This was only a suggestion, not fully developed, and he included some assumptions which were contrary to experience. He never undertook any research to apply his suggestion. His suggestion was not the direct ancestor of the travel cost method of valuing outdoor recreation, as that subsequently was developed, but he did indeed have the first glimmering of the idea.

The first successful and quantitative application of the travel cost method was in 1959. This was a crude model, using data collected for other purposes, but it was the beginning of a large field of research. This first model was refined and extended in a subsequent book.[14]

The travel cost method of valuing outdoor recreation starts with the recognition that the experience of visiting an area must be viewed as a whole. That is, there are five phases to the experience: anticipation or planning, travel to the site, on site activities, travel back, and recollection. Each phase has three dimensions: physical, economic, and psychological. Our concern here is with the economic dimension, but the others must be kept in mind. All parts of the experience are essential, each has its cost (monetary, time, and sometimes psychological). Although each has its cost, and each may have positive value, they are not really separable. There is uncertainty about how to value time spent in travel; often it is merely cost, to be endured to get to the site, but some people drive for pleasure, hence it has a positive value for them.

Zones of origin of visitors are established, average costs per visit from each zone estimated, and the number of visits from each related to the basic population in each zone. The result is a schedule of quantities and prices— the basic factors in a demand curve. From this, by established methods of economic analysis, may be calculated such items as the effect of an increased user fee upon the volume of use, or the user fee which will produce the maximum net revenue, or the value of the site for recreation purposes, and others. In all of this, once the demand curve has been estimated, outdoor recreation involves no unusual features of economic analysis.

The President's Commission on American Outdoors, in its Literature Review in 1988, lists several dozen American studies which utilize this

This family enjoys an outdoor recreation experience—one that economic researchers try to quantify using the "travel cost method."
(AMERICAN FOREST INSTITUTE PHOTO)

method of analysis.[15] The travel cost approach to economic studies of out-door recreation is, by now, a well-established research procedure, widely used by many researchers and planners. It is completely a growth of the past half century, and indeed is one, perhaps the only one, of the ideas not covered by the Duerr/Vaux book.

To my personal knowledge, this method has also been employed in a number of studies in countries of Western Europe, in the British Isles, and in Nigeria. We got some hint that it had been used in the USSR also—but, of course, one could never be sure about what was done there, especially if the results were not published.

A parallel approach to measuring the value of outdoor recreation is the contingent value approach. By means of interviews, the stated intentions of defined populations to pay for recreation opportunities are obtained. One strength of this approach is the ability to estimate values of alternative sites, facilities, and methods of management. One weakness of this approach is that what people say may be different from what they are willing to back up with expenditures.

Decisions on recreation policy, such as areas of land to be acquired, investments to be made, fees to be charged, and the like rarely depend on economic analysis. Other factors are dominant. But the careful decision-maker should consider the economic costs and economic values of any decision. As with other estimates of value, they pertain to the future, and hence many uncertainties inevitably attach to any estimates.

RESOURCES FOR THE FUTURE

Resources for the Future (RFF), founded in 1952, is a private nonprofit research and educational organization concerned with natural resources and the environment. Its research is conducted by its staff and by grants to other nonprofit organizations, especially to universities. Since it is concerned with all natural resources, it has naturally included forests. Several of its early studies included material on forests. The Duerr text and the Mead study, described earlier in this chapter, were each made possible, in part, by a small grant from RFF.

In 1976 RFF announced the inauguration of an expanded program of research on forest economics and forest policy. The next year it conducted a symposium on the subject, the results of which were published.[16] Some forty-seven scholars, from universities, federal agencies, and private firms met with several RFF employees, to discuss a number of papers prepared in advance. In many ways, this was an update of the Duerr/Vaux study of twenty-five years earlier. Zivnuska was the only person contributing to each study. Passage of time had led to the replacement of nearly all of the earlier participants, but the later group was the more explicitly trained and devoted to forest economics. The RFF symposium did not explicitly inventory all the studies in this field, as had the Duerr/Vaux study.

In 1977 RFF also appointed Roger A. Sedjo to direct its expanded program of research in forest economics and forest policy. He has remained at RFF ever since. In addition to studies about forests in the United States, Sedjo and his associates have given substantial attention to forests around the world, including the less developed countries and including forest plantations made in several countries.

Over the years, RFF has had a number of researchers who conducted major research projects on economics of forestry: John A. Krutilla, Michael D. Bowes, William F. Hyde, Samuel Radcliffe, Roger A. Sedjo, Marion Clawson, and their cooperators. They have produced a number of books.[17] A considerable part of the total literature on forest economics during the past twenty years has come from this group. In his 1987 book, *Resource Economics for Foresters*, Gregory includes an extensive bibliography; twenty-three of the items listed were written by RFF personnel and twenty-four of the items were written by Forest Service personnel.[18]

RFF has not been the only nonprofit organization concerned with forests, but it has been the largest (in terms of output) and perhaps the most innovative one. The outsiders (outside of the public agencies, that is) have some advantages and some disadvantages in conducting research in any field, including forestry. They avoid the institutional biases of managing agencies, but of course individual researchers have viewpoints (biases) of their own. It is possible for the outsider researcher "to tell it like it is" more easily than can persons employed by forest managing agencies and firms. But the outsider may lack access to important information and may not fully understand the considerations which led to various decisions and actions.

THE LAST TWENTY YEARS

Forest resource economics and policy research has evolved strikingly over the past forty years in a number of fundamental respects. For example, the size of the effort devoted to such research has grown substantially. The research methods available for use have become much more sophisticated. And the array of problems needing and receiving attention has multiplied in number and become much more complex."[19]

Vaux and Josephson thus succinctly and accurately describe the great changes which have occurred. Later in the same book, Ellefson says: "In sum, the total number of individuals working as researchers in the field of forest economics in all employments probably exceeds 500."

His definition of forest economists may not be identical with the definition I used when I estimated there were but nine men trained in economics and in forestry in 1953, as stated above. Hence exact comparisons may not be possible, but the contrast in numbers is in the right direction and probably of about the right magnitude. By any standards, forest economics grown greatly in recent years.

Under these conditions, a comprehensive yet reasonably brief survey of the field today is impossible. There has not been, and could not well be, any report in recent times which has dominated the professional scene as did the Duerr/Vaux report of 1953. But there have been two published reports which do in some measure describe the field of forest economics today.

Ellefson in 1989 edited the book noted above, with twenty-five chapters by twenty-seven contributors. A few men wrote more than one chapter and a few chapters had more than one author. The authors include Henry Vaux and H. R. Josephson, who had participated in the Duerr/Vaux study of thirty-six years earlier. The authors also include very few men who had participated in the RFF study of about twenty years earlier. But for the most part, the authors were a new crowd—evidence of the changing and growing field, among other factors.

The Ellefson book includes more or less standard subjects, such as demand, supply, and forest management. It includes more emphasis on the non wood outputs of the forest, upon environmental and institutional considerations, and upon international forestry, than had the earlier studies. Basically, the book is an outline of desired and probably effective future research. The subtitle of the book is: "Strategic Directions for the Future," and this well describes the emphasis throughout. As such, it is necessarily based upon the present but it describes the present only briefly, and its treatment of history is still shorter. The book does include many references at the end of each chapter. No hint appears in any chapter of the special research problems likely to arise as a result of forest ecosystem management.

Hyde et al. in 1991 surveyed forest economics as applied to the developing countries.[20] They used international sources and examples from other countries to an extent not found in the earlier comprehensive surveys. Their analysis is not directed at the United States but at other countries, and it is research only in the sense that it brings research and other information into a consistent and reasonably comprehensive picture. But their study does present forest economics of today in an integrated fashion.

Forest economics research in recent years has given major attention to the international aspect of forestry.[21] Earlier studies of the economics of forestry in the United States typically had a brief concluding section discussing some of the international situations and problems. By the recent decade, analyses frequently dealt with the world situation or at least some major regional situations. Wood supply and demand are now more commonly dealt with in global terms.

These past twenty years or more have seen a good deal of legislation relating to forests in the United States. Two items have special importance for forest economics: the Forest and Range Land Renewable Resources Planning Act of 1974, and the National Forest Management Act of 1976. These acts were more concerned with planning than with research, but the planning necessarily rested on research done in the past or for the purpose of these acts. Each emphasized all the outputs of the forest, not wood alone. Each required some consideration of values and costs but each explicitly rejected a precise monetary comparison, emphasizing other values as well. They have been characterized as including some economics, but not too much. They did require the assembly of data. Building models was often the best way to develop plans. The planning process has been carried out by the federal agencies but also by private groups, which often did not accept the agencies' findings.

All of this planning work has been greatly facilitated by the development of computers, including personal computers on which the planner could carry out his own analyses. The computers can very quickly make analyses which would have been slow or impossible by methods common fifty or more years ago. The effect of different assumptions can quickly be ascertained. Some modelers indeed seem more entranced by the machine and by its processes than interested in the results. Models are clearly no stronger than the data and the assumptions on which they rest. Some of this modeling has left some of the old timers in forestry subdued and isolated.

Forest economics and the analysis of forest policy have clearly matured by now, into well recognized and solidly established professional fields. The results of economic analysis have not always been followed, of course. Many forest economists today are practitioners as much as they are researchers or innovators.

CONSERVATION AND ECONOMIC EFFICIENCY

The activity groups generally considered as being conservationist/preservationist/wilderness-lovers have generally rejected economic analysis for the formation and defense of their policies. They have rested their support on other values—"inherent values"—which are not measurable in monetary terms. And many of them have considered costs as irrelevant to their programs.

In recent years, however, some persons in such groups have discovered the fact that the Forest Service was selling timber when the cash revenues did not equal the cash costs of making the timber sale—below-cost timber sales, in the current terminology. These sales have then been attacked by these persons on the grounds of their economic inefficiency; but the real basis of the objection was that they wanted to stop timber harvest on these areas. Many resource economists, including the present writer, would also like to stop timber sales under these conditions, but solely on the grounds of economic inefficiency. This example is a relatively small one but it does illustrate very well a general principle: economic analysis is a method of examining management policies and is not an end in itself.

EFFECTIVENESS OF RESEARCH IN FOREST ECONOMICS AND FOREST POLICY

What difference has the research on forest economics and forest policy made to forest landowners, to the wood processing industry, to other users of forests, and to the nation as a whole?

Appraisal of the effectiveness of research is always difficult and the results are rarely unequivocal. The researcher typically assembles data, makes analyses, and draws conclusions with either explicit or implicit policy consequences. If a new policy is instituted or an old one changed, to embody the suggestions of the research, this might suggest that the research has indeed been effective. But it is at least possible that the same changes in policy would have occurred had there been no research. On the other hand, if the suggested changes in policy were fully not made, this might seem to indicate that the research had not been effective. But, again on the other hand, the policies and actions taken might have been affected in some degree by the research.

These relationships apply to private as well as to public forests, as they also apply to many fields other than forestry. The most careful analysis of the effectiveness of research is generally desirable but at the best its conclusions are tentative or provisional.

My own and necessarily subjective conclusions about the effectiveness of research on forest economics and forest policy are:

1. In many instances, specific research has led to effects, on such matters as optimum age of timber harvest, economic level of investment and management, and valuation of the nonpriced outputs of the forest. I find it impossible to be precise and quantitative, but I think these results have been considerable.

2. Perhaps more important, the development of forest economics as a professional field has introduced economic thinking, at least to some degree, into much decision-making, both public and private. It has also led to the development of a new generation of foresters with more concern for the economics of their management.

3. But it must be recognized that forest economics has been but one force, and not always the most important force, in establishing forest policies, both public and private, and in carrying out those policies.

4. Forest economics has been a positive force in the nation, but also and necessarily a somewhat limited one.

NOTES

1 Martin Faustmann wrote in German in 1849. His book was translated into English in 1968: *Martin Faustmann and the Evolution of Discounted Cash Flow*, by M. Gane, and published as Commonwealth Forestry Institute Paper No. 42, University of Oxford.

2 John A. Zivnuska, *Business Cycles, Building Cycles, and Commercial Forestry* (New York: Institute of Public Administration, 1952).

3 William A. Duerr and Henry J. Vaux, editors, *Research in the Economics of Forestry* (Washington, D.C.: Charles Lathrop Pack Forestry Foundation, 1953).

4 Henry C. and Anne Dewees Taylor, *The Story of Agricultural Economics in the United States, 1840–1932* (Ames: Iowa State College Press, 1952).

5 William A. Duerr, *Fundamentals of Forestry Economics* (New York: McGraw-

Hill Book Company, 1960); Albert Worrell, *Principles of Forest Policy* (New York: McGraw-Hill, 1970); G. R. Gregory, *Forest Resource Economics* (New York: Ronald Press, 1972).

6 William A. Duerr, Dennis E. Teeguarden, Neils B. Christiansen, and Sam Guttenberg, editors and authors, *Forest Resource Management-Decision-making Principles and Cases* (W. B. Saunders Company, 1979).

7 M. Mason Gaffney, *Concepts of Financial Maturity of Timber and Other Assets*, A. E. Information Series No. 62 (Raleigh: Department of Agricultural Economics, North Carolina State College, 1960).

8 Walter J. Mead, *Competition and Oligopsony in the Douglas Fir Industry* (Berkeley: University of California Press, 1966).

9 The several reports by the Forest Service were; 1920, *Timber Depletion, Lumber Prices, Lumber Exports and Concentration of Timber Ownership*, Report on Senate Resolution 311, 66th Congress, 2nd Session,(Capper Report); 1933, *A National Plan for American Forestry*, Senate Document No. 12, 73rd Congress, 1st Session (Copeland Report); 1941, *Forest Lands of the United States*, Senate Document 32, 77th Congress, 1st Session; 1948, *Forests and National Prosperity*, U. S. Dept. Agric. Pub. No. 668; 1958, *Timber Resources for America's Future*, Forest Resource Rep. No. 14, U. S. Forest Service; 1965, *Timber Trends in the United States*, Forest Resource Rep. No. 17, U.S. Forest Service; and 1973, *The Outlook for Timber in the United States*, FRR-20.

10 Marion Clawson, "Forests in the Long Sweep of American History" *Science* (June 13, 1979).

11 Henry Clepper, *Professional Forestry in the United States*, Published for Resources for the Future (Baltimore: Johns Hopkins Press, 1971).

12 President's Materials Policy Commission, *Resources for Freedom*, (Paley report) (Washington, D.C.: Government Printing Office, 1952).

13 Samuel Trask Dana, *Forest and Range Policy—Its Development in the United States* (New York: McGraw-Hill Book Company, 1956).

14 Marion Clawson, *Methods of Measuring the Demand for and Value of Outdoor Recreation*, Reprint No. 10, Resources for the Future, 1959. Marion Clawson and Jack L. Knetsch, *Economics of Outdoor Recreation*, published for Resources for the Future (Baltimore: Johns Hopkins Press, 1966).

15 The President's Commission on Americans Outdoors, *A Literature Review*, (Washington, D.C.: Government Printing Office, 1988).

16 Marion Clawson, editor, *Research in Forest Economics and Forest Policy*, Research Paper R-3 (Resources for the Future, 1977).

17 In addition to publications noted above, staff at Resources for the Future have produced the following books and working papers, which have been published by or for Resources for the Future: Michael D. Bowes and John V. Krutilla, *Multiple-use Management: The Economics of Public Forestlands*, 1989; William F. Hyde, *Timber Supply, Land Allocation, and Economic Efficiency*, 1980; John V. Krutilla, editor,

Natural Environments: Studies in Theoretical and Applied Analysis, 1972; John V. Krutilla and Anthony C. Fisher, *The Economics of Natural Environments: Studies in the Valuation of Commodity and Amenity Resources*, 1975; Roger A. Sedjo, *The Comparative Economics of Plantation Forestry: A Global Assessment*, 1983; Roger A. Sedjo and Samuel J. Radcliffe, *Postwar Trends in U. S. Forest Products Trade: A Global, National, and Regional View*, 1981; Roger A. Sedjo and Kenneth S. Lyons, *The Long-Term Adequacy of World Timber Supply*, 1991. Marion Clawson, Research Paper R-4, *Decision Making in Timber Production, Harvest, and Marketing*, 1977; Marion Clawson, Working Paper EN-6, *The Economics of National Forest Management*, 1976; and Marion Clawson, *Forests for Whom and for What?*, 1975.

18 G. Robinson Gregory, *Resource Economics for Foresters* (New York: John Wiley & Sons, 1987).

19 Henry J. Vaux and H. R. Josephson, "Development and Accomplishments of Research Programs" in Paul V. Ellefson, ed. *Forest Resource Economics and Policy Research: Strategic Directions for the Future* (Westview Press, 1989).

20 William F. Hyde, David H. Newman, and Roger A. Sedjo, *Forest Economics and Policy Analysis, An Overview*, World Bank Discussion Paper (The World Bank, 1991).

21 In addition to the reports listed above, see 1) Hans M. Gregerson and Arnold H. Contreras, *Economic Analysis of Forestry Projects* (FAO Forestry Paper 17, 1979); and 2) Jan G. Learman and Roger A. Sedjo, *Global Forests-Issues for Six Billion People* (New York: McGraw-Hill, 1992); and extensive bibliography in latter.

BRUCE ZOBEL AND JERRY SPRAGUE

DEVELOPING BETTER TREES BY GENETIC MANIPULATION

THE PRELUDE TO AN ERA

The application of genetics in operational forestry is often considered a recent development, although Japanese redwood (*Cryptomeria japonica*) has been developed for centuries and reproduced as rooted cuttings for a long time. Some of the earliest published results relative to genetics of forest trees date from the 1300s in Japan. Few intensive genetic studies occurred world-wide until the late 1800s when a number of studies were conducted on several species in different parts of the world.

Some of the earlier work has been outlined in Table 1; however, references to the work done in Asia or the southern hemisphere are not included. Some excellent forest genetic work was done in Asia, South America (Brazil especially), South Africa, Australia, and New Zealand. Although these early studies were not coordinated, they represent a body of information utilized by forest geneticists. A lack of knowledge of the basic forest tree biology and of the use of statistical methodology, however, has greatly restricted application of the early results. In light of these limitations some of the early accomplishments remain truly remarkable, and the results obtained are now known to be correct.

Applied genetics in forestry had a slow and difficult start partially because of the foresters themselves. Although they recognized how genetics had been used successfully in agriculture, horticulture, and animal production, they

Table 1. Some of the Early Forest Geneticists and Their Areas of Interest[1]

1717	Bradley (England)	Importance of seed origin
1760	Duhamel de Monceau (France)	Inheritance: oak
1761	Koehlreuter (Germany)	Hybridization
1787	Bursdorf (Germany)	Plantations for seed production
1840	Marrier de Boisdhyver (France)	Vegetative propagation
1840	de Vilmorin (France)	Fir hybrids
1845	Klotzsch (Berlin)	Intraspecific hybrids; oak, elm, and alder
1904	Cieslar (Austria)	Provenances: larch and oak
1905	Engler (Switzerland)	Elevation differences of species: fir, pine, spruce, larch and maple
1905	Dengler (Germany)	Provenance tests: fir and spruce
1906	Andersson (Sweden)	Vegetative propagation
1907	Sudworth, Pinchot (USA)	Breeding nut and other forest trees
1908	Oppermann (Denmark)	Straightness: beech and oak
1909	Johannsen (Sweden)	Elite stands
1909	Sylven (Sweden)	Self-pollination: Norway spruce
1912	Zederbauer (Austria)	Crown form: Austrian pine
1918	Sylven (Sweden)	Seed orchards
1922	Fabricius (Austria)	Plantation for seed production
1923	Oppermann (Denmark)	Seedling seed orchards
1924	Schreiner (USA)	Poplar breeding
1928	Burger (Switzerland)	Pine selection
1928	Bates (USA)	Seed orchards
1930	Larsen S. (Denmark)	Controlled pollination: larch
1930	Heikinheimo and M. Larsen (Finland)	Curly grain: birch
1930	Nilsson-Ehle, Sylven, Johnsson, Lindquist (Sweden)	Pine and aspen breeding
1935	Nilsson-Ehle (Sweden)	Triploid aspen

somehow thought forest trees would not respond to genetic manipulation. When the use of genetics was proposed for forestry, many foresters showed little or no enthusiasm. Their primarily ecological training led to the belief that tree size and quality were determined only by environmental conditions and that parentage had little, if any, effect.

Most foresters also believed that tree breeding would take so long that it could not possibly prove economically feasible. Because the geneticists had little knowledge of tree genetics and the potential gain from using it, they could not effectively argue otherwise. They could demonstrate, however, from the research already conducted that forest trees responded to genetic manipulation, probably more so than other organisms. Without the initial "Period of Great Faith" discussed later, the use of genetics in operational forestry would have been very slow to develop.

Gain from tree breeding can be roughly defined as $G = h^2 \times SD$ ($h^2 =$

heritability, SD = selection differential). In the early stages of tree improvement, the variations of many important characteristics were not known, but it soon became evident that they varied tremendously, thus allowing for large selection differentials through intensive selection. Forest trees have nearly twice the variability of any other organism. The presence of such variation becomes understandable when one visualizes the greatly varied environmental conditions under which trees survive and grow over their long life spans. The requirements for genetic gain could be met once the geneticists learned what causes the variations to exist (genetics or environment) and if the genetic variation (additive or non-additive) was of a type that could be utilized in a seed or vegetative propagation regeneration program. Although the following is an oversimplification, the job of the forest geneticist was to:

> Determine if variation exists in the species of interest
> Determine the amount of the variation
> Assess the causes of the variation
> Package the desired characteristics into usable genotypes
> Mass produce the package that had been developed

In addition, it became necessary to determine which tree characteristics should be emphasized in a genetics program; this determination depends both upon the genetic control of the characteristic and its economic value. Initially, some knowledge existed about the value of different characteristics, and those such as growth rate, that could be easily measured and had high value, were emphasized. Although quality characteristics, such as bole straightness or small limb size, intrinsically have considerable value, it proved most difficult to put monetary values on them. Thus, the success of the early tree improvement programs tended to be assessed on growth improvement alone, although it is now known that the value of quality improvement, usually proves equal to, or greater than, volume gain. Thus, the early value of tree improvement was underestimated when based only on growth characteristics strongly influenced by environmental factors. Many early workers did not appreciate the importance and magnitude of the genotype x environment interaction.

THE ERA OF MAXIMUM DEVELOPMENT

The fifty year period from 1940 to 1990 proved one of active development of forest genetics, as well as good cooperation among researchers. Starting in the 1950s, operational programs used some of the existing information, and a surge of new research occurred. Although the application of conventional plant breeding methodology marked much of this period, toward the end, new developments in biotechnology, tissue culture, and molecular biology became prominent. Table 2 illustrates the breadth of activity during just the decade of the 1950s. The list is, of course, incomplete, including only a small sample of the active scientists. It remains, however, of interest to note the number and locations of the countries involved. Although the application of genetics occurred in many areas, the major emphasis in operational programs took place in the southern United States, Australia, New Zealand, and the Scandinavian countries. Use of genetics in forestry then spread rapidly to many other areas of the world and became an integral part of well-based silvicultural programs.[3]

Table 2. A Sampling of Authors Who Published on Applied Tree Improvement and Forest Genetic Activities During the 1950s.[2]

Author	Date	Country	Type of Article
Barner	1952	Denmark	General tree breeding
Bouvarel	1957	France	General tree improvement
Buchholz	1953	Russia (Germany)	Review of Soviet activities
Duffield	1956	USA	Breeding approaches
Fielding	1953	Australia	Variation studies
Fischer	1954	Germany	General tree breeding
Greeley	1952	USA	History of Institute of Forest Genetics
Haley	1957	Australia	Status of tree breeding in Queensland
Heimburger	1958	Canada	General tree breeding
Hellinga	1958	Indonesia	General tree improvement
Hyun	1958	Korea	General tree breeding, hybridization
Johnsson	1949	Sweden	Results of breeding
Langner	1954	Germany	General tree breeding
Larsen	1951	World (Denmark)	General tree breeding
Matthews	1953	Great Britain	General tree breeding
Pauley	1954	USA	General tree breeding, poplars
Perry & Wang	1958	USA	Value of genetically improved seed
Rao	1951	India	General tree improvement
Schreiner	1950	USA	General tree breeding
Thulin	1957	New Zealand	General tree breeding
Toyama	1954	Japan	General tree breeding
Wright	1953	USA	General tree breeding
Zobel	1952	USA	improving wood quality

Coordination and Cooperation

Although good cooperation has always existed among research scientists, the Era of Maximum Development included optimal coordination among companies, universities, and governmental agencies rarely witnessed in major production organizations. Although most company mills remained very proprietary about their operations (some would permit no photographs or note taking), the same companies cooperated freely and exchanged information, research results, and even plant materials with one another relative to tree improvement. People in other professions have commented with disbelief how such extensive cooperation could be possible. It appears to result from the long developmental nature of forestry and an understanding that if progress were to be achieved, the organizations needed to help one another. Current findings would not be of immediate financial benefit, so no reason existed to keep results proprietary. Also, the scarcity of technological expertise in forestry made the maximum use of trained people essential. Only toward the end of the 1980s did the word proprietary become common, and attempts occurred to develop patents or in other ways to keep tree improvement findings secret.

At the beginning of the 1950s, little was known about genetic gains in forestry. The term "The Period of Great Faith" illustrated that millions of dollars were expended during this time on tree improvement research and operations without any proof of economic returns. Without this attitude of "faith without proof," little applied genetics would have been done on forest trees because of the long time it takes to obtain definitive results.

Many examples of "faith without proof" could be given but we shall use only two illustrations here. When the first applied tree improvement cooperative began in Texas, no knowledge existed of the payout, although the members were promised a five percent gain in volume from the application of genetics in silviculture. Fifteen hundred miles east of the Texas Cooperative work area, Union Bag Paper Company, under the direction of Mr. Gunnar Nicholson, asked to become a member of the cooperative, paying its fair share. Union Bag had no opportunity to use the improved trees that were developed, because its forest ecosystem was so different from that further west. When asked after many years of membership in the cooperative, "Why did you spend money when no payout was possible?" Nicholson said, "I knew nothing about genetics or silviculture but the potentials sounded good. We in forestry need to support activities that may help in the future."

Another example of this great faith occurred when a cooperative member in Louisiana heard that his contribution was less than his competitor in Texas. He called the director of the cooperative and demanded to pay as much as his Texas competitor. In addition, he gave the cooperative an air-conditioned greenhouse and a twenty-five foot tall greenhouse for studying large trees. The desire of some industry leaders for recognition as the most

Cone collection in a loblolly selection seed orchard at the Crown Zellerbach Tree Improvement Center, in Bogalusa Louisiana.
(FOREST HISTORY SOCIETY PHOTO)

progressive in forestry provided one reason for such an attitude. As a result, real competition existed among them, but it contributed to the general cooperation because each wanted to know what the others were doing.[2]

In many parts of the world, the formation of cooperatives provided the key to cooperation in tree improvement efforts. These began in the southern United States, but then rapidly spread to other areas of the world.[3] That only a few highly technically trained people were required to guide and advise the activities for a number of organizations made the cooperatives attractive. The employees of the cooperatives conducted the technical activities such as design, analysis, wood assessment, and operational recommendations, while each cooperative member executed the actual work of tree selection, seed orchard establishment, and progeny testing.

This system required a minimal contribution of funds directly to the cooperatives, so that the bulk of the financial outlay for tree improvement (often ten times as much as went to the cooperatives) was spent on the lands of the cooperators actually implementing the tree improvement programs. It proved a most efficient use of funds. Further, each member was much more willing to use its own funds on its own lands with its own personnel than to send money to some research institute over which it had minimal control. A research institute or a university usually housed the headquarters of the cooperatives. During this time the industries were generally suspicious of the efficiencies and motives of government organizations. A couple of cooperatives even ruled out participation of government organizations. Government institutes, however, initiated some cooperatives, of which some succeeded and others failed. The key to organizing a successful cooperative was for the members to need (want) help, organize, and then ask a neutral institution, like a university, to administer the cooperative. It never worked to have an industrial member of the cooperative also serve as its administrator.

A cooperative was considered successful when its members viewed the activities of the coops as their own and referred to "our selected trees, our seed orchards, or our progeny tests." This feeling of pride became frequently evident in the public relations area with advertisements, television programs, articles, and field days where a cooperative member could showcase its progressive accomplishments. Comments such as "Look at my selected tree—isn't my seed orchard beautiful?" indicated the success of a cooperative program. Although the members of a cooperative exchanged information and plant material with other members, the plants developed and results obtained were the property of each member to use as desired.

Tree Improvement

During the "Period of Maximum Development," the genetic methodologies initially used were simple and straightforward. Not enough information had become available to employ sophisticated breeding and testing systems. Furthermore, in most programs, time proved a key element because, with the establishment of large plantations, the improved trees were needed right away. This urgency led to the use of shortcuts, with the result that many of the early programs did not prove the best from a scientific point of view. Using this approach, however, the cooperative members obtained partial, but good, gains while the more scientific approaches were being developed. When feasible, the early tree improvement programs "borrowed" methods used in agriculture and, especially, the breeding programs employed by large animal producers. Of course, the methods had to be adjusted and adapted for efficient application to forest trees. Compared to the knowledge of the 1990s, the early programs appear slow, cumbersome, and inefficient as a result of the lack of genetic knowledge and the urgency immediately to use what gain was possible. The inheritance patterns (heritabilities) for most important tree characteristics were not yet known, so intuitive estimates had to be made. Two good examples of this occurred in the area of wood properties and tolerance to certain diseases.

Most foresters believed that the wood properties of a tree predominantly resulted from the site on which the tree grew as well as the growth pattern of the tree. Parentage was not considered important, and those who proposed to include wood properties in tree breeding programs received ridicule. Yet, later results indicate that many wood properties have a high heritability plus much variation, so the gain from using them in genetic programs is good. Similarly, some pathologists thought that breeding for disease tolerance could not be productive and discouraged including this in tree improvement programs. In fact, tolerance to a number of important diseases of forest trees has eventually been found. In the southern United States, the genetic improvement in tolerance to fusiform rust on pines alone has more than paid for all the cost, time, and effort of all tree improvement activities in the area. Similar results have been found elsewhere such as breeding for tolerance to *Dothistroma* in radiata pine in New Zealand and Australia. Breeding for tolerance to disease, initially largely ignored in tree breeding, became a major component of tree improvement.

The need for basic knowledge prompted activities in two directions. The first required maintaining interest in genetics, so that the major contributors would continue to support tree improvement. Results were needed that could be incorporated into operational programs, leading to an economic advantage in forestry operations. It takes, however, many years to obtain meaningful results from a genetics program, so the early tree improvers had a problem. Many of them kept interest of the contributors by obtaining results such as in wood variation or certain silvicultural actions. This strategy kept the financiers interested and pleased until measurable solid genetic

J. G. Hoffman grafts a scion from a superior tree
onto root stock in a North Carolina seed orchard.
(AMERICAN FOREST INSTITUTE PHOTO)

improvement could be obtained. Although many programs initially "charged ahead" not really knowing whether their actions were scientifically sound, a knowledge of the species, assessment of agricultural methods, and intuition resulted in more positive genetic results than had been expected. One example comes from the North Carolina State University-Industry Cooperative in which an initial estimate of five percent gain in volume from the use of genetics was presented to the members as an obtainable goal. The results, however, indicate that even with the crude methodologies used at first, about a fifteen percent gain was achieved.

The second need required obtaining more basic genetic information about forest trees to permit implementation of more sophisticated genetic programs. To do this, several organizations established major genetic analytical studies that supplied information on the size, extent, and kind of variation present and the genetic patterns of the important forest tree characteristics. This approach enabled the packaging of the desired characteristics into individual genotypes. Along with this, the development of methods to mass produce good genotypes on an operational basis both with seed and by the use of vegetative propagation proceeded rapidly.

Toward the middle and end of the period, some sophisticated breeding methodologies were developed that in many ways equaled agricultural breeding, despite the long life-cycle, slow flowering, and difficulties in working with large organisms like trees. Progress proved nearly miraculous; much of the credit for this must go to the fabulous cooperation among scientists, both nationally and internationally. The numerous meetings and symposia where results were reported and information exchanged made possible the worldwide application of forest genetics. The world network in forest genetics was probably the best of any scientific area. Most scientists personally knew the others and had access to the work and methodologies developed by the others.

Better Utilization of Genetic Potential

An expanding development, mostly in its infancy (but well known for a few species), was the use of vegetative propagation for operational planting. Although in widespread use for genetic studies, seed orchard establishment, and clone banks, it was not generally used for commercial planting. Most geneticists consider vegetative propagation an absolute necessity if maximum gains are to be obtained from the use of genetics in forestry. Certainly, vegetative propagation is not a breeding method - its contribution is to mass produce the best genetic material available. It also enables the efficient capture and use of the non-additive genetic variances not easily exploited through a seed orchard program. For important tree characteristics with low heritabilities, like growth rate, greatly improved yields become possible using vegetative propagation compared to seed regeneration. Uniformity in tree morphology, silvicultural handling, and wood properties provides another major advantage of vegetative propagation. Some forest geneticists consider wood uniformity to be the most significant contribution that genetics can provide for the industry.

Vegetative propagation proves relatively simple for those species that sprout and where the sprouts are juvenile, so that they are easy to root and will grow normally and not plagiotropically (grow like a branch). The poplars, eucalypts, and several hardwoods, including some tropical hardwoods, follow the pattern of sprouting and normal growth. A few conifers, such as redwood (*Sequoia sempervirons*), *Cryptomaria japonica*, and *Pinus oocarpa* sprout in the same way. But most conifers, especially the pines, are more difficult to root unless juvenile material is used. By the time, however, a tree has grown long enough to show its potential, it will no longer root satisfactorily. This has become a serious barrier to widespread use of vegetative propagation for conifers (See "The Present and The Future").

Adaptability

As forest practices expand, opportunities for planting often become relegated to marginal sites on which normal trees do not grow satisfactorily. In order to use such marginal and degraded sites, special "land races" must be developed.[4] For many tree improvement organizations, development for adaptability was a major objective. Luckily, characteristics needed for marginal sites, such as drought tolerance, cold tolerance, tolerance to excess moisture, and nutrient deficiencies have proven very responsive to genetic manipulation.

Breeding for adaptability became a major activity after 1980. Results proved dramatic, and millions of hectares by the mid-1990s were producing economically viable forest crops on sites previously unproductive. Fortunately for the tree breeder, usually little or no genetic correlation exists between adaptability and economically important forest tree characteristics. Thus, it is possible to produce drought-tolerant trees that are straight or crooked, or cold-hardy trees that have small limbs or large limbs. The large variation in morphological characteristics and in adaptation, along with their lack of genetic relationship to each other, enables the tree breeder to "tailor make," land races both suitable for the degraded sites and also those of good economic value.

Development of adaptable land races was especially needed in the tropics where most planting occurs on degraded sites. Land management often followed the pattern of harvesting forests, farming, and grazing followed by abandonment. When this sequence occurs no chance exists for natural regeneration of these sites because no residual seeds or sprouts are available. Further, the sites are frequently grass covered, making plantation establishment very difficult and costly, because of root competition between grass and young trees and the allelopathic characteristics of grass (production of poisons from the roots). Developing adaptable land races for the degraded grasslands requires intensive breeding, mostly with the hardy exotic species. Another problem is that usually the past treatment of the site, such as compaction, erosion, or nutrient deficiency, makes these sites unsuitable for the species growing there originally. Without the ability to develop adaptive land races, much of the degraded land in the tropics will remain unproductive. Genetics in forestry has been responsible for a remarkable restoration of productivity to large areas of little use before the establishment of forest tree plantations.

Environmental Considerations

During the Era of Maximum Development, intense activity occurred relating to forestry and the environment. Environmentalists expressed concern about many things, some directly related to forest genetics. One major concern was the monoculture that, to the layman, results from planting large areas with a single species. Yet, planting of a single species does not necessarily prove bad if the genetic base used is kept broad. More serious is the use of clones as rooted cuttings, where genetic diversity can be seriously reduced if too few clones are planted. But if sufficient, genetically different clones are used and deployed correctly, no fear should exist of a dangerous monoculture. Some non-justifiable restrictions have been made on operational application of clonal forestry in several countries, such as Germany and Sweden. Generally, the forest geneticist is very sensitive to environmental issues and attempts to incorporate suitable safeguards into breeding plans.

Geneticists have a major advantage in working with forest trees because the populations remain mostly still natural. The stands have much variation and are little changed by generations of breeding, such as occurs with agricultural crops. Most important characteristics of forest trees, whether economic or adaptive, are genetically independent from each other. This makes possible breeding for a broad adaptability while at the same time reducing variation in economic characteristics. This approach is not possible in most agricultural crops, where the genetic base of variation has been reduced. Breeding for adaptability and to avoid environmental extremes has provided a major success for forest tree geneticists.

Economics of Forest Tree Improvement

At the early stages of applied forest tree improvement, no one knew how profitable the long term application would be. In recent years, various organizations have made in-depth analyses of the value of genetics to forestry. For programs of sufficient size, all analyses have found that tree improvement, despite its delayed payout and long term nature, is economically profitable. A key indication of the value of genetics occurs when the companies shift the tree improvement activities from research to operations.

The complete value of tree improvement often remains unknown because some data are considered proprietary. For example, what is the value of an increase in wood density for an organization wanting high-density wood? Generally, the value of higher density wood is assessed based upon greater yields in pulping or greater strength as boards. Added density also makes harvesting and hauling less expensive, handling in the mill — such as chipping and pulping — more efficient, and the quality of the final product improved. When all these factors are included, the economic effect of wood density becomes much greater than its effect on yield. As several large organizations have indicated, "tree improvement gives us greater returns on our investment than any other forestry activity."

Where was the Action?

As the Period of Maximum Development began, a limited number of organizations were involved in forest genetics studies. That was no longer true at the end of the period, when no specific location could be designated as outstanding in applying genetics to forestry. The organizations from North America and Europe proved active from the start. Perhaps the greatest gain in applying genetics happened in the tropical area, especially in South America. In the Southern Hemisphere, a good deal of intensive genetic work has been done in New Zealand, Australia, and South Africa, in addition to that in South America. Some excellent applications have also occurred in Africa and Asia. In Asia a number of countries are involved in forest genetics, especially Indonesia and China. Activities using genetics in forestry are now worldwide, making a major contribution to progress in forestry. The movement to make a greater use of vegetative propagation, along with advanced generation breeding, became prominent at the end of the period.

THE PRESENT AND THE FUTURE

Activities in forest genetics are currently at a high level, and it appears this will continue into the future. The *kind* of activities rapidly changes with greater emphasis on vegetative propagation and biotechnology and less on what might be called conventional breeding. Efficient advanced generation breeding techniques and the development of better quality and more uniform wood have received increased attention.

Problems

One major obstacle to forest tree breeding which will likely become worse is the lack of financing, especially of practical (operational) breeding and for developmental research (basic research to further operational programs). Resistance to financing research occurs despite the fact that all in-depth studies of tree improvement have shown it to be the best economic investment in forestry.

Three reasons explain the financial problems:

(1) The forest industry as a whole has been in financial difficulty for several years. During attempts to reduce costs, the first move usually is to curtail biological research. A curtailment of the intensity of forest management (silvicultural) activities, making the genetics research appear to be less necessary, usually occurs along with this. The 1990s "fad" involved an emphasis on methods of natural regeneration, tried for many years and found inefficient. (This includes things such as seed tree regeneration, shelterwood forestry, and selective logging. The latter almost always degenerates into "high grading," taking out the good trees and leaving the poor). These systems, often called "new forestry," nominally occur for environmental (biodiversity) reasons but, in fact, are done because they are less expensive in the short term, even though they can spell disaster in the long term. Regardless, the trend toward natural regeneration reduces emphasis on operational

types of tree improvement research. Money available for biotechnological and environmental work frequently causes the shifting of funds out of applied genetics studies into the currently favored areas.

(2) There is a strong move toward privatization of forestry activities. When this happens, governmental sources of funds for research are reduced or removed. Since the forest industry is generally not noted for its research support, the organizations involved end up searching for funds wherever they can be obtained, much from consulting activities or from granting agencies. Assurance of long term, well-funded research so desperately needed for successful tree improvement research disappears, and usually the research undertaken is short term with quick payout to satisfy the funding organizations.

By the mid-1990s numerous very valuable, long term research activities were being curtailed or canceled. The results of this fallacious action were not too evident as the century drew to a close, but the long term effects will prove serious and produce a major reduction of the potential value of forestry. Many governmental organizations operate with the attitude that "industry makes the profits, let it shoulder the costs involved in growing timber." In years past, the attitude was that forest research was needed for the good of the country and both government and industry contributed.

(3) Foresters are faddists, and research activities follow the trends where the money is available. Currently, and for the foreseeable future, money is most easily obtainable for environmental studies (especially for biodiversity) and for studies in biotechnology and tissue culture. Numerous researchers have gravitated toward these areas in order to continue the level of funding needed for their organizations. Although nothing is wrong with research in these areas, the limited financing frequently results in an imbalance of funding, reducing support for standard forest genetics activities. The administrators in charge are apparently not aware that most of the activities of the biotechnologists relate to efforts to mass produce desired genotypes. Those improved genotypes are produced by conventional breeding, and, if this work is curtailed, the biotechnological discoveries will have much less utility.

The Supply of Wood

Wood is a natural, renewable resource, good for many uses. Despite the current trends to use plastics in place of wood, or steel beams in construction, a need for wood will always exist. The alternative materials create more pollution problems and are made from non-renewable resources. The question is not "Is wood needed?" but "Where will it come from?"

Genetic manipulation of forest trees is mandatory to supply society with wood at a reasonable price. The quantity and quality of the wood resource must be improved so its use will efficiently satisfy needs. Increasing land pressures make less land available on which trees can be grown successfully. To expand the productive forest base requires several intensive genetic activities.

(1) An increase in forest land requires the development of special land races more adapted to the marginal and degraded sites available. The most effective uses of genetics in forestry have been in developing adaptability to marginal sites and in tolerance to pests, which results from an intensive application of forest genetics. Most of the available lands for forest production in the tropical areas are degraded and not suited to the forest tree genotypes currently available. If the expansion into these lands is to continue and be economically successful, intensive breeding programs are necessary. If enough area is to be planted in trees to supply the world's need for wood, *very skilled and intensive activities are needed to develop the necessary land races*, as well as to learn how best to handle silviculturally the trees on the marginal sites.

(2) Predictions are common of wood inventories relative to the tropical rain forests with the assumption that they will be a major source of wood supply for the future. *This assumption is wrong!* The tropical rain forests will produce only a small volume of wood for future needs, and, therefore, the bulk of the wood must come from degraded lands. This approach will not be successful without intensive breeding programs. Although the reasons for the relative unimportance of the tropical forests in the world wood supply are many and complex, some are very briefly summarized below:

(a) *Lack of infrastructure.* It often is not economic to log and transport wood to the market.

(b) *Variability of wood.* Most tropical forests have many species of trees on a hectare with greatly differing wood properties. Mixing the various quality woods is usually not successful for final products of high utility.

(c) *Wood properties are unknown.* Some of the tropical hardwoods have the best quality wood in the world. But they are not accepted

385

Georgia Pacific's research forester, Angelo San Fralelo, checks the "super pine." This loblolly strain grew six feet in six months. This strain can supply seed for reforesting marginal lands.
(GEORGIA PACIFIC PHOTO, 1971)

in the forestry trade because the buyers are not familiar with them.

(d) *Establishing plantations is too difficult.* After harvesting tropical hardwoods a huge amount of limbs, cull trees, large stumps and other debris is left on the land, making plantation establishment difficult and costly.

(e) *Growth of competition is severe.* After harvesting, the growth of competing grass and vines in the tropics is enormous. Cleanings may be required as much as twelve times per year if the planted trees are to survive and grow successfully.

(f) *Environmental pressures are severe.* Many tropical areas are off limits to large harvesting operations.

(g) *Natural regeneration is possible but not economical:* Efficient production of timber in the tropics requires the use of the millions of hectares of degraded lands. To do this requires intensive breeding activities to develop suitable land races with desired wood.

(3) Wood quality improvement is required. The greatest need is for wood uniformity along with development of the best wood for each desired product. Genetic manipulation can achieve much of this objective. More intensive breeding relative to wood will be required in the future than has been done in the past.

Environmental Influences

Although tree breeders have been aware of the need to maintain a broad genetic base, it is becoming essential to make this fact known to the general public. This is particularly true with the use of vegetative propagation and the development of specific genotypes for unique environments. Breeding plans must be carefully made and understood to prevent enactment of rules and laws lacking biological justification. Intensive studies are required to produce data showing the safest and best method of deploying clones.

The general public has developed the attitude that any genetically improved forest tree is dangerous to the environment and biodiversity, even though all breeding methodologies attempt to avoid such a problem. The public generally does not understand that forest trees have over twice the variability compared to other organisms and that most important characteristics are genetically independent of each other, especially those related to adaptability and economic characteristics. Thus, one can have straight trees that are drought tolerant or drought susceptible, and that are tolerant or susceptible to disease. Forest tree breeders have all the advantages over other breeders because the variation and genetic patterns enable them to "tailor make" trees that are well adapted and also have maximum economic value. Forest tree breeding in the future must make maximum use of this wonderful potential.

NOTES

1 From Sziklai and Katompa, *Forest Tree Improvement* (Budapest, Hungary: Mezogazdsasagi Kiado, 1981).

2 Bruce Zobel and Talbert, *Applied Tree Improvement* (New York: Wiley & Sons, 1984).

3 Bruce Zobel and Jerry Sprague, *A Forestry Revolution—The History of Tree Improvement in the Southern United States* (Durham: Carolina Academic Press and the Forest History Society, 1993).

4 For fuller details and numerous stories about the cooperative era see Zobel and Sprague, *A Forestry Revolution.*

5 See Zobel and Sprague, *A Forestry Revolution,* pp. 45–66, for the story of the development of cooperatives in the southern United States.

6 A land race is a group of trees well adapted to a given site which has the qualities desired by the forester.

instruction began to diminish, pushed on by the influence of the early forest soil scientists. Over time forest soils became a valid area in the study of forest resources and worthy of study and research in its own right. Most forestry students in the 1990s, therefore, experienced direct contact with the subject matter of forest soils and how to apply forest soil science to forest land management. Many students also choose to specialize in forest soils at the graduate level.

CONTRIBUTION OF HILGARD AND JENNY

E. W. Hilgard and Hans Jenny proved especially effective in advancing our understanding of forest soils. Although Hilgard from Germany and Jenny from Switzerland received their initial backgrounds and showed considerable influence from European science, their contributions to the field of forest soils were distinctly North American. Most likely they would not have had the considerable impact they did on the understanding of forest soils outside of the pristine natural laboratories they found in North America.

Hilgard, who grew up in a German colony near Belleville, Illinois, developed his career and made his early reputation as state geologist for Mississippi from 1855 until 1870. His 1860 report on the geology and agriculture of the State of Mississippi brought him considerable fame. The first half dealt primarily with geology of the state, but the second half was devoted almost entirely to soils. This part established his reputation. His many observations about the nature of soils and how to manage them for agriculture as well as for forestry established many principles still used more than a century later. In recognition of his contributions he was made professor of experimental and agricultural chemistry at the University of Mississippi during the final years of his work there. In 1870 he moved to the position of professor of geology and natural history at the University of Michigan. After two years there he moved to a professorship in soils at the University of California at Berkeley, remaining there until his death in 1916.

In the course of his extensive field work in Mississippi, Hilgard observed that tree species distribution and general forest vigor and productivity have a direct relationship to the occurrence of different soils. During many years of field and laboratory work in California he extended his observations and began to mention some of the relationships in his writings. Hilgard also recognized that factors other than geology appeared to determine the characteristics of a particular soil in a particular location.

Hilgard also established the concept of soil layers (later called soil hori-

zons) as being natural components of soils and that consistent differences exist between surface soil and subsoil. In 1895 he took these ideas a step further and published a paper that discussed particular characteristics of soil by depth. In Mississippi, Hilgard recognized low calcium status of soils by the presence of pines and absence of legumes. His observations regarding distribution of trees and soils were typically quite specific. For instance, one quote from his writing states, "It is very essential, however, to take into account not only the species of trees, but also their mode of growth. The black-jack and post oak especially, as species, characterize the poorest as well as the richest upland soils, but their mode of development is very different in each case." Such observations of soil effects on forest growth are abundant in his writings.

When a new soils map for the state of Mississippi was created in 1942, Hilgard's visionary work of the 1850s provided the basis for the new map and demonstrated his accuracy in describing soil properties. In fact, Professor Vanderford commenting on the 1942 map wrote to Dr. Jenny that, "Many of the lines first drawn by Hilgard are still used on the present soil area map." Dr. Jenny paid tribute to Hilgard by publishing a special volume in 1961 entitled *E. W. Hilgard and the Birth of Modern Soil Science*.

It remained for the keen mind of Jenny to put the total picture of soil formation and distribution into sharp focus with his book *Factors of Soil Formation* in 1941. A number of early soil scientists had previously recognized that several different factors resulted in what came to be called a "soil profile" forming in a particular spot. No one, however, had put the observations together and developed a unified theory about the formation of all soils until Jenny. His teachings and writings clearly established the role of each factor he considered to be of importance in soil formation. He developed his theory with a strong mathematical foundation and illustrated the role of each factor with many field examples and studies. His teachings and theoretical development of this important concept became widely accepted across the world. Soils professionals, as well as most foresters, have some concept of how soils develop under the influence of the five factors of climate, topography, organisms, different parent materials, and time.

Due to the assimilation of previous ideas with new concepts introduced by Jenny, soil scientists have a basis to think of soils as a product of these soil-forming factors. Foresters and all those interested in soils are thus able to evaluate what soil characteristics and problems to expect in a given part of their landscape. Stated simply, the effort and contributions of Jenny en-

able us to actually put some order into the study and classification of soils. His contribution to soil science has been recognized through a special publication by the Soil Science Society of America resulting from a 50th Anniversary Symposium in 1991.[7]

ORGANIZATIONS RELATED TO FOREST SOILS

During the initial stages of the development of forest soils as a study area, forest soil scientists remained heavily dependent for technical and instructional material on associates who were primarily interested in using soils information to manage agricultural lands. Forest soil scientists also depended upon the organizations they had developed to provide a basis for transfer of information and research discussions in the agricultural arena. Forest soil scientists still used these original organizations in the 1990s for some aspects of their information transfer needs. The detailed history of those organizations and their relationship to forest soils is available in papers by Lutz and Stone.[8]

The American Society of Agronomy (ASA), composed of three divisions including the Crop Science Society of America, the Soil Science Society of America (SSSA), and the Agronomy Society of America (commonly referred to as the "tri-societies"), initially sponsored much of the activity in forest soils. The American Society of Agronomy was organized in 1907 with sections of crops and soils specifically to further interest in agronomy programs at Land Grant colleges. Many changes in organization and incorporation of associated groups have occurred over the years, and they continue. Important events have included the incorporation of the American Soil Survey Association (ASSA) into the organization in 1925. The first soil survey had started in 1889 with considerable expansion during 1900 to 1922. In all of this time the ASSA offered well developed annual technical meetings. Early activity in forest soils focused on soil surveys, so this group had a special relationship to the needs of forest soils even though forest and range soil surveys as such were not initiated until later. For example, when the Weyerhaeuser Company decided to make forest soil information a part of their land management base in the 1950s, it initially employed Dr. Eugene Steinbrenner to survey the soil of their forest land and afterward developed site quality information based on soil survey.

The ASSA demonstrated its sincere interest in forest soils by sponsoring a "Forest Soil Symposium" at its 1936 Annual Meeting. The papers given appeared in the January 1937 *Journal of Forestry* with abstracts in the Official Report of the Soil Science Society of America for 1936. The titles and authors of papers given at this early symposium were:

> "Some Soil Characters Influencing the Distribution of Forest Types and Rate of Growth of Trees in Arkansas" —Lewis M. Turner.
>
> "Contrasts Between the Soil Profiles Developed under Pines and Under Hardwoods" —P. R. Gast.
>
> "Physical and Chemical Studies of Two Contrasting Clay Forest Soils" —Roy L. Donohue.
>
> "Distribution of Forest Tree Roots in North Carolina Piedmont Soils" —T. S. Coile.
>
> "The Effect of Frequent Fires on Profile Development of Longleaf Pine Forest Soils" —Frank Heyward.
>
> "A Study of Certain Calcium Relationships and Base Exchange Properties of Forest Soils" —Robert F. Chandler, Jr.
>
> "The Effect of Forest Litter Removal upon the Structure of the Mineral Soil" —Herbert A. Lunt.
>
> "Nomenclature of Forest Humus Layers" —S. O. Heiberg

Some if not all of these names remain familiar to those who have followed forestry and forest soils work during this century. John Auten, who did much to advance the study of forest soils, was chairman of the Forest Soils Committee at that time.

The combination of the Soils section of the American Society of Agronomy and the independent American Soil Survey Association formed the Soil Science Society of America. The current organization named Soil Science Society of America was proposed and voted on at a joint meeting of the American Society of Agronomy and American Soil Survey Association in 1935 and became official in 1936. Professor Emil Truog had the foresight to plead not to start the organization name with "American," but to start with "Soil." He proposed "Soil Science Society of America" as the official name of the new organization. That modern name was thus adopted, and the new society avoided an embarrassing acronym. At the time of its formation, the Soil Science Society had six official subject matter sections:

I. Soil Physics
II. Soil Chemistry
III. Soil Microbiology

IV. Soil Fertility

V. Soil Genesis, Morphology, & Cartography

VI. Soil Technology

Although some of the early workers in forest soils were members of ASA and SSSA, their field of activity was not formally recognized as such in either society. Forest Soils, however, was organized as Division V-A of SSSA, under Soil Genesis, Morphology and Cartography. Stone lists R. F. Chandler, Jr. as the first chair of the Forest Soils Section.[9] At the 1954 Annual meeting of the SSSA the president announced that a mail ballot had approved divisional status of S-7 with the title "Forest and Range Soils." John C. Retzer chaired the Division at that time.

The structure of SSSA in the 1990s recognized the original six divisions (with the names of S-3, S-4, S-5 and S-6 changed), as well as three added official divisions and one recent provisional division:

S-1. Soil Physics

S-2. Soil Chemistry

S-3. Soil Biology & Biochemistry

S-4. Soil Fertility & Plant Nutrition

S-5. Soil Genesis, Morphology, and Classification

S-6. Soil & Water Management & Conservation

S-7. Forest and Range Soils

S-8. Nutrient Management & Soil & Plant Analysis

S-9. Soil Mineralogy

S-10. Wetland Soils (Provisional)

A strong international focus and activity has always existed in forest soils, first seen by an influx of European thought into the study of forest soils, and, during the late twentieth century, by a high amount of information flow in all directions both into and out of North America. It is important to consider international soil science activities and their impact on the study of forest soils in North America. The first International Conference of Soil Scientists took place in Hungary in 1909 with primary interest in soil classification and agricultural problems.

The fifth International Conference of Soil Scientists occurred in Washington, D.C. in 1927. Remarkably, this fifth conference was officially termed the "First International Congress of Soil Science," sponsored by the International Society of Soil Science, ASA, and ASSA. The presentations at this early conference did show considerable interest in forest soils with a range of papers related to forestry. From the standpoint of the creation of an organization with a widespread, methodical approach to the study of forest soils with continental interest and distribution, this conference could actually be considered as the beginning of the organized study of North American forest soils. In 1933 three independent soils organizations existed in the United States: (1) the Soils Section of ASA, (2) ASSA, and (3) the American Section of the International Society of Soil Science. In 1936 ASSA merged with the Soil Science Society. In 1937 the American Section of the International Soil Science Society became part of the SSSA. Thus, this one organization came to represent major activities in both national and international relationships in soils. North American forest soil scientists have a number of other avenues for international activities such as the International Union of Forest Research Organizations, but these will not be discussed further in this short history.

EARLY DEVELOPMENTS IN FOREST SOILS

Organizations subsequently developed which are more specific to forest soils.[10] The senior author, the late Stanley Gessel, was appointed to the first faculty position dealing specifically with forest soils at the University of Washington beginning in September of 1948. At the time no formal local organization concerned with forest soils existed in North America.

With the help and support of many segments of the forest industry and Dean Gordon Marckworth at the University of Washington, the first action relative to development of a forest soils organization came with a conference held in Portland, Oregon, on September 15, 1948. This meeting brought together representatives of both private and public forests who had an interest in what information about forest soils could contribute to the management of their forest land. The conference quickly disclosed that little information existed about forest soils in the Northwest, few people were working in the area, and that programs needed to be established in order to develop interest, information, and a forest soils "network" of professionals. Specific study areas that needed research priority were identified:

1. Productivity—Site classification
2. Forest tree physiology, nutrition, and water in relationship to soils
3. Nursery soils
4. Forest management activities and soils
5. Physical, chemical, and biological properties of forest soils

A more long-lasting action of this conference, however, was to organize and support "The Forest Soils Committee for the Douglas-Fir Region." A regular schedule of two major meetings a year was established, one a winter business and review meeting and the second a summer field trip for the purpose of reviewing forest soils research or problems. This schedule continued to be followed in the 1990s. This kind of an organization proved very useful and efficient, and now has counterparts in Northeastern and Southeastern United States as well as in California and Canada. The committee has since changed its name to "The Northwest Forest Soils Council" and has sponsored many special short courses and conferences on various aspects of forest soils and published a text entitled *Forest Soils of the Douglas fir Region.*[11]

In the early years of the committee, it offered regular short courses on forest soils at the University of Washington and at Charles Lathrop Pack Forest to give practicing foresters a working knowledge of the subject matter of forest soils, with special discussions relative to soil damage, watershed management, and productivity. As information about forest tree nutrition and forest fertilization developed, these subjects were also covered. These kind of short courses became the forerunner of the "Silvicultural Institute" at the University of Washington and Oregon State University, which typically teaches thirty to thirty-five professionals per year on various aspects of forest management, including two modules on forest soils.

Another organization which has proven very effective in promoting forest soils and disseminating information about current management of forest soils is the North American Forest Soils Conference, designed to bring the major producers of soils information together with the users. The first meeting, held in 1958 at Michigan State University, received limited sponsorship but covered a wide variety of subjects. Thirty-one papers were given with the principal address by Charles Kellogg.[12] In 1962 the SSSA, the Society of American Foresters, the Canadian Soil Science Society and the Canadian Institute of Forestry took over sponsorship.

The eighth conference convened in 1993 at the University of Florida. Proceedings of the conferences have all been published in individual book form so that eight volumes of research results along with discussions about application to forest problems were available in the mid-1990s. Organizers have attempted to direct each conference to highlight a particular subject matter. For example, the second conference spent considerable effort on Soil Surveys of Forest and Range Lands, as that subject was of major concern in the 1960s. The seventh conference concentrated on Productivity Problems and Questions, and the proceedings are titled *Sustained Productivity of Forest Soils*.

The professional forestry society in the United States, the Society of American Foresters (SAF), has also moved to recognize the importance of soils through the establishment of a soils working group. This has been in operation for a number of years, with elected officers and a program at the annual national meetings as well as at local levels. For example, the program at the 1994 annual meeting in Alaska was, Sustaining Long-Term Productivity—A Soils Based Approach.[13] The soils working group of the SAF regularly publishes a newsletter, becoming an increasingly important communication group for forest soils professionals.

FOREST SOILS IN CANADA [14]

Although a number of forest soils researchers in British Columbia have been members of the Northwest Forest Soils Council and worked closely with U.S. soil scientists, a divergence exists, especially in relationship to teaching, management oriented research, and the use of soils information. Canadian forest soils research and teaching centers on site classification. This was brought about after a rather large administrative and scientific debate during 1950 to 1970 that concerned the approach Canada should take to classifying forest soil productivity. The site approach won out because forests in Canada basically consist of a band of boreal conifer forests arranged in a somewhat consistent way across a glaciated landscape with strong physiographic and climatic control. In Canada, the great diversity found in forest lands across the United States does not exist, which makes a national approach to classifying forest site productivity simpler than in the U.S. Most of the land is publicly owned, and there has been no national soil survey as in the U.S. Therefore soil surveys of forest areas basically do not exist. The common denominator between Canadian and American soils studies exists only in pure soils research. The writings of Hills and Rowe explain the Canadian approach.

University of British Columbia Professor Gordon Weetman lists some of the reasons the U.S. and Canada have gone separate ways:

1. Forest soils teaching in the U.S. was often based in agricultural institutions and facilities. They therefore tended to follow the agricultural lands and methods.

2. One of the initial goals of the U.S. Soil Conservation Service as set by H. H. Bennett was to map soils in the entire land area of the U.S., and foresters had an incentive to use available mapped soil information to classify forest productivity, and not site.
3. Soil was seen as a separate item for study and not a part of site.
4. There is great ecological/climatic and ownership variation in the U.S., but not in Canada.

Weetman observed that "forest soils," as taught in Canadian schools, has to distinguish between "soil" as a legitimate object of study and its field application in provincially approved forest site systems. He also believes that the teaching is more influenced by European concepts than in the United States. He observes that forest site classification is finding more use and is providing a solid base for management. This kind of a base could lead to better political decisions about forest land use than we in the United States are making.

ACHIEVEMENTS IN FOREST SOILS

A major achievement of forest soils study has been to get the soil recognized as one of the major resources in forestry and one which must be carefully managed.

Soil Surveys

The first step in the wise use and management of a natural resource consists of knowing what exists and where. In the case of soils this means a soil survey. The American Soil Survey Association has existed since 1899, and its members were actively conducting soil surveys before that time. Such activity, however, was largely directed to lands either at that time under, or destined for, agriculture until about 1950. One of the first North American papers calling attention to the need to survey and map soils on forest areas was that of Lunt and Swanson. [15] In fact in this paper the authors substantiate the need by the statement, "At a forest soils conference held in Portland, Oregon, last September [1948] it was agreed that the setting up of a soil survey program for forest land was first in importance for the Douglas-fir region."

Soil survey activity on forest land in the U.S. has gone very well, and large areas of all classes of land ownership are now mapped. In the early years, large land owners such as Weyerhaeuser employed their own surveyors for both western and southern forests. The regular survey programs of the U.S. Soil Conservation Service (later the U.S. Natural Resource Conservation Service) expanded to gradually include all forest land, so that most small forest land owners can expect to find maps and soil descriptions in the updated standard county surveys. In future years, the NRCS hopes to enlist the aid of many professionals in the effort to develop better soil interpretations for forestry. Larger public land owners such as the U.S. Forest Service, the Bureau of Land Management, and state land agencies have also carried out surveys on their lands. The same is true for many of the forested Indian reservations. The adoption of Geographical Information System methods and more broad scale land management planning is rapidly changing the whole field of soil survey.

Soil Factors Which Affect Tree Growth

Soil, in concert with climate, proves of primary importance in determining the potential for forests to be present in a particular area as well as the potential productivity or growth rates of those forests. Though soil itself is not specifically a necessary requirement for trees, soil supplies many of the growth requirements, including water and nutrients. In addition, soil offers a support medium for growing trees and a buffer against the extremes of temperature. All soil properties, therefore, which relate to any of these are important to tree growth. These are commonly expressed as physical, chemical, and biological properties. Forest soils researchers have worked on all of these and have advanced our understanding of each one in relationship to tree growth and health. The research effort has become quite specialized with educational institutions, research centers, and researchers typically known for their contributions to specific areas, as well as overall forest management.

Water availability is one of the primary controlling factors for the presence of trees as well as vigor and growth rates. The problem with water on most forest areas is the presence of either too little or too much. In the case of too much water, flooded and wetland conditions often exist. The role of water is assuming greater importance in overall management of forest lands. Many researchers have made significant contributions to our understanding of the water relationships of forest soils. The work of Zahner has been very useful. [16] The symposium "Soil Moisture—Site Productivity," held in

1977 at Myrtle Beach, South Carolina, reviewed the entire field of forest soil water relationships and provides a good reference point for this aspect of forest soils.

Many of the other soil physical properties have been studied in detail throughout the forest areas of North America, so that we now have a good information base on the range of values to expect for such properties. Values have been used to predict tree growth rates, or site productivity, while others are used to evaluate the effects of forestry operations such as harvesting.

Forest soils provide habitat for large numbers of organisms from those very small and relatively immobile, such as bacteria and other microbes, to larger mobile insects and earthworms, to relatively large mammals, amphibians, reptiles, and even birds. Although many studies have been conducted relative to these organisms, and an extensive literature exists, our knowledge of the different kinds and the roles they play in the forest ecosystem remains rather incomplete, particularly as they relate to soil. Some organisms, such as the major decomposers, nitrogen fixing, or disease causing organisms of large economic importance, are well known and highly studied. Forest soil scientists, for instance, initially advanced the study of mycorrhizae. In this particular case, the agricultural soil scientists have followed the lead of forest soil scientists. Knowledge sometimes includes how we can affect population numbers and activities, but most studies have large degrees of uncertainty attached to them. Some organisms, such as the edible mushrooms, are becoming increasingly important in the management of forest areas as nontraditional sources of revenue, often with high value per area of forest land.

Chemical properties of forest soils have also been rather thoroughly investigated and described for most forest areas of North America, and a very extensive literature exists. We therefore know what level of acidity to expect in soils of most of the forest areas and how this has or has not changed under the influence of man's activity. We also generally know the chemical composition of most of our soils and the status of the elements as to available, non-available, or exchangeable, even though we may not know in an extremely practical way how to interpret the information relative to forest growth. Again an extensive literature exists on these aspects of forest soils for all areas of North America, and many researchers have made significant contributions.

Productivity-Site

Although these terms have been historically used somewhat interchangeably in forestry, we can review the contribution of forest soils to each. Site has been used to convey some concept of overall growing conditions and also specific productivity. A system of site classification called "site index," which has been developed for the major wood producing forest species, uses height of dominant or codominant trees at a specific age as an index. Many different concepts about productivity exist but in this review we will use the traditional biological context as stated by Hansen where productivity is "the total quantity of organic material produced in a given period by an organism." [17] For forest tree species foresters have condensed this to merchantable or total volume and express it as cubic feet or cubic meters per unit area of land per year.

Forest soil researchers have contributed a great deal to our understanding of the factors which determine site. As a result we now recognize the role that some, but not all, soil physical, chemical, and biological properties play in determining site. This information has been put to practical use through the development of soil-site indices for given species in local areas. These contributions are well documented in an extensive literature specific to the local areas and species.

Productivity evaluation has assumed an increasingly important role in discussions about the management of forest land. Some believe that forest harvest, and other uses by man, will reduce or destroy forest productivity, while others believe that both normal forest use and increased future production can occur on a given area of land. Obviously, determining what will happen on any given area of land would involve understanding what factors determine productivity and whether and how much we change these factors in use and management. Forest soils research has contributed substantially to our understanding of soil factors which are important in productivity and therefore can be used to predict or quantify changes due to land management.

The political and policy ramifications of the maintenance of forest productivity will be with us for a long time. Although many forest productivity studies have been completed, environmental attention is now focused on what is termed "sustained productivity," and a general literature is developing on how forest productivity can be destroyed. A major program in long term evaluation of forest productivity in different areas of the United States was in the early developmental stage by the U.S. Forest Service in the 1990s, and practically every forest research organization had some kind of a forest productivity program. Maintenance of soil and site quality is written into the management plan of each of the national forests, with monitoring plans designed to detect losses in forest productivity of fifteen percent or greater being designed and implemented. [18] The potential for loss of inherent site quality and the ability to grow productive forests remains on the forefront of forest soil research, probably second only to biodiversity as a central issue in the management of North American forests.

Forest soil researchers will have ample opportunity for further contributions. Earl L. Stone made a major contribution to this area with his 1973 report on "The Impact of Timber Harvest on Soil and Water." Several new symposia on the general subject of maintaining forest productivity occur each year. The demand for forest products has been increasing both in per-capita and absolute terms, but the area of land devoted to forest production has decreased over that same time frame. Interest has grown, therefore, in increasing the productivity of land for wood production through intensive management of forests. Intensive forest management basically increases wood production in two ways. The first method is by improving tree genetic potential and applying appropriate stocking, thinning, harvesting, and other silvicultural improvements. The second method involves improvement of soil properties that partially determine site. Improvement of site for forest management is probably best illustrated by the effort made in forest nutrient management and forest fertilization, which served as the underpinning for early forest soils research and continues to be a major soil manipulation that can improve site and thus forest productivity.

NUTRIENT MANAGEMENT AND FOREST FERTILIZATION

Extensive early research work throughout all of North America has clearly shown that elemental deficiencies exist in many specific forest areas. Trees growing on those sites, therefore, cannot make full use of the other growth factors of the sites. In addition to reduced growth rates trees may also be suffering decline in vigor and eventual death from nutrient deficiencies.

Research activity has not only described these nutrient deficiencies but also shown how corrective action can be taken to eliminate them through forest fertilization, or the adding of nutrients, designed to add elements either lacking or in low supply. To be successful, fertilization must be very specific to the needs of a given forest area and to the economics of forest management. In North America, forest fertilization is generally well-developed where the following three basic conditions exist:

1) Forests respond to fertilization with significant increases in growth rates,
2) high demand in that region makes the price of raw wood expensive, and
3) the infrastructure for buying, moving and applying fertilizers exists.

In the future, increasingly stringent environmental considerations may restrict the ability of forest land managers to apply fertilizers, even where these three conditions make application profitable and attractive.

An extensive literature has been developed to cover this subject, and many reports of special symposia and conferences exist. The development of the information base for good programs of forest fertilization in all of the major forest areas of North America has been an important contribution of forest soil research and the development of the relation of forest soils to tree growth.

Fertilization research and practice has also extended into the application of municipal waste to forest land. In many areas of North America programs for use of "biosolids" (previously called sewage sludge) are well underway. This use on forest land provides an important contribution to recycling and environmental protection goals of society. The program will expand in the future. Again an extensive literature covering all aspects of such use exists. One of the earliest books dedicated to the subject of application of wastes to forest land was published by Sopper and Kerr and coined the term "living filter" for the high ability of forests to retain nutrients added as waste materials. [19]

For many areas, researchers have developed nutrient cycles, which include the sizes of discrete forms or locations of nutrients (termed nutrient pools) and the relative movement of nutrients from one pool to another (termed nutrient fluxes). There is justified concern that one form of management or harvest may remove more of a particular element than the area can replace or sustain over a particular time period. If the harvest is repeated at frequent intervals then nutrient deficiencies may occur over time. Harvest at long intervals and only partial crop removal constitute less of a potential problem.

Research remains badly needed in the area of the effect of nutrients on long-term productivity of forests to address claims (with little or no data provided) that on some forest soils as few as three harvest cycles would totally destroy productivity. As a result we are now in an era of expanded discussion and research on the question of "sustainable forestry" and what role forest soils play as sources of nutrients for forest growth. A great deal of need exists for continuation and expansion of the research that early pioneers in forest nutrient cycling research conducted as forest management systems change.

OTHER RESOURCES

The increasing attention to the use of forest land, especially that in public ownership, for purposes other than wood production has expanded the need for a soils information base to serve these uses. Alternative uses of forest land will continue to expand with increasing population and changing desires for the use of that land. Forest soil research has responded by incorporating these alternative uses into research plans. and information development and communication are well underway.

Very important uses are water catchments and water supply for agriculture, municipalities, or in some cases, for wildlife such as salmon. In local areas this use of forests may have first priority, and management practices shifted to reduce water use by forest trees while still maintaining quality. An extensive literature to which a large number of researchers have contributed exists in the area of water quantity and quality effects of forest management.

Many challenges remain that will need to be addressed in the future by forest soil scientists in order to utilize and protect the soil resource for the benefit of man and nature. The final section of this chapter will review important needs, realizing that many of the original questions asked several decades ago remain largely unanswered, and many compelling new questions have been asked since then.

EVALUATION AND CHALLENGES

Perhaps the most important accomplishment comes from the general recognition that soil is an independent natural body which plays an important role in the functioning of all natural terrestrial systems. Knowledge of the soil provides a critical part of our understanding of these systems. At the same time, forest soils have achieved universal recognition as a legitimate object of study at all levels and become the subject of much research effort and reporting.

One of the greatest challenges to researchers and educators will be to keep on the right track in their research and instructional efforts in the face of many charges of universal soil degradation and claims of inability to practice forestry while at the same time protecting forest soils. Forest soils are not magical entities, though their properties often prove complicated and their form and functions fascinating to layman and advanced scientist alike. Some soils are very resilient and not highly subject to degradation by even the most intense forest management, while other soils remain very fragile and should be highly protected from management that may degrade soil quality. Some of the concerns and charges of universal soil degradation have developed because so little attention was paid to forest soils for so long in forest management. Clearly soil research must identify specific cases where soils can and cannot be easily degraded and make that information available to forest management professionals as well as the public.

Expanding human populations over the past two hundred years places pressure on land resources and increases the need to maintain soils in the best possible condition in order to maintain productivity. The 1994 Annual Report of the German publication *Prisma* focuses on the threat to soils and especially soil degradation. The advisory council to this group came up with twelve syndromes for depicting the causes of soil degradation. One of these, the so-called "Sarawak Syndrome," attributes the degradation to conversion or over-exploitation of forests that have delicate soils. [20]

Another recent publication takes a broader look at the subject by considering what is termed "Soil/Resilience and Sustainable Land Use." [21] A number of papers in this volume deal with degradation of soils from many types of uses, including forestry. The concept of soil resilience and recovery, however, is introduced and described. Considerable emphasis, therefore, falls on the ability of the soil system to recover from degradation.

Forests and forest land have been used for many purposes by man. Native peoples in forested areas, generally, but not necessarily few in numbers per unit area of land, used the forest mainly as a source of food and shelter and typically without a great impact. In many areas, however, indigenous people used fire to bring about certain desired changes in conditions for grazing and agriculture and had widespread impacts on the land. Hilgard refers to this in his early writings about soils in Mississippi.

Reason exists to give priority to managing land resources to maintain or sustain basic productivity. To do this requires, however, an understanding of the factors which control forest productivity and how these may be affected by human action. Soil properties do play a central role in forest productivity, but so do many other factors. Soils also vary greatly from place to place both on the earth's surface and locally. The effect, therefore, of one forest crop harvest every forty years may be very insignificant on one soil, but cause negative impacts on another, even over small distances. For instance, if we are concerned with essential elements as the negative factor, we must recognize that the amount removed in a given harvest may be quite small in relationship to either the total or available storehouse. Some forest soils are deficient in one or more elements for good growth of trees under virgin or natural conditions. Although the practice of forest fertilization may not mesh well with some concepts of "sustained productivity," problems of nutrient deficiency can be corrected by applying the necessary essential elements.

Much of the agricultural land in parts of North America or other continents was developed by removing the forest cover, frequently by total area burning and then intensive agriculture. In the early years of development this typically occurred without the benefit of modern fertilizer and management practices. If total soil destruction could occur, we should certainly expect it under these drastic conditions. The extremely large loss of topsoil due to soil erosion from agriculture in the southeastern United States during the 1800s provides a good case in point. At the end of the twentieth century this region encompassed some of the most productive land with the highest forest growth rates in the U.S. Some of the highest growth rates measured in the region occurred on "old fields," once considered so degraded as to be no longer acceptable for agriculture and thus available for forestry.

To meet some of the increased demands on forest resources and furnish raw materials, trees must be harvested. Man has of course devised many different methods for harvesting and removing trees from forest areas but, depending upon how they are applied, various kinds and degrees of impacts can be made on the forest system. For large areas of the world, and particularly in North America, the purpose of tree removal has been conversion to other land uses, especially agriculture. Unfortunately, many of these efforts have been described as forestry in the popular literature. The blame, therefore, for much of the deforestation associated with land clearing for agriculture falls on the practice of "forestry" and, for maximum effect on the public, "multinational companies." Conversely, in many cases, watchdog environmental organizations appear reluctant to blame small farmers for defor-estation, even when it is warranted. Widespread research evidence shows that forest management in general is sustainable from the standpoint of nutrient cycling in most situations that have been studied. Two of the im-portant roles that forest soil scientists can play concern identifying situations where a given level of forest management would or would not be sustain-able and providing the range of possible corrections that would offer sustainability for all forest management.

Natural systems are all resilient to a certain extent, with soils typically very much so. Forest land can and does recover from mismanagement, though mismanagement of the soil resource may result in a period of time that a given area of land is unproductive. Recovery of a natural system is typically a certainty, provided that the necessary organisms that might have been eliminated can enter the system. A good case in point is the frequent total destruction of forests in the Pacific Northwest by mudslides and vol-canic eruptions, and the widespread incidence of other severe disturbances in many other areas. For instance, the example of the regeneration of for-ests on and near Mt. St. Helens offers a very good example of the resilience of "totally destroyed" forest systems. Such patterns of total destruction of forest ecosystems and their rebuilding over time dominate the landscape of many parts of North America. Damage to forest ecosystems by improper forest management prove typically much less destructive than severe natu-ral disturbances. No valid reason exists to believe that damage will be longer in duration just because it is man-induced and not natural.

The arguments for total destruction of the soil from present day forest harvesting appear to have been developed primarily to support the environ-mental movement's programs against any harvest of forests termed "old growth." While recognizing the high ecological value of these resources, many of the methods used to provide old growth protection, including uti-lizing the myth of permanent ecosystem destruction due to normal forest management, stand open to question. Existing old-growth forests often regenerated from previously disturbed systems, and the age class distribu-tion of many old-growth stands in the Pacific Northwest indicate that dis-turbance may have been the original event that set those forests in motion.

CONCLUSIONS

Progress began from the early times of forest soil science as simply recogniz-ing that forest soils are related to the type and productivity of a forest that grows on that soil. This initiated classification and mapping of forest soils in order to estimate potential forest productivity and concentrate management on productive areas. Later work showed that soil was a primary reservoir of critical nutrients for forest growth, and recognized that soil nutrients often limit forest growth, and that nutrients can be depleted or increased by forest management practices. The concept of the forest nutrient cycle provides a framework for evaluating the potential impacts of forest management on the nutrient capital of the forest ecosystem. In addition, methods of nutrient enhancement by fertilization or culture of certain species or tree/organism symbiotic associations provide a means of increasing nutrient availability and possibly forest productivity.

All of the earlier findings related to forest soils remain valid and useful and still active fields of study. The concept of forest soil as a component of a delicate forest ecosystem where constant linkages between all parts of the forest ecosystem are critical to long-term forest health is also not recent. In fact, the conservationist Aldo Leopold mentioned in his earliest writings that it is important not to throw away any parts of an ecosystem, or we couldn't put it back together in a way that would function, saying, "The first rule of intelligent tinkering is to save all the cogs and wheels." The widespread acceptance, however, of the forest ecosystem as a functioning "organism-like" unit remains a subject of considerable debate, a concept by no means universally accepted. Over short time scales all parts of a natural forest ecosystem are not necessary for forests to "function." Some organisms may not serve a critical function, and some may be redundant for critical processes.

Instances certainly exist of specific, widespread disasters where dominant species have been eliminated from forests such that the forest ecosystem changes but continues to function. Examples include the blight of the American Chestnut and the Dutch elm disease. Such systems are certainly different, and in most ways less desirable than the original forest system, but they do function with many of the properties of the original forest due to the resiliency of the forest. The authors are unaware of any instance where elimination of one or more "cogs" of an ecosystem caused a complete nonfunctioning of a future forest ecosystem. It may be telling that the only instances of widespread continual deforestation in local sites that come to mind involve destruction of the soil resource. Thus, the challenges to future forest soil scientists include developing knowledge of forest soils both as an entity worthy of study in its own right, but particularly as a critical component with myriad linkages to the whole forest ecosystem. It is also clear that the primary importance of forest soil research in the future will be to assure and demonstrate that forest management be sustainable in the long term. Otherwise, forest management in North America may largely cease to be an acceptable land use option politically due to real or perceived losses in forest productivity.

NOTES

1 Due to the death of Professor Stanley P. Gessel during the writing of this chapter, and before acknowledgments were written, it is not possible to recognize each person that provided significant help to him. The junior author would like to thank Chuck B. Davey, Earl L. Stone, Dick E. Miller, Gordon Weetnam, and J. Stan Rowe for valuable inputs to this chapter.

2 E. Meeker, *Seventy Years of Progress in Washington* (Seattle, 1921).

3 S. A. Wilde, *Forest Soils: Their Properties and Relation to Silviculture* (New York: The Ronald Press, 1958); "Cotta's Preface," *Newsletter of The Appalachian section of The Society of American Foresters* (Winter 1992).

4 S. A. Wilde, "Forests and Letters" *The Land* 12 (1953):328–330.

5 S. A. Wilde, "Glinka's Later Ideas on Soil Classification" *Soil Science* 67 (1949):411–413.

6 T. S. Coile, "Soils And Tree Growth" *Advanced Agronomy* 1 (1952).

7 "Factors of Soil Formation: A Fiftieth Anniversary Retrospective" SSSA. Special Publication 33.

8 J. F. Lutz, "History of the Soil Science Society of America" *Soil Science Society of America Journal* 41 (1977):152–173; E. L. Stone, "Some highlights of Division S-7, Forest and Range Soils" *Soil Science Society of America Journal* 50 (1986): 1094–1095.

9 Stone, "Some Highlights."

10 We begin by stating that the review will not be uniform for North America and readers may find coverage of their area is not included. We apologize for this but point out that contents of this chapter must come from local knowledge and personal contacts, since regularly published materials are so few on this subject. Because our careers are limited in geographical area, information available to us to write this chapter is incomplete for some areas. We hope that our effort will eventually result in the production of a more complete history of Forest Soils in North America as readers with more complete information become aware to us, and we encourage them to contact us with that information.

11 P. Heilman, editor, *An Introduction to Forest Soils of the Douglas-Fir Region of the Pacific Northwest* (1957).

12 *First North American Forest Soils Conference Proceedings* (East Lansing, Michigan: MAES, 1958).

13 "Program of the Soils Working Group at the Society of American Foresters National Convention, Anchorage, Alaska," Society of American Foresters, (1994).

14 The authors do acknowledge that the coverage is principally related to the United States. We acknowledge that the basic information for a section specifically related to Canada was supplied by Professor Gordon Weetman of the University of British Columbia.

15 H. A. Lunt and C. L. W. Swanson, "Mappable Characteristics of Forest Soils" *Journal of Soil and Water Conservation* 4 (1949): 5–44.

16 W. E. Balmer, editor, *Proceedings Soil Moisture—Site Productivity Symposium, Myrtle Beach, South Carolina* (1978).

17 H. D. Hansen, *Dictionary of Ecology* (Washington, D.C.: The Catholic University of America, 1962).

18 R. F. Powers, D. H. Alban, G. A. Rourk and A. E. Tiarks, "A Soils Research Approach to Evaluating Management Impacts on Long-Term Productivity" in W. J. Dyck and C. A. Mees, editors, *Impact of Intensive Harvesting on Forest Site Productivity*. F.R.I.. Bull 159. (Rotorua, New Zealand: FR.I., 1990): pp. 127–145.

19 W. E. Sopper, and S. N. Kerr, *Utilization of Municipal Sewage Effluents and Sludge on Forest and Disturbed Land* (University Park: Pennsylvania State University Press, 1979).

20 German Advisory Council, "World in Transition: Threat to Soils" *Prisma* 5 (1994):10–17.

21 L. R. Oldeman, "The Global Extent of Soil Degradation" in D. J. Greenland, editor, *Soil Resilience and Sustainable Land Use* (Tucson: Cab International, 1994): pp. 99–118.

STEPHEN G. BOYCE AND CHADWICK D. OLIVER[1]

THE HISTORY OF RESEARCH IN FOREST ECOLOGY AND SILVICULTURE

INTRODUCTION

Extensive forest management began in the United States in the late 1800s and early 1900s at a time of increasing concern about conservation of natural resources. Specific techniques to improve forests—thinning, regeneration, and others—were first studied and applied systematically in the United States during this time. Referred to as silvicultural techniques, these were based on European silvicultural experience.[2] During this same time, ecology developed as a science. Ecological research should test the theoretical basis for silviculture; and, in turn, silvicultural research should test ecological theories. Often the two fields have interacted this way; different influences on the two fields, however, have caused elements of them to clash at times. This paper will describe the changes in forest ecology and silviculture research trends in the United States, rather than trying to trace individual personalities or origins of schools of thought. Research will be used in the broad sense to include experiments, observations, and syntheses of these experiments and observations into generalities which allow predictions and explanations of phenomena.

Tracing those implicit assumptions which the scientists of each period took for granted, but are not as obvious now, proves more difficult than tracing the trends of explicit scientific thought. A significant assumption has generally been that the primary needs in silvicultural research are biological, and so silviculture has interfaced strongly with ecology and physiology. Silviculturalists have acknowledged a need for site-specific decision making, economic and technical efficiency, and enfranchising the workers and public, but have rarely done research on these aspects of modern management.[3]

Ecology

Ecology emerged in the late nineteenth century in a scientific atmosphere of great progress in holistic theories. These held that the complex whole of nature produces interrelated properties that can be explained in relatively simple laws.[4] The Laws of Thermodynamics, Darwin's theory of evolution, Einstein's Special Theory of Relativity, the peneplane concept in geology, and Marx and Engel's social theories all appeared during the late nineteenth and early twentieth century, at the same time early ecological theories developed.

Some ecologists sought a holistic theory of ecology and developed the concept of succession and climax communities.[5] Other ecologists sought more mechanistic explanations to observed phenomena and used explanations based on the fundamental physiological, edaphic, climatic, and other processes involved. For several decades, the two approaches to ecology—the holistic and mechanistic—developed synergistically. Eventually the approaches diverged.[6]

Early silviculture research looked to ecological theory to understand how forests develop so they could be sustained through silvicultural systems. Many researchers worked as both ecologists and silviculturists, often making contributions in both fields. The close association of silviculture and ecology continued in the late twentieth century, with most silviculturists contributing to ecological research.

Silviculture

Silviculture appeared in the United States at the same time as ecology. Silviculture has been defined as "the art and science of controlling the establishment, growth, composition, health, and quality of forests and woodlands to meet the diverse needs and values of landowners and society on a sustainable basis."[7] It applies the theories of more fundamental biological and social sciences to managing forests. Silvicultural research calibrates these theories and applies them to specific conditions. When the theories can not explain phenomena observed by silviculturists, silviculture research provides feedback to the other disciplines and helps with research in these disciplines.

Silviculture also provides theories used by even more applied disciplines such as policy and management.

Silviculture can roughly be divided into two parts. *Silvicultural operations* includes individual activities done at discrete, short times to alter the stand structure and development. The other part consists of the *coordination of specific operations over time* to achieve the desired long-term changes in stand structure and development. These coordinations have (perhaps awkwardly) been referred to as "silvicultural systems." The term "silvicultural systems" came from nineteenth century Europe, where forestry was based on each forester's perception of what trees to harvest and how large to make canopy openings to obtain regeneration and to encourage growth of young trees.[8] Fernow, a forester from Germany and early chief of the United States Division of Forestry, described three controls for organizing a forest:

1. rates of timber harvest, measured as the proportion of total canopy removed per year, or some equivalent measure.
2. sizes of canopy openings formed to encourage regeneration and growth of young trees.
3. kind of regeneration encouraged by sizes of canopy openings, removal of competing plants, or planting desired species.[9]

He designated different combinations of these three controls—"harvest systems," "regeneration systems," or "silvicultural systems." Schenck described different "harvesting" types (analogous to silvicultural systems) to obtain "natural seed regeneration." He also described a series of forest operations (e.g., weeding, thinning, and pruning) but added that it "is of little importance in America at the present time since there are no wood crops at hand which might be profitably tended." Hawley acknowledged that silvicultural systems are sometimes used synonymously with regeneration methods. Troup emphasized that silvicultural systems include tending as well as tree removal and regeneration.[10]

Although people had been managing forests in one form or another in North America for many centuries, silviculture was introduced from Europe as a specialized discipline during the environmental movement of the late nineteenth and early twentieth century. Concern about the sustainability of soil productivity, water quality, and wood products led to studies of ways to reforest deforested areas and to sustain the values on existing forests. Agriculture-like plantation operations reforested areas degraded by erosion, overgrazing, nonsustainable agriculture, wildfires, and logging. Existing forests were to be managed in ways which would not reduce their sustainability.

The greatest area of uncertainty in early silviculture was not in individual operations, organization, or socio-economic considerations; rather, it was in understanding how forests developed, so that management practices could be done in harmony with natural processes. The overall objective was to provide the various commodities to society; consequently, the social science basis for silviculture appeared to be agreed upon by foresters and the public. Most early foresters were familiar with agriculture or engineering and had knowledge of such individual forest operations as seeding, planting, weed control, and harvesting. Less well known was what silvicultural systems to select—how the various operations would be organized and applied to an existing stand over time to achieve the objectives.

PRE-1920

Ecological Research

Research in ecology developed about 1900 at a number of colleges and universities, including Nebraska, Chicago, Columbia, Minnesota, Michigan, North Carolina State, Yale, and Harvard. Much of the field research addressed specific silvicultural problems and followed a mechanistic approach. This research was related to competition, succession, and regeneration in forests or on lands being reforested. Silviculture textbooks by Toumey, Hawley, and Baker summarized much of this research. The mechanistic research consisted of short term observations of permanent plots, but was limited by the rudimentary nature of statistical tests.[11]

Development of holistic theories of ecology can be traced to the European philosophers G. W. F. Hegel and Herbert Spencer, and the biologist Ernst Haeckel, who argued for viewing natural communities as a kind of biological organism in which the whole was more than the sum of its parts. Charles Bessey, Rosco Pound, and Frederick Clements, biologists at the University of Nebraska, expanded this philosophy to the perspective that plant communities were co-evolved, super-organisms and that each plant community had a behavior and life of its own. It was hypothesized that, as plant succession proceeded, plants occupying a site modified the substrate, enabling the next group of plants (sere) to establish. The deterministic succession of seres culminated in a self-perpetuating, stable state termed "climax." Vegetation comprising the climax consisted of the most shade-tolerant, perennial species for a given geographical and climatic region. Although a few early ecologists acknowledged that disturbances and competition were important factors, the teleological view of the holistic community was espoused by most influential ecologists of that time and persisted well into the twentieth century.[12]

These ecologists and most of their colleagues believed that the natural or normal vegetation for any geographic region was its climax, a self-perpetuating steady-state or equilibrium condition. Disturbances, from their viewpoint, were extrinsic to the natural system and generally of human origin. The emerging conservation movement to control overcutting, overgrazing, excess burning, and frequent erosion and to reforest abandoned farmland probably masked the importance of natural disturbances and added to the perception that disturbances were human-caused.[13]

Much of this holistic theory was inferred from analogies among different grassland, shrub, and forest communities as well as from chronosequence observations, short term permanent plot observations of grasslands, and studies which had inferred small trees were necessarily younger than larger trees in the same stand, and so represented later invading "seres."[14]

Silvicultural Research

In forestry schools, much of the emphasis fell on engineering and wood products. Mensuration studies developed ways to estimate stand volume and site productivity, and early attempts to develop yield tables were made for the few species considered valuable. Both national forest and large estate planning involved management studies. The "estate approach" to forestry reflected the European education of many early American foresters. Only later were forest management approaches considered for small landowners. Forest biological research centered on describing the biological requirements of species, as well as studying seeding habits and insect and disease pests.

Fernow directed the first North American silvicultural studies.[15] From this source, European philosophy and methods of research in forestry and ecology were linked, with modifications, to forestry in the United States. Research methods consisted mostly of short term remeasurements of permanent plots, observations of results from unreplicated field practices, and chronosequence observations. The body of scientific knowledge increased rapidly.

Silvicultural systems were poorly developed for the different forest types,

*This permanent sample plot of sugar maple and white pine on the
Biltmore estate in western North Carolina is marked for thinning.*
(USDA FOREST SERVICE PHOTO FROM A LANTERN SLIDE)

and forestry guides were promoted as temporary solutions until more
knowledgeable silvicultural solutions could be found.[16] In 1903, the Cornell
University Forestry School lost its political support when Dean Fernow
proposed managing some of its lands by clearcutting. Gifford Pinchot, who
replaced Fernow as chief of the Division of Forestry, favored selection cutting in a 1903 research article.[17] Although favoring selection cutting was not

scientifically documented, it established a management approach which could be tested. This approach was compatible with much of the logging, where the large diameter trees of a few species were removed with the rest left standing. Much silvicultural research examined stand responses to these partial harvests. Responses such as growth and amount of decay of residual trees, and susceptibility to windthrow proved common, immediate concerns. Selection harvesting was politically pragmatic from Pinchot's perspective (based on such events as the Cornell University experience), and it was operationally pragmatic for establishing relations with the logging companies who found partial cutting (actually, more high grading than selection cutting) to be economically feasible. It was scientifically credible since it was compatible with the succession/climax ecological theory of the day.

Light burning to reduce future conflagrations was opposed in print in the late 1800s. This helped give conceptual design to a Division of Fire Control created in 1915. The socioeconomic perspective (or theory) of providing conditions where forest commodities could be sustained for the American public drove silviculture during the late nineteenth and early twentieth century. Organization and management theory was just being defined as a discipline suitable for research in the late 1800s and early 1900s. Further research in management concentrated largely on efficient, top-down ways to coordinate harvest of forests to sustain timber flows.

By 1920 a large number of rule-based silvicultural systems had been described in North America. The systems, however, were primarily used as a teaching tool. Practicing foresters in Europe and America had found that rule-based harvest and regeneration practices limited the ability to adapt to the myriad of field situations, and by 1911, the USDA Forest Service abandoned rule-based harvest and regeneration systems. Over the years, many experienced foresters continued to warn against formalized, largely textbook concepts that gave poor results.[18]

1920–1940

Ecological Research

Ecology and silviculture became more formally linked in research and education by accreditation of forestry schools in the 1930s. In the previous decade, concerns arose about the quality of forestry education and research in universities. After many studies, discussions, and publications, the Society of American Foresters launched accreditation of forestry schools in 1936.

Specific courses were not defined; however, areas of instruction required for accreditation included ecology or equivalent material in forest biology, silviculture, and related subjects. One intent (and effect) of accreditation was to ensure close linkage of ecology and silviculture and to stimulate ecological studies of forests.[19]

In contrast to the framework and unity of direction for research in forest ecology and silviculture, no generally accepted direction or unity existed for research in general plant ecology. Differences in philosophy about organization of natural communities erupted in the ecological community in the 1920s and affected ecological and silvicultural research for many decades. Of the many schools of thought, the two most prominent in 1935 were the "holistic" and the "mechanistic."[20]

The Holistic School

One school followed Clements' holistic philosophy, which included viewing succession as a function of superorganisms and progress toward a climate-linked climax. Professors of this school used the textbook *Plant Ecology*. The idea of communities as superorganisms was later linked to the holistic philosophy of Smuts, who opposed Darwinian selection theory and operational research methods. Clements and his followers linked holism and the two thousand-year-old belief that nature was orderly and purposeful. The terms "balance of nature" and "unity of nature" appealed to many people, including ecologists and silviculturists.[21]

Proponents of the superorganism philosophy believed individuals were linked in mutual relationships to form systems that exhibited properties greater than the sum of the properties of individuals. Communities followed predetermined cycles of development, such as ecological succession, competitive exclusion of species, and emergence of predetermined dominants, that led to a predictable climax for a given climate. Not aimlessly changing aggregations of individuals, natural communities were directed toward a goal-- a climax. Included was the supposition that organisms cooperated to produce systems greater than individuals and, thus, must somehow incorporate genetic material to control this greater system. Under this philosophy, plants and animals that invaded disturbed communities changed soils and other physical properties to direct ecological succession toward an ecologically determined climax.[22]

The philosophy of superorganisms had many followers, especially those who liked phrases modified from Spencer, such as "the whole community of organisms is greater than the sum of the parts," those who liked to think reductionism led to destructive exploitation of forests, and those who believed holism led to preservation and stability. Mysticism, implied in the superorganism philosophy, appealed to many people who believed everything had to have a purpose. The idea of a stable climax as the purpose of succession was satisfying to many people, including some biologists. Clements' followers, especially John Phillips, pushed the superorganism philosophy into agriculture and forestry. The word "climax" continued to be found in some literature during the 1980s. The philosophy of communities and forests functioning as superorganisms extends to the present in many forms and phrases, such as mutualism, steady-state, animal and plant strategies, habitat types, relative stability, sustainable state, "ancient forests", and some definitions of "old growth."[23]

Clements' superorganism philosophy was more closely tied to the writings of Spencer and Smuts than to any romantic "back to nature" proto-environmentalism. In the 1920's, most ecologists, including Clements and his followers, were concerned with how people interact with "nature" to produce goods and services for growing human populations.[24]

The Mechanistic School

The mechanistic school searched for operational answers in ways fundamentally different from those who supported the superorganism philosophy. Cowles and the botanists working with Fernow intended to develop field studies with procedures others could repeat, evaluate consequences, and use to adjust, support, or discard theories. The method was scientific and operational. Outcomes from these mechanistic methods led a number of ecologists and silviculturists to challenge the superorganisms philosophy in the 1920s.[25]

The strongest challenge came in 1926, when Henry A. Gleason presented evidence for the dynamics of natural communities to be understood only as the sum of the dynamics of individuals.[26] The paper stimulated research to disprove the superorganism philosophy and to present experimentally collected and statistically evaluated evidence to support the individualistic concept. The next five decades produced many publications in ecology about the superorganism versus the individualistic theory of natural community organization.

Comparison of the schools

The Ecological Society of America organized a meeting to discuss and publish papers on these two philosophies in 1935. At this meeting, Tansley used the word "ecosystem" to attempt to establish a scientific framework and unity of direction for research in ecology, basing the concept on the emerging perspective of analyzing systems. Tansley presented reasons for rejecting the terms "complex organism" and "biotic community", which were used at the time to imply holism, unity of nature, and balance of nature. Tansley wrote: "The whole method of science, as H. Levy has most convincingly pointed out, is to isolate systems mentally for the purpose of study." In this context, ecosystems "form one category of the multitudinous physical systems of the universe, which range from the universe as a whole down to the atom." As presented in his earlier papers, Tansley was referring to an ecosystem as a mentally isolated organism or community and the associated environments identifiable for study, experimental research, and classification in the same way that Levy had isolated physical systems for study and management. The intent of Tansley's paper was to turn research and the philosophy of ecology toward scientific procedures, such as those described by Levy.[27]

The 1935 meeting ended with neither a unity of philosophy nor a consensus on research direction. Even as the debate proceeded, many scientists of the mechanistic school published studies which enhanced the ability of silviculturists to implement silvicultural operations. H. J. Oosting wrote most of his 1956 book *Studies of Plant Communities* as lecture notes in the 1930s and used them to educate many student foresters at Duke University in the 1940s.[28]

Silvicultural Research

Research in silviculture continued at a slow pace until after World War II. Several textbooks, however, which merged ecological information and silviculture appeared in the 1920s and 1930s. These books established a common framework and direction for research in forest ecology and silviculture. Toumey and Hawley published textbooks about trees and the environment and suggested ways to apply the information. In the 1930s, Korstian worked with Toumey and expanded his book into *Foundations of Silviculture upon an Ecological Basis*. Hawley's original book enjoyed eight revisions. In 1934, Baker completed *Theory and Practice of Silviculture*, while two years later, Westveld published *Applied Silviculture in the United States*. Every edition of these textbooks emphasized operational procedures. These books taught students procedures to solve scientifically a diversity of management problems.[29]

Silvicultural research began to have additional questions about the appropriateness of selective harvesting. Foresters maintained a goal of sustaining a timber supply to the public; therefore they viewed what is currently called "old growth" as "overmature" forests, where decay was occurring more rapidly than growth. To maintain a timber supply, it was considered best to remove this timber before it rotted further and replace it with young, fast growing trees. Foresters began asking the specific question: would a timber supply be better sustained by allowing residual trees to grow after selective cutting, or by allowing new trees to grow after clearcutting all trees in the previous forest? Research plots were established to determine the effects of selective harvesting in many parts of the country. Guidelines for diameter distributions for selection cutting developed.[30] Meanwhile, research stud-

Silvicultural research began to address questions concerning the benefits of selective cutting versus clearcutting. This image shows selective cutting of fifteen acres of young Douglas fir, Clackamas County, Oregon.
(SOIL CONSERVATION SERVICE PHOTO BY BRANSTEAD, 1944)

ies on clearcutting, shelterwood cutting, and seed tree cutting as alternatives were conducted in many regions.

Increasingly, research showed that selective harvesting caused a shift toward more shade tolerant species. While this shift was in keeping with the contemporary ecological theory of succession, the shade intolerant species were considered the most valuable for timber. Foresters began to realize they wanted to maintain these intolerant species, rather than let the forest grow to a "climax" condition.

Another emerging concern was the competition of undesirable "weed" species with seeded or planted crop species (usually conifers) after harvesting "old field" plantations. The initial "old field" plantations consisted of "pioneer" conifers, and so behaved according to the "succession/climax" ecological theory. According to this theory, harvesting the pines should reset "succession" to the initial stages, in which the pioneering conifers again had a competitive advantage. After the harvest, the unexpectedly intensive competition with hardwoods and other "late-successional" species after the harvest began to cause silviculturists to ignore the "succession/climax" holistic theory, considering it impractical for management. With the emphasis on manipulating the forest for certain values (usually timber), the theory was not so much questioned as ignored.

While problems emerged with selective harvesting of natural stands, plantations established on old fields and cut over lands grew well. Early experiments with planting seedlings proved promising enough that nurseries were established in many places for large-scale tree planting. Scientists began to question the wisdom of excluding all fires, as researchers were showing longleaf pine forests in the Southeast and ponderosa pine forests in the West grew well with fires.[31]

Research began to consider economic competition rather than regulation as a way to maintain timber supplies. On national forests, apparently limitless timber inventories masked the dichotomy between forest harvest regulation, based on centrally planned harvest allocations, and silvicultural practices, based on site-specific decisions.

1940–1960

The successes of science in the war increased funding for research following World War II and created many changes. More faculty began training graduate students, who did increasingly more of the research. At the same time, availability of large sums of money independently in silviculture and forest ecology, strong competition for the money, a growing dependence of universities on overhead from this money, and control of the money by federal agencies led to more aggressive efforts by ecological and silvicultural scientists to obtain funding and produce results. By 1960, more than one thousand technical and scientific papers, theses, and dissertations flowed from schools of forestry and research institutions every year. Possibly half of these publications could be classified under silviculture and forest ecology.

The increased pressure of academicians to obtain funding gradually shifted the decision of what was important silvicultural research from university professors to the Forest Service. Also, in 1955 the Forest Service required all of its research papers to be approved by a statistician. Those qualifying as statisticians were sometimes limited in their statistical creativity. Consequently, research in silviculture became restricted to subjects which could be researched in statistically acceptable ways. The Forest Service established formal research publication outlets, and, in 1955, the Society of American Foresters began publishing a peer-reviewed journal, *Forest Science*.

Ecologists developed new areas of research and explored new sources of funding, no longer depending on the Department of Agriculture—especially the Forest Service—as the primary source. The National Science Foundation, formed in 1949, became a primary source of funds to support ecological research, especially in universities. Additional funding came from the Atomic Energy Commission and other federal agencies for nontraditional research, as well as from the military, chemical companies, and various industries. Research interest, objectives, and even procedures shifted away from the statistical rigor and the management-oriented concerns of silviculturists.[32]

Ecological Research

Mechanistic ecological research continued in the 1940s and 1950s. At first, the research was interpreted in the perspective of the classical, holistic "succession/climax" theory, which did not appreciate the dominant role of disturbances. By the 1950s ecologists openly questioned the "steady state" and "succession" concepts of the classical holistic theory. A study at the Harvard Forest by Earl P. Stephens, which reconstructed the history of a stand, revealed that disturbances had periodically disrupted the stand and that most species regenerated following these disturbances, not in a "successional sere" manner. Although the major portion of the study was not published until 1977, the unpublished evidence influenced numerous scientists who had visited Harvard Forest, such as Raup, Spurr, and Henry and Swan, to question the holistic, classic "succession/climax" theory.[33]

Ecologists and silviculturists rarely used the word "ecosystem" until the 1950s, when it was revived with new definitions and connotations. Oosting used "ecosystem" in the sense of a whole system, a unit for research and management as described by Tansley. Yet Oosting elsewhere linked belief in the unity and balance of nature by claiming natural communities are groups of species adapted to living together, and the organisms are in balance with potential productivity of the environment. He described how elimination of a species or introduction of a new species may disrupt this balance. Thus, the term "ecosystem" was used to imply a kind of progressive holism, which was not the intent of Tansley in 1935.[34]

The new definitions for "ecosystem", embodied in the phrase "ecosystem ecology", contributed to the broader environmental consciousness that later developed strong links between ecology in the academic communities and environmentalism. The linkage expanded ecology into one of the most active areas of biology and later increased the status of ecology among environmentalists; however, the linkage with environmentalism decreased the scientific and authoritative status of ecology among scientists in the academic community.[35]

The numbers of university-trained ecologists tripled between 1945 and 1960, and doubled between 1960 and 1970, but lacked unity as a discipline. Degrees in ecology were offered in departments of botany, zoology, wildlife, history, anthropology, and forestry, as well as new departments of ecology and environmental sciences and schools of medicine, engineering, agriculture, and law. The holistic theory allowed students to grasp an overview of ecology and link it with other fields without undertaking the rigorous studies of the mechanistic approaches to ecology.[36]

Silvicultural Research

During the 1940s and 1950s, textbooks which updated and assimilated knowledge in silviculture, related fields of silvics, and anatomy and tree physiology continued to be published.[37] These texts kept a broad focus to research.

Throughout the 1940s and 1950s research studies began to reveal increasing problems with uneven-aged management. Problems included high costs, slow growth rates, insect attacks, decay, ice breakage, and shifting to shade tolerant species. At the same time, even-aged management from plantations showed successes. Trials of shelterwood and seed tree even-aged systems also proved promising. As a result of this research, a general shift occurred from uneven-aged to even-aged systems. Silviculturists still generally accepted the succession/climax ecological theory; however, where they were trying to manage for intolerant, "early successional" species, the theory suggested (and their research showed) that even-aged silvicultural systems were better.

In shade tolerant species and in mixed species stands - such as hardwoods in the southeastern United States, selection cutting was still considered appropriate. Shade tolerant species, considered "late successional" or "climax," could be perpetuated in this way. In mixed species stands, the smaller trees were assumed to be younger "late successional" species and would respond to removal of the other trees.

The general, but not universal, acceptance of the suitability of even-aged management, both when establishing plantations on abandoned farmlands and when managing an existing forest, meant efforts to make this management more efficient rapidly increased. Much research was done on operational treatments to enhance even-aged stands. These enhancements included site preparation, planting and other forms of regeneration, weed control, pruning, pre-commercial thinning, commercial thinning, and fertilization. To the credit of silviculturists, the focus fell primarily on regenerating native species within their natural range. The few cases where species had been moved out of their range or introduced from abroad had been unsuccessful, reinforcing the practice of relying on native species.

Silviculturists increasingly studied the mechanisms of regeneration, leading forest scientists to accept that stands originated in many ways and could develop along many pathways. Ecologists who were searching for holistic patterns and ignoring human-initiated stands generally overlooked this realization. More mechanistic ecologists, however, began to recognize these variations. Some began to challenge the conventional holistic theories.[38] Emphasis on these enhancements and the expanded research funding led many young scientists to specialize in genetics, soils, and physiology for the production of seedlings and plantation establishment.

The need increased to identify site productivity, both to determine what species to regenerate and to determine whether and how much to invest in regeneration. In the Southeast, much of the regeneration occurred on abandoned agricultural land, and site productivity relationships were based on soil conditions. In the Pacific Northwest, many forests had been clearcut and often burned in the early 1900s. By the late 1950s and 1960s they had regrown to even-aged stands; and site productivity could be determined by measuring site index of standing trees, although some efforts to determine productivity by soil conditions were made. In the inland West, fires and selective cutting created uneven-aged stands that could not easily be aged for site index, and the varied topography made studying soil conditions difficult. Consequently, estimates of productivity were based on vegetation classification, which assumed key plant species indicated certain site conditions and the holistic succession/climax ecological theory.[39]

This Forest Service timber survey in the inland West shows an unevenaged stand. This large western white pine had a 46-inch DBH.
(USDA FOREST SERVICE PHOTO, POWELL RANGER DISTRICT,
CLEARWATER NATIONAL FOREST, IDAHO, 1951)

Silvicultural focus began to shift to the most efficient way to grow timber volume, with relatively little emphasis on tree size and quality. The acceptance of even-aged management and the concentration on present net worth analyses enhanced the focus on efficiency at the stand level. The focus on specialization led to separation of objectives and means to achieve the objectives. Silvicultural research began to fragment into many specialties—with research specialists being hired to replace the silvicultural generalists of previous generations.

At the same time, economic or growth and yield calculations had to be done by slide rules or desk calculators, and so remained slow, crude, and simplistic. Often, these analyses did not consider changes in tree sizes and wood quality—only total volume. Advances during the 1940s and 1950s led foresters to believe that technology would overcome problems of smaller, more knotty wood grown in the plantations on short rotations.

1960–1975

Environmentalism became attractive to many young ecologists who believed management of natural resources for industrial use was destroying environmental amenities. Many silviculturists who had sought scientific assistance from ecologists found themselves being criticized instead. Since the ecologists had not been trained in management, economics, or policy, they used a biological basis for their concern. The large amount of high quality old growth timber being harvested from the western United States and Canada masked future issues of wood quality. Consequently, silviculture emphasized growing volume.

"Top-down" legislation such as the Wilderness Act, National Environmental Policy Act, and Endangered Species Act created a major change in funding directions for silviculture and forest ecology. The U.S. Forest Service supported high levels of research in silviculture and specialized subfields and developed a large organization of scientists. These Forest Service stations were often located on university campuses near forestry colleges to ensure a "critical mass" of research scientists. The method of research consisted of writing a research proposal, having it approved, accomplishing the task, and publishing the research. Scientists who emerged from the International Biome Project (IBP) program or inhouse U.S. Forest Service research became adept at writing such proposals—often adjusting their area of specialty to coincide with funding emphases.

Public funding of research continued, increasing in basic ecology much more than in applied fields such as silviculture, since applied research was considered a subsidy of private industry. The shortage of this funding by the federal government led to private industry support of much silvicultural research—primarily oriented toward timber production. Industry and the U.S. Forest Service increased research cooperatives, which focused on narrow areas of research such as tree breeding, fertilization, or growth and yield. To obtain the expertise of universities, to avoid any appearance of collusion, and to ensure a long-term continuity of the projects, these cooperatives were often located at universities; and Forest Service research stations and universities often became active in these. The *Canadian Journal of Forest Research* began publication in 1971, thus establishing another, independent avenue for publication of silvicultural and forest ecology research.

Ecological Research

Many ecologists continued mechanistic studies, which paralleled and enriched many of the efforts done in the 1940s. New editions of the early silvics texts of Toumey and Korstian were not published after the 1950s. Texts giving insights into the mechanistic approach to ecology primarily concerned physiological and anatomical subjects, such as those by Kozlowski, Wilson, Zimmerman and Brown, and Esau, and a forest ecology text by Spurr with later editions coauthored by Barnes. A process-based ecological model was built on a computer which recognized the response of forests to disturbances.[40]

The search for holistic laws caused many ecologists to overlook differences in behaviors of plants and animals. A conceptual paper by G. E. Hutchinson had led animal and plant ecologists to write many theoretical papers on diversity and stability. Although the term "stability" was defined variously, some ecologists associated "old growth" forests with the most "stable" and "diverse" since they had experienced the longest time since a disturbance.[41]

The large amount of research in these old forests and the concept of their "stability" led environmentalists to object to human activities (especially clearcutting) which "disrupted" their condition. At the same time, some scientists still based their work on the classical, holistic "succession/climax" theory, often doing research which described "old growth" patterns and processes, rather than testing the classical theory. Several textbooks and studies of vegetation were still based on the "classical" theory. Mathematical models of ecological processes reinforced stability and so supported the "holistic" perspective of ecology.[42]

At the same time, research papers began to give evidence that forests undisturbed by humans did not remain in the holistic, climax condition of the classical "succession/climax" theory. Some ecologists reinterpreted the "benign" perspective of the classical theory to encompass small disturbances which kept the forest in a dynamic steady state.[43]

As research and direct and indirect evidence of large-scale natural disturbances accumulated, scientists slowly began to abandon the "holistic" perspective of forests existing in a "steady state." Using large research funds from the National Science Foundation, one study actually tested the hypothesis that natural forests were in a delicate balance and subject to degradation by clearcutting. The study disproved the hypothesis and showed that the forest was very resilient to disturbances much more intense than clearcutting. Various conferences focused on the role of natural disturbances. Research by Harper from England began to be appreciated by ecologists in the United States. Since he often studied plant populations on obviously disturbed agricultural fields and pastures, his work eroded many implicit assumptions that only undisturbed ecosystems could give insights into natural ecological processes.[44]

Silvicultural Research

Research and development in forestry-related machinery declined in the United States. Efficient power saws and mechanized harvesting equipment began being imported from Scandinavian countries. By contrast, research and development of forest chemicals for weed and insect control increased. Public funding for wood product research began declining, which restricted training of new scientists in these fields. The accumulation of plot data and the emerging capabilities of the computer enabled mensurationists to develop yield information and models for stands managed by various silvicultural treatments. Often, the mensurationists became the "statisticians" whose approval was necessary for U.S. Forest Service research proposals and papers.

Forest management research began to stress that the selection of objectives for stand management was a management and policy decision, and that silvicultural decisions primarily concerned how to achieve the objectives most efficiently. Management scientists highly criticized silvicultural research and papers if value judgments appeared. Forest management research focused on economic efficiency and planning. Using quantitative methods, forest managers had difficulty incorporating non-commodity values such as aesthetics, wildlife habitats, or biological diversity. Forest management did not study the emerging new approaches of "quality control" or "flexible systems," which moved away from top-down planning and encouraged "site-specific" decisions compatible with advances in silviculture.[45]

Silvicultural research generally accepted that even-aged management was both an ecologically natural way for forests to grow and appropriate for regrowing the valuable, "early successional" species of trees. Results of earlier research plots confirmed this perspective. Since some logging companies had interpreted "clearcutting" as cutting only merchantable trees, silviculturists developed the terms "cleancut" or "silvicultural clearcut" to indicate that unmerchantable trees also had to be felled, so they would not compete with new regeneration. Smith and others suggested that mixed species stands and stands of shade tolerant species were commonly even-aged and grew after an initial disturbance. Such statements implicitly questioned the holistic "succession/climax" theory.[46]

Silviculturists studied how forests changed, developing ways both to regrow forests after natural and human disturbances and to disturb forests to create the desired conditions. They examined a variety of forest structures both through their research and through transects resulting from timber cruising and land surveying. The close association between silvicultural practitioners and researchers gave the scientists access to a variety of tests of their research findings. In contrast, many ecologists focused on old forests and processes free of recent disturbances in an attempt to study the "stable" condition.

The different approaches led to differing interpretations of forests. Silviculturists noted the many interactions among species and the soil, climate, and other aspects of the forest and interpreted the complexity as resilient and not directional (or unidirectional). Their interpretation of the resiliency was based on experiences in recovering degraded lands through reforestation, in studying natural disturbances—fires, winds, floods, insects, and diseases—and in often failing to obtain desired results through management. Ecologists were looking for general, holistic "laws" of diversity and stability and often assumed old forests were the most stable and diverse.

Although researchers never agreed on a single view of "stability," the focus on "undisturbed" forests led environmentalists to argue that human disturbances (especially clearcutting) were "destroying" stable, diverse eco-

systems. Silviculturists generally ignored, rather than challenged, the holistic, stability trend of forest ecologists, since such ideas seemed to apply to
select "old growth" forests and did not help with the immediate tasks of
determining how to grow and manage forests. The issue did become visible
enough that President Nixon asked a panel of silviculture and mechanistic
forest ecology scientists (among others) to review the issue of clearcutting.
In this process, they gave scientific evidence supporting the resiliency of
forests.[47]

Specialists conducted most silvicultural research of this period, and it
involved ways to make even-aged management more efficient. Consequently, large advances occurred in tree breeding, nursery and seedling handling, chemical control of weeds, control of browse by animals, and pre-
commercial thinning—fields easily studied by known statistical techniques.
Little pruning research was done, since it was considered uneconomical.[48]

Quantitative growth relationships were studied and developed to determine the relation of growth to growing stock. More thinning trials were
established on a broad basis, and computerized growth models were developed. By necessity these trials took place in older natural stands or plantations; however, the results were expected to apply to new plantations. The
statistical emphasis of forestry studies resulted in thinning trials and growth
models more statistically precise than silviculturally creative, often concentrating on empirical results rather than what was possible by studying the
processes.

As forestry schools hired specialists to meet the demand for technical
information, scientists with an overview of silviculture began to disappear,
creating a general decline in thinking and research in silviculture systems.
The only general silvicultural texts published between 1960 and 1979 were
Smith's *Practice of Silviculture* (the seventh edition of the text begun by
Hawley in 1921) and a synthesis of regional practices of silviculture, *Regional
Silviculture*, edited by Barrett.

Unlike fields such as ecology, which deepened and expanded as books
and research increased, silviculture did not expand as a cohesive, in-depth
discipline. Specialists, who concentrated on individual fields or operations
such as physiology, genetics, soils, tree improvement, planting, or weed
control, filled silvicultural research and teaching positions. They focused less
on the full array of silvicultural systems which could be examined.[49]

When the public objected to unsightly clearcutting, environmental
groups often used biological arguments based on the holistic "succession/

climax" theory to discredit the U.S. Forest Service and private forestry. Research findings that openings were necessary for regeneration of many of the intolerant commercial species led many foresters to argue that large clearcuts were a biological necessity. In fact, the biological arguments that large clearcuts should or should not be made were unfounded. The most valid argument against the large clearcuts was their unsightliness, and the most valid argument in favor of large clearcuts was economic efficiency. Some research foresters pointed out the errors of basing the extremes on biological sciences.[50]

In both the U.S. Forest Service and private industry, the silviculturist usually occupied a subservient position to the person deciding to harvest timber—if there were a separate silviculturist at all. Consequently, the role of silviculturist increasingly became one of implementing recovery from timber harvesting operations, rather than deciding how (or if) the harvesting should be done.

The earlier dichotomy between forest management science focusing on centralized planning (allowable cut, etc.) and silvicultural sciences focusing on site-specific decisions by local foresters began to reveal problems as timber harvesting became highly visible. Either through mandate or from a lack of budget, harvest and silvicultural operations were frequently simplified to "clearcut and regenerate with a single species." When conflicts did arise between management practices and the public, however, management specialists relied on silviculturists to give biological justifications for what were in fact non-biological actions.

1975–1995

Universities became increasingly dependent on overheads from grants. Silviculture research had benefited from growth in research programs in the Department of Agriculture during the 1950s. Since the 1970s, however, funding for applied forest research in silviculture and forest ecology had declined relative to financial aid for environmental studies.[51] Funding increased for research in environmental issues, such as global warming, air pollution, water pollution, soil contamination, endangered species, and acid rain. University and Forest Service scientists became increasingly adept at obtaining funds from these sources. Research shifted from manipulating the forest to produce goods and services to documenting effects of acid rain, designing models to predict effects of global warming, predicting carbon dioxide cycles in the forest, and acidification of the soils. The U.S. Department of Agriculture established a competitive grants program to fund more applied research than the NSF, but it never became very large because of federal budget constraints.

The number of refereed journals publishing articles in forestry increased with *Forest Ecology and Management, New Forests, Journal of Sustainable Forestry*, and regional Society of American Foresters journals emerging. In addition, the Ecological Society of America began *Ecological Applications*, a journal which allowed ecologists to publish articles commenting directly on management, instead of the more traditional pattern of biological scientists giving technical input to those trained in management. Many symposia proceedings were also published, as conferences exposed professionals to new research information.

Ecological Research

Dramatic advances occurred in ecological theory at the same time. Whereas past ecological studies had examined physiological, nutrient cycling, and similar processes which occurred over short times, new studies began to examine changes over long times. These studies gave ecologists a perspective similar to the silvicultural one, which incorporates changes over time. Paleobotanical studies began to show that present "old growth" communities actually consisted of species which had not been together long enough to have coevolved. "Reconstruction" research techniques, using methods similar to archeology, became accepted in ecology and also showed that presumably stable communities were not only impacted by severe disturbances, but were actually dependent on them for their species composition.[52]

These studies, and recognition of the importance of large natural disturbances, led to much scientific questioning of the ecological paradigm of classical "succession/climax."[53] By 1990, it was generally accepted by ecologists that the classical, holistic, stability theory was no longer valid, as reported at a session of the Annual Meeting of the Ecological Society of America:

> Ecologists have traditionally operated on the assumption that the normal condition of nature is a state of equilibrium, in which organisms compete and coexist in an ecological system whose workings are essentially stable . . . A forest grows to a beautiful, mature climax stage that becomes its naturally permanent condition.

This concept of natural equilibrium long ruled ecological research and governed the management of such natural resources as forests and fisheries. It led to the doctrine, popular among conservationists, that nature knows best and that human intervention is bad by definition.

Now an accumulation of evidence has gradually led many ecologists to abandon the concept or declare it irrelevant, and others to alter it drastically. They say that nature is actually in a continuing state of disturbance and fluctuation. Change and turmoil, more than constancy and balance, is the rule. As a consequence, say many leaders in the field, textbooks will have to be rewritten and strategies of conservation and resource management will have to be rethought.[54]

Many papers emerged in light of the accepted shift of paradigms. Oliver and Larson detailed a "conceptually unified" perspective which explained behaviors of both managed and unmanaged forests while avoiding such terms as "succession," "seral stage," and "climax." Botkin developed a more universal theory for explaining biological phenomena without reverting to the "steady state" perspective of classical ecology. Over time, these perspectives will probably be modified, accepted, or rejected.[55]

Ecologists and silviculturists became actively involved in policy making, heading teams to develop management alternatives for policy makers, testifying at hearings, or developing the scientific basis for certain concerns.[56] In addition, private timber companies began to incorporate the dynamic landscape perspective into plans to manage their forests for both commodity and non-commodity values.

Not all scientists shifted away from the holistic, "stability" approach immediately. Scientific papers described general "old growth" structural attributes, as if "old growth" were a general state, rather than a classification dependent on the purpose of classification. The concepts of "island biogeography", developed under the premises of diversity and stability of the 1960s, were used to develop ecological theories for landscapes. At times, these theories reverted to the classical holistic stability concepts by emphasizing the importance of not fragmenting "old growth," although scientists elsewhere pointed out the negative consequences of fragmenting openings as well.[57]

Some very good studies still described interactions, rather than changes with time; however, the large amount of these done in "old growth" forests may have perpetuated the impression that "old growth" forests contain more things of value than other forests. Some forest ecologists, as well as scientists in fields peripheral to forest ecology (e.g., wildlife sciences, economics, and physiology), used the stability concepts of classical, holistic ecology to justify large "reserves" where silvicultural operations were essentially prohibited.[58]

Silvicultural Research

Despite the concerns of forest machinery's impact on soil compaction and sustainability, forestry educators placed relatively little emphasis on research and training scientists to develop new machinery. Teaching and training of wood products specialists received even less emphasis. With advancements in computers, mensuration research in growth and yield models expanded. These strongly empirically based models contrasted to the JABOWA model developed by ecologists or a later silvicultural model. Management sciences continued to emphasize economic efficiency and developed models emphasizing optimization and linear programming. One such model, FORPLAN, was used to develop plans for national forests, and so to comply with the Forest Management Act of 1976, although the model's author later admitted it was poorly suited to watershed management purposes.[59]

Forest entomologists began to recognize the social, technical, and biological undesirability of pure insecticide or other "single-technology" approaches to managing insect pests. They developed an "integrated pest management" approach, which involved monitoring and various condition-specific decisions. This strategy contained many elements of modern management theory and set a precedent for more general forest management approaches developed later by silviculturists.[60]

A significant number of researchers in silviculture began to return to broader questions of systems for managing stands over long times. Well trained in statistical techniques, these researchers could use statistics creatively, rather than be encumbered by experimental designs designated by statistical specialists. These scientists worked with relatively little research money compared to their colleagues doing ecological research, and they developed research techniques which required little money for equipment. Rapid advances occurred in the development of silvicultural techniques for mixed species stands, stand density and thinning opportunities, uneven-aged management, management relative to disturbances, and early tending of stands.[61]

These same researchers worked both in silviculture and in mechanistic approaches to forest ecology, contributing to the changing theories there.

Unlike earlier silvicultural specialists who had largely ignored holistic ecological theories, these scientists were also well versed in holistic ecological
theory, forest management, and physiology. Textbooks appeared which summarized the status of knowledge in silviculture and forest ecology, often
combining the two subjects.[62] These scientists felt comfortable addressing
the two fields since they had studied the "holistic" succession/climax theory
and its limitation of studying primarily undisturbed forests. Much of their
research used the mechanistic perspectives of J. W. Toumey, C. F. Korstian,
H. J. Lutz, and J. L. Harper and human-manipulated stands to gain knowledge about ecological processes.

At the same time, specialized research into operations to make even-aged
management more efficient continued with advances in genetics, seedling
handling, fertilization, weed control, and spacing. Concern also grew about
wood quality, and so pruning and other methods were investigated to improve the quality.[63]

Studies on the relationship of stand density, growth, and thinning addressed both theoretical relationships of population ecology and analyses of
thinning studies begun in previous decades. These relationships were incorporated into various thinning guidelines. Various growth models developed
by mensurationists also included these relationships, although silviculturists sometimes pointed out that these strongly empirical models did not
account for many silvicultural possibilities—such as insect epidemics or
increased growth through close attention to stand structures while thinning.[64]

Empirical studies showed what silviculture scientists had been suggest-
ing for many years: that mixed-species stands and stands of very shade tolerant species could develop as even-aged stands, and so selection silvicultural
systems were not necessarily most appropriate to perpetuate these stands.
Empirical studies also demonstrated how selection systems could be accomplished—and not accomplished—if they were desired.[65]

Despite the focus on both individual operations and systems, silviculturists remained relatively unified in their approaches and underlying theories.
In fact, many silviculturists did research on both systems and individual
operations. Silviculture instructors from all universities continued their
annual meetings begun in the 1960s. Between 1975 and 1995, this group
developed a unified set of silvicultural definitions—despite heated arguments over technical terms. A biennial silvicultural conference began in the
southeastern United States with its proceedings published. In the mid-1980s
a series of articles looking at the future of silviculture appeared in sequen-

tial issues of the *Journal of Forestry*, and scientists collaborated on two new editions of the *Regional Silviculture* text.[66]

The U.S. Forest Service instituted a "silviculture certification" program in the 1970s to ensure professionally updated silviculturists were making condition-specific decisions in national forests. This program gave impetus to many mid-career updating courses at universities attended by state, private, and federal foresters. These programs emphasized the decentralized silvicultural approach of Frothingham, Baker, Smith, and others wherein site-specific silvicultural decisions replaced those mandated through rigid rules.

Silviculturists began to work with wildlife ecologists, economists, and others to develop ways to maintain both commodity and non-commodity values.[67] These studies suggested a shift to concentration at the landscape level, in contrast to the stand level of previous decades. This research focused on the ecological, social, and economic aspects of this management and emphasized site-specific decisions carried out by an enfranchised professional and technical infrastructure, rather than centrally planned and mandated approaches. These scientists began to realize that the obstacles to managing across landscapes were not always biological and thus began to expand their research to organizational, management, and policy questions.

Unlike early in the century, silvicultural knowledge generally outpaced silvicultural practices. Much of the information on the advantages of mixed species stands, the possibilities of selection systems, wide spacings, thinning, and management for wood quality was not extensively applied. With some exceptions, management of both government and private forest lands became more centrally planned and controlled—with forest plans on national forests, forest practice regulations on private lands, and expected flows of commodities from both lands. The mandated flows and budgets left little latitude and resources for applying creative, site-specific treatments.

THE FUTURE OF RESEARCH IN
FOREST ECOLOGY AND SILVICULTURE

Research in forest ecology and silviculture is beginning to converge again, as silviculturists look for diverse ways to manage forests, and more ecologists accept that forests constantly change—and so do not perceive managed forests as innately outside the realm of study. Ecologists, silviculturists, forest managers, and the public are also beginning to realize that not all perceived environmental problems and solutions are purely biological. There are biological, economic, social, organization/management, and technological components to the solution.

Concerns of the 1990s about productivity and sustainability of forests are the same which stimulated much of the ecological and silvicultural research of the 1890s. In addition, the concerns of biodiversity and aesthetics are added. Now, however, a robust, biological understanding exists of how forests grow and can be manipulated to resolve these and future, unforeseen concerns. Max Plank and Charles Darwin both noted that acceptance of a changed scientific paradigm may not occur until the older generation of scientists gives way to a new generation that has internalized the new paradigm. A similar lag may be necessary before the current understanding of the dynamic nature of forests and roles of silviculture are accepted by the American public. The lag may be longer if various non-scientific groups continue to emphasize a "steady-state" perspective of nature. It is unclear how the forests will be managed while this understanding is being accepted. It is very possible that research in public education, decision-making, and organization and management can expedite this understanding.

Funding for research for the past few decades has emphasized basic biological research, in hopes this would lead to resolution of environmental problems. The realization that pure research can be pursued to infinite detail on any subject and that many of the forest-related problems have non-biological solutions means that funding for basic ecological research could increase dramatically while these problems remain unresolved. Applied research seeking solutions may be more effective. Properly managed, applied research will channel funds to those elements of basic research which will help solve society's concerns. An integration of research with management using concepts of adaptive management, quality improvement, and flexible systems may provide the funding, long term stability, and feedback to practitioners which will stabilize support for applied silvicultural research.[68]

In addition, centrally planned approaches to management—for either commodity or non-commodity values—will not resolve the forest environmental problems. To date, they have simply reduced management of forests and added to stresses on the global environment, biodiversity, the economy, and society's well-being. Forest ecologists and silviculturists now have the biological knowledge to begin managing forests for commodity and non-commodity values. It remains unclear, however, the extent to which this knowledge and management approaches will apply to different types of land ownership in the United States.

It is very possible that forest management for both commodity and non-commodity values will not be done in many forests of the United States, with associated declines in the economic, social, and environmental health of the country. Given the many advantages to managing forests and the knowledge in forest ecology and silviculture which has accumulated over the past century, some countries probably will utilize the accumulated knowledge to become leaders in forest management, even if the United States chooses not to do so. Other systems, such as the quality control system which made Japan economically strong, have been adapted and used by other countries, but ignored in the United States where they were developed.

NOTES

1 The authors are grateful for reviews of earlier drafts of this paper—Professor David M. Smith (retired) of Yale University School of Forestry and Environmental Studies, Dr. Robert Curtis of the USDA Forest Service Olympia Laboratory, Dr. Harold K. Steen of the Forest History Society, Inc., Professors W. Dwight Billings and Kenneth R. Knoerr of the Duke University School of the Environment, and Mr. Patrick Baker and Mr. Jeremy Wilson of the Silviculture Laboratory, University of Washington College of Forest Resources.

2 S. H. Spurr, "German Silvicultural Systems," *Forest Science* 2 (1956): 75–80; J. V. Thirgood, "The Historical Significance of Oak," in *Oak Symposium Proceedings*, D. E. White and B. A. Roach, chairs, (Upper Darby, PA: USDA Forest Service NE Forest Experiment Station, 1971).

3 E.g., A. V. Feigenbaum, *Total Quality Control*, (New York: McGraw-Hill Book Company, 1951, 1983); W. E. Deming, *Out of the Crisis*, (Cambridge: Massachusetts Institute of Technology Center for Advanced Engineering Study, 1982); and R. H. Reich, *The Next American Frontier*, (New York: Time-Life Books, 1983).

4 D. Worster, *Nature's Economy: A History of Ecological Ideas*, (New York: Cambridge University Press, 1985).

5 A. Chase, *In a Dark Wood: The Fight Over Forests and the Rising Tyranny of Ecology*, (New York: Houghton-Mifflin Company, 1995).

6 Worster, *Nature's Economy*; D. Worster, *The Wealth of Nature: Environmental History and the Ecological Imagination*, (Oxford: Oxford University Press, 1993).

7 Definition of the Silviculture Instructor's Subgroup, Silviculture Working Group, Society of American Foresters, 1994; for other definitions see R. M. Burns, editor, *The Scientific Basis for Silvicultural and Management Decisions in the National Forest System*, (Washington, D.C.: USDA Forest Service General Technical Report WO-55, 1989).

8 Spurr, "German Silvicultural Systems;" Thirgood, "The Historical Significance of Oak."

9 B. E. Fernow, "The Forest as a National Resource," *Proceedings of the American Forestry Association, 1890*, (1891): 36-53.

10 C. A. Schenck, *Biltmore Lectures on Sylviculture*, (Albany, NY: Brandow Printing Company, 1907); C. A. Schenck, *The Art of the Second Growth or American Sylviculture*, (Albany, NY: The Brandow Printing Company, 1912); R. C. Hawley, *The Practice of Silviculture*, Second, Third, Fourth, and Fifth editions, (New York: John Wiley and Sons, 1929, 1935, 1937, 1946); R. S. Troup, *Silvicultural Systems*, New York: Oxford University Press, 1928).

11 J. R. Troyer, "Bertram Whittier Wells (1884–1978): A Study in the History of North American Plant Ecology," *American Journal of Botany* 73 (1986): 1058–1078; R. C. Tobey, *Saving the Prairies. The Life Cycle of the Founding School of American Plant Ecologists*, (Berkeley: University of California Press, 1981); D. G. Sprugel, "A 'Pedagogical Genealogy' of American Plant Ecologists," *Bulletin of the Ecological Society of America* 61 (1980): 197-200; R. L. Burgess, "Sources of Biographical Information on American Ecologists," *Bulletin of the Ecological Society of America* 58 (1977): 236-255; F. N. Egerton, editor, *History of American Ecology* (New York: Arno Press, 1977); C. D. Oliver, and B. C. Larson, *Forest Stand Dynamics* (New York: McGraw-Hill, 1990); C. D. Oliver, and B. C. Larson. *Forest Stand Dynamics*, Revised Edition (New York: John Wiley and Sons, 1995); J. W. Toumey, *Seeding and Planting* (New York: John Wiley, 1916); J. W. Toumey, and C. F. Korstian, *Foundation of Silviculture Upon an Ecological Basis,* Second Edition (New York: John Wiley and Sons, 1937); R. C. Hawley, *Forest Protection* (New York: John Wiley and Sons, 1921); F. S. Baker, *Theory and Practice of Silviculture* (New York: McGraw-Hill, 1934).

12 Herbert Spencer, *A System of Synthetic Philosophy*, 10 volumes (London: Williams and Norgate, 1862-1896); Chase, *In A Dark Wood*; F. E. Clements, *Plant Succession: An Analysis of the Development of Vegetation* (Washington, D.C.: Carnegie Institute Publication 242, 1916); F. E. Clements, "Nature and Structure of the Climax," *Journal of Ecology* 24 (1936): 252-284; E. L. Braun, *Deciduous Forests of Eastern North America* (New York: Macmillan, 1950); E. P. Odum, "The Strategy of Ecosystem Development," *Science* 164 (1969): 262-270; Worster, *Nature's Economy*; W. S. Cooper, "The Climax Forest of Isle Royale, Lake Superior, and its Development," *Botanical Gazette* 55 (1913): 1-44; H. A. Gleason, "The Individualistic Concept of the Plant Association," *Bulletin of the Torrey Botanical Club* 46 (1926): 7-26.

13 Worster, *Nature's Economy*.

14 For a sequence of references which perpetuated this inference without rigorously testing it, see Oliver and Larson, *Forest Stand Dynamics*, Revised Edition, p. 238.

15 B. E. Fernow, *Report Upon the Forestry Investigations of the U.S. Department of Agriculture, 1877-1898*, U.S. House of Representatives Doc. 181, 55th Congress, 3rd Session (Washington, D.C.: GPO, 1899).

16 e.g., E. E. Carter, "The Silvicultural Results of Marking Timber in the National Forests" Society of American Foresters *Proceedings* 3 (1908): 18-28, cited from A. P. Mustian, "History and Philosophy of Silviculture and Management Systems in Use Today," in *Uneven-aged Silviculture and Management in the Western United States: Proceedings of an In-service Workshop, October 19-21, 1976*, (Redding, CA: Timber Management, USDA Forest Service, 1976).

17 Gifford Pinchot, "A Primer of Forestry," *Farmers' Bulletin* No. 173 (1903).

18 R. S. Troup, *Silvicultural Systems*, Second Edition, edited by E. W. Jones (New York: Oxford University Press, 1952); Fernow, "The Forest as a National Resource"; C. A. Schenck, "Forest Finance"; *Proceedings of the American Forestry Association, Special Meeting in Asheville, North Carolina, and Nashville, Tennessee, September 17-22, 1897* 12 (1897): 124–133; See reports in preface to USDA Forest Service Resource Report No. 23; F. S. Baker, "Notes on the Composition of Even aged Stands," *Journal of Forestry* 21 (1923): 712–717; F. S. Baker, *Theory and Practice*; Burns, *The Scientific Basis*; F. S. Baker, *Principles of Silviculture* (New York: McGraw-Hill, 1950); E. H. Frothingham, "Ecology and Silviculture in the Southern Appalachians: Old Cuttings as a Guide to Future Practice" *Journal of Forestry* 15 (1917): 343–349; J. N. Spaeth, *Twenty Years Growth of a Sprout Hardwood Forest in New York: A Study of the Effects of Intermediate and Reproduction Cuttings* (Ithaca, New York: (Cornell) Agricultural Experiment Station Bulletin 465, 1928); E. H. Frothingham, "Ecology and Silviculture in the Southern Appalachians: Old Cuttings as a Guide to Future Practice" *Journal of Forestry* 15 (1917): 343–349; J. N. Spaeth, *Twenty Years Growth of a Sprout Hardwood Forest in New York: A Study of the Effects of Intermediate and Reproduction Cuttings* (Ithaca, New York: (Cornell) Agricultural Experiment Station Bulletin 465, 1928); R. M. Burns, technical compiler, *Silvicultural Systems for the Major Forest Types of the United States* (Washington, D.C.: USDA Forest Service Agricultural Handbook 445, 1983); S. G. Boyce, "Management of Eastern Hardwood Forests for Multiple Benefits (DYNAST-MB)" USDA Forest Service Research Paper SE-168, 1977; W. H. Carmean, and S. G. Boyce, "Hardwood Log Quality in Relation to Site Quality," USDA Forest Service Research Paper NC-103, 1973; D. W. Peterson, "Pisgah Working Circle, Plan Period 1969–1978" Files, National Forests in North Carolina, USDA Forest Service, Asheville, NC, 1968; J. R. Runkle, "Patterns of Disturbance in Some Old-growth Forests of Eastern North America" *Ecology* 62 (1982): 1533–1546; D. M. Smith, "The Scientific Basis for Timber Harvesting Practices" *Journal of the Washington Academy of Science* 67 (1977): 3–11; D. M. Smith, *The Practice of Silviculture*, Eighth Edition. (New York: John Wiley and Sons, 1986); G. W. Wood, L. J. Niles, R. N. Hendrick, J. R. Davie, and T. L. Grimes, "Compatibility of Even-aged Timber Management and Red-Cockaded Woodpecker Conservation," *Wildlife Society Bulletin* 13 (1985): 5–17.

19 H. S. Graves, "Education in Forestry" *Journal of Forestry* 23 (1925): 108–125; H. S. Graves and C. H. Guise, *Forestry Education* (New Haven, CT.: Yale University Press, 1932); F. H. Kaufert, "Improving Professional Forestry Education Through Accreditation" *Journal of Forestry* 73 (1975): 464–469.

20 Worster, *Nature's Economy*; J. B. Hagen, *An Entangled Bank: The Origins of Ecosystem Ecology* (New Brunswick, NJ: Rutgers University Press, 1992).

21 Clements "Nature and Structure;" John Phillips, "Succession, Development, the Climax and the Complex Organism: An Analysis of Concepts" *Journal of Ecology* 22 (1934–1935); see also Chase, *In A Dark Wood*; J. E. Weaver and D. E. Clements, *Plant Ecology*. Second Edition. (New York: McGraw-Hill, 1938); J. C. Smuts, *Holism and Evolution* (London: Macmillan Press, 1926); C. J. Glacken, *Traces on the Rhodian Shore: Nature and Culture in Western Thought From Ancient Times to the End of the Eighteenth Century* (Berkeley: University of California Press, 1967); M. Oelschlaeger, *The Idea of Wilderness: From Prehistory to the Age of Ecology* (New Haven: Yale University Press, 1991).

22 Clements, *Plant Succession*; Clements "Nature and Structure"; Phillips, "Succession, Development, the Climax and the Complex Organism"; Weaver and Clements, *Plant Ecology*; F. B. Golley, *Ecological Succession* (New York: Halsted Press, 1977); F. B. Golley, *A History of the Ecosystem Concept in Ecology* (New Haven: Yale University Press, 1993).

23 Spencer, *A System of Synthetic Philosophy*; Smuts, *Holism and Evolution*; Phillips, "Succession, Development, the Climax and the Complex Organism;" E. P. Odum, *Basic Ecology* (New York: Saunders College Publishers, 1983).

24 P. J. Bowler, *The Norton History of the Environmental Sciences* (New York: W. W. Norton and Co., 1992).

25 A. G. Tansley, "The Classification of Vegetation and Concept of Development," *Journal of Ecology* 8 (1920): 118–144; A. G. Tansley, "Succession: the Concept and its Values" *Proceedings of the International Congress of Plant Sciences 1926* (1929): 677-686; H. C. Cowles, "The Ecological Relations of the Vegetation on the Sand Dunes of Lake Michigan," *Botanical Gazette* 27 (1899); Gleason, "The Individualistic Concept;" B. W. Wells, and I. V. Shunk, "The Vegetation and Habitat Factors of the Coarser Sands of the North Carolina Coastal Plain: An Ecological Study" *Ecological Monographs* 1 (131): 465–520; P. W. Bridgman, *The Logic of Modern Physics* (New York: Macmillan Publishing Co., 1927); Troyer "Bertram Whittier Wells"; Golley, *Ecological Succession;* Oliver and Larson, *Forest Stand Dynamics*.

26 Gleason "The Individualistic Concept."

27 Tansley, "The Classification of Vegetation and Concept of Development"; Tansley, "Succession: the Concept and its Values"; A. G. Tansley, "The Uses and Abuses of Vegetational Concepts and Terms" *Ecology* 16(1935): 284-307; A. N. Whitehead, *Science and the Modern World: Lowell lectures, 1925* (New York: Free Press, Macmillan, 1925); Phillips, "Succession, Development, the Climax and the Complex Organism"; H. Levy, *The Universe of Science* (London: The Century Company, 1923).

28 J. R. Roughgarden, R. M. May, and S. A. Levin, *Perspectives in Ecological Theory* (Princeton: Princeton University Press, 1989); "Ecosystem management of the National Forest and Grasslands. Memorandum from the Chief of the Forest Service to Regional Foresters and Station Directors. June 4, 1992." USDA Forest Service File 1330-1, Washington, D.C.; Toumey and Korstian, *Foundation of Silviculture*; Baker, *Theory and Practice*; Hawley, *The Practice of Silviculture*, 1929, 1935, 1937; R. H. Westveld, *Applied Silviculture in the United States* (New York: John Wiley and Sons, 1939); H. J. Oosting, *The Study of Plant Communities;* (San Francisco: W. H. Freeman and Company, 1956).

29 Toumey, *Seeding and Planting*; J. W. Toumey, *Foundation of Silviculture Upon an Ecological Basis* (New York: John Wiley and Sons, 1928); Toumey and Korstian, *Foundation of Silviculture;* Hawley, *Forest Protection*; Hawley, *The Practice of Silviculture*, Second, Third, Fourth, and Fifth Editions; R. C. Hawley and D. M. Smith, *The Practice of Silviculture*. Sixth Edition. (New York: John Wiley and Sons, 1954); D. M. Smith, *The Practice of Silviculture*, Seventh Edition. (New York: John Wiley and Sons, 1962); Smith, *The Practice of Silviculture*, Eighth Edition.

30 H. A. Meyer and D. D. Stevenson, "The Structure and Growth of Virgin Beech-Birch-Maple-Hemlock Forests in Northern Pennsylvania" *Journal of Agricultural Research* 67 (1943): 465–484; C. E. Fiedler, "The Basal-Area-Maximum Diameter-q (BDq) Approach to Regulating Uneven-Aged Stands" in K. L. O'Hara, editor, *Uneven-Aged Management: Opportunities, Constraints, and Methodologies. Proceedings of a Workshop* (Missoula: School of Forestry, The University of Montana, 1995).

31 H. H. Chapman, "Fire and Pines" *American Forests* 50 (1944): 535; S. J. Pyne, *Fire in America* (Princeton: Princeton University Press, 1982).

32 Chase, *In A Dark Wood*; Golley, *A History of the Ecosystem*; Hagen, *An Entangled Bank*.

33 A. F. Hough and R. D. Forbes, "The Ecology and Silvics of Forests in the High Plateau of Pennsylvania" *Ecological Monographs* 13 (1943): 299–320; Earl P. Stephens, "The Historical-Developmental Trend of Determining Forest Trends (Ph.D. diss., Harvard University, 1955); E. P. Stephens, "Research in the Biological Aspects of Forest Production" *Journal of Forestry* 53 (1955): 183–186; F. E. Egler, "Vegetation Science Concepts: I. Initial Floristic Composition: A Factor in Old-Field Vegetation Development" *Vegetation* 4 (1954): 412–417; C. D. Oliver and E. P. Stephens, "Reconstruction of a Mixed Species Forest in Central New England" *Ecology* 58 (1977): 562–572; H. M. Raup, "Vegetation Adjustment to the Instability of Site," in *Proceedings and Papers, Sixth Technical Meeting of the International Union, Conservation of Natural and National Resources*, (Edinburgh: the International Union, Conservation of Natural and National Resources, 1957): 36–48; H. M. Raup, "Some Problems with Ecological Theory and Their Relation to Conservation" *Journal of Ecology* 52 (1964, supplement): 19–28; S. H. Spurr, *Forest Ecol-*

ogy (New York: Ronald Press, 1964); J. D. Henry, and M. A. Swan, "Reconstructing Forest History from Live and Dead Plant Material—An Approach to the Study of Forest Succession in Southwest New Hampshire" *Ecology* 55 (1974): 772–783.

34 Oosting, *The Study of Plant Communities*, pp. 29, 228; Tansley, "The Uses and Abuses;" Hagen, *An Entangled Bank* ; Worster, *The Wealth of Nature.*

35 A. Jamison, "A Tale of Two Brothers," *Science* 261 (1993): 497–498; Hagen, *An Entangled Bank.*

36 Worster, *Nature's Economy.*

37 Hawley, *The Practice of Silviculture*. Fifth Edition; Hawley and Smith, *The Practice of Silviculture*. Sixth edition; R. H. Westveld, *Applied Silviculture in the United States*. Second Edition. (New York: John Wiley and Sons, 1949); Toumey and Korstian, *Foundation of Silviculture*; Baker, *Principles of Silviculture*; *Plant Anatomy* (New York: John Wiley and Sons, 1953); P. J. Kramer and T. T. Kozlowski, *Physiology of Trees* (New York: McGraw-Hill, 1960).

38 H. J. Lutz, *Trends and Silvicultural Significance of Upland Forest Successions in Southern New England* (New Haven: Yale University School of Forestry Bulletin Number 22, 1928); H. J. Lutz, *Disturbance of Forest Soil Resulting From the Uprooting of Trees* (New Haven: Yale University School of Forestry Bulletin Number 45, 1940); H. J. Lutz, *Aboriginal Man and White Man as Historical Causes of Fires in the Boreal Forest, With Particular Reference to Alaska* (New Haven: Yale University School of Forestry Bulletin Number 65, 1959); Egler "Vegetation Science Concepts."

39 T. S. Coile, *Relation of Soil Characteristics to Site Index of Loblolly and Shortleaf Pines in the Lower Piedmont Region of North Carolina* (Durham: Duke University School of Forestry Bulletin 13, 1948); C. W. Ralston, "Some Factors Related to the Growth of Longleaf Pine in the Atlantic Coastal Plain" *Journal of Forestry* 49 (1951): 408–412; J. E. King, *Site Index Curves for Douglas-fir in the Pacific Northwest.* Weyerhaeuser Forestry Paper 8. (Centralia, WA: Weyerhaeuser Company Research Center, 1966); E. C. Steinbrenner, "Soil-site Relationships and Comparative Yields of Western Hemlock and Douglas-fir" in W. A. Atkinson and R. J. Zasoski, editors, *Western Hemlock Management*. Institute of Forest Products Contribution Number 34. (Seattle: University of Washington College of Forest Resources, 1986): 236–238; R. Daubenmire, "Forest Vegetation of Northern Idaho and Adjacent Washington and its Bearing on Concepts of Vegetation Classification" *Ecological Monographs* 22 (1952): 301–330; R. Daubenmire, "Vegetation: Identification of Typical Communities" *Science* 151 (1966): 291–298.

40 P. Marks, "The Role of Pine Cherry (*Prunus pennsylvanica L.*) in the Maintenance of Stability in Northern Hardwood Ecosystems" *Ecological Monographs* 44 (174): 73–88; D. G. Sprugel, "Dynamics Structure of Wave-Generated *Abies balsamea* Forests in the Northeastern United States" *Journal of Ecology* 64 (1976): 889–912; T. T. Kozlowski, *Growth and Development of Trees.* Volumes I and II. (New York: Academic Press, 1971); B. F. Wilson, *The Growing Tree* (Amherst: University of Massachusetts Press, 1970); M. H. Zimmerman and C. L. Brown, *Trees: Structure and Function* (New York: Springer-Verlag, 1971); K. Esau, *Plant Anatomy* (New York: John Wiley and Sons, 1965); K. Esau, *Anatomy of Seed Plants* (New York: John Wiley and Sons, 1977); S. H. Spurr, and B. V. Barnes, *Forest Ecology*, Second Edition (New York: John Wiley and Sons, 1973); S. H. Spurr and B. V. Barnes, *Forest Ecology*, Third Edition. (New York: John Wiley and Sons, 1980); Spurr, *Forest Ecology*; D. B. Botkin, J. F. Janak, and J. R. Wallis, "Some Ecological Consequences of a Computer Model of Forest Growth" *Journal of Ecology* 60 (1972): 948–972.

41 G. E. Hutchinson, "Homage to Santa Rosalia, or Why are there so Many Kinds of Animals?" *American Naturalist* 93 (1959): 145–159; Brookhaven National Laboratory, *Diversity and Stability in Ecological Systems* BNL 50175 (C-56), Biology and Medicine-TID-4500, (Upton, NY: Biology Department, Brookhaven National Laboratory, 1969); E. P. Odum, "Relationships Between Structure and Function in Ecosystems" *Japanese Journal of Ecology* 12 (1962): 108–118; E. P. Odum, *Fundamentals of Ecology*, Third Edition. (Philadelphia: Saunders College Publishers, 1971); R. Margalef, "On Certain Unifying Principles in Ecology" *American Naturalist* 97 (1963): 357–374; M. L. Cody and J. M. Diamond, editors, *Ecology and Evolution of Communities* (Cambridge: The Belknap Press of Harvard University Press, 1975); R. H. Whittaker, *Communities and Ecosystems*. Second Edition. (New York: Macmillan Publishing Company, 1975); R. H. May, "The Evolution of Ecological Systems" *Scientific American* 239(3) (1978): 160–175; D. B. Botkin, "A Grandfather Clock Down the Staircase: Stability and Disturbance in Natural Ecosystems" in *Forests: Fresh Perspectives from Ecosystem Analysis. Proceedings of the 40th Annual Biology Colloquium* (Corvallis: Oregon State University Press, 1979): 1–10.

42 F. H. Bormann and M. F. Buell, "Old-Age Stand of Hemlock-Northern Hardwood Forest in Central Vermont" *Bulletin of the Torrey Botanical Club* 91 (1964): 451–465; C. C. Grier, D. W. Cole, C. T. Dyrness, and R. L. Frederiksen, *Nutrient Cycling in 37- and 450-year old Douglas-fir Ecosystems*. Coniferous Forest Biome Bulletin 5, (Seattle: University of Washington, 1974); C. C. Grier, and R .S. Logan, "Old-Growth *Pseudotsuga menziesii* (Mirb.) Franco Communities of a Western Oregon Watershed: Biomass Distribution and Production Budgets" *Ecological Monographs* 47 (1977): 373–400; W. B. Leak, "Age Distribution in Virgin Red Spruce and Northern Hardwoods" *Ecology* 56 (1975): 1451–1454; J. F. Franklin, K. Cromack, Jr., W. Denison, A. W. McKee, C. Maser, J. Sedell, F. Swanson, and G. Juday, "Ecological Characteristics of Old-growth Douglas-fir Forests" USDA Forest Service General Technical Report PNW-118, 1981; Odum, *Fundamentals of Ecology*; R. Daubenmire, *Plant Communities: a Textbook of Plant Synecology* (New York: Harper and Row, 1968); Chase, *In A Dark Wood*; D. B. Botkin, *Discordant Harmonies: a New Ecology for the Twenty-first Century* (New York: Oxford University Press, 1990).

43 Raup "Some Problems;" W. H. Drury, and I. C. T. Nisbet, "Succession" *Jour-*

nal of the Arnold Arboretum 54 (1973): 331–368; D. B. Botkin and M. T. Sobel, "Stability in Time-Varying Ecosystems" *The American Naturalist* 109(970) (1975): 625–646; Botkin "A Grandfather Clock;" M. L. Heinselman, "Restoring Fire to the Ecosystems of the Boundary Waters Canoe Area, Minnesota and to Similar Areas" *Tall Timbers Fire Ecology Conference* 10 (1970): 9–24; O. Loucks, "Evolution of Diversity, Efficiency, and Community Stability" *American Zoologist* 10 (1970): 17–25; See also Brookhaven National Laboratory, *Diversity and Stability*.

44 F. H. Bormann, G. E. Likens, D. W. Fisher, and R. S. Pierce, "Nutrient Loss Accelerated by Clear-cutting of a Forest Ecosystem" *Science* 159 (1968): 882–884; F. H. Bormann, and G. E. Likens, *Pattern and Process in a Forested Ecosystem* (New York: Springer-Verlag, 1981); Heinselman, "Restoring Fire"; M. L. Heinselman, "The Natural Role of Fire in the Northern Conifer Forests" *Naturalist* 21 (1970): 14–23; M. L. Heinselman, "Fire in the Virgin Forests of the Boundary Waters Canoe Area, Minnesota" *Journal of Quaternary Research* 3 (1973): 329–382; J. L. Harper, *Population Biology of Plants* (London: Academic Press, 1977).

45 Barney Dowdle, "The Role of the Society of American Foresters in Formulating Public Policy" in *Proceedings of the Society of American Foresters Annual Meeting, September 12–15, 1966* (Bethesda, MD: Society of American Foresters, 1966): 31–33; Deming, *Out of the Crisis*; Reich, *The Next American Frontier.*

46 Smith, *Practice of Silviculture* Seventh Edition; R. J. Wilson, Jr. "How Second-growth Northern Hardwoods Develop After Thinning" USDA Forest Service Northeastern Forest Experiment Station Paper Number 62: (1953); Raup, "Some Problems;" J. W. Johnson, "Silvicultural Considerations in Clearcutting" In R. D. Nyland, editor, *A Perspective on Clearcutting in a Changing World. Proceedings of the Winter Meeting of the New York Section of the Society of American Foresters* (New York: New York Section of the Society of American Foresters, 1972): 19–24; Drury and Nisbet, "Succession."

47 Chase, *In A Dark Wood*; J. P. Kimmins, *Forest Ecology* (New York: Macmillan Publishing Company, 1987); F. A. Seaton, R. D. Hodges, Jr., S. H. Spurr, M. Clawson, and D. Zinn, *Report of the President's Advisory Panel on Timber and the Environment.* Submitted April 30, 1973 (Washington, D.C.: GPO, 1973).

48 E. W. Shaw and G. R. Staebler, *Financial Aspects of Pruning* (Portland, Oregon USDA Forest Service, Pacific Northwest Forest and Range Experiment Station, 1950); K. L. O'Hara, "Uneven-aged Management: Opportunities, Constraints, and Methodologies," Proceedings of a Workshop. School of Forestry, The University of Montana, Missoula, 1995.

49 C. D. Oliver, "The 2030 Forest: Directions of Silvicultural Research" in *Proceedings of the Fifth Biennial Southern Silvicultural Research Conference. November 1–3, 1988.* USDA Forest Service General Technical Report SO-74, 1989: 15–22.

50 D. A. Marquis, *An Appraisal of Clearcutting on the Monongahela National Forest* (Warren, PA: USDA Forest Service, Northeastern Forest Experiment Station, 1973);

J. F. Franklin and D. S. DeBell, "Effects of Various Harvesting Methods on Forest Regeneration" in R. K. Hermann and D. P. Lavender, editors, *Even-Age Management* (Corvallis: Oregon State University School of Forestry Paper 848, 1973): 29–57.

51 USDA Forest Service, *Natural Resources, Federal Spending and Resource Performance, 1940–1989* (Washington, D.C.: Office of Budget and Program Analysis, 1993).

52 M. B. Davis, "Quaternary History and the Stability of Forest Communities" in D. C. West, H. H. Shugart, and D. B. Botkin, editors, *Forest Succession: Concepts and Application* (New York: Springer-Verlag, 1981): 132–153; P. A. Delcourt, H. R. Delcourt, D. F. Morse, and P. A. Morse, "History, Evolution, and Organization of Vegetation and Human Culture" in W. H. Martin, S. G. Boyce, and H. C. Echternacht, *Biodiversity of the Southeastern United States: Lowland Terrestrial Communities* (New York: John Wiley and Sons, 1993): 47–80; L. B. Brubaker, "Vegetation History and Anticipating Future Vegetation Change" in J. K. Agee and D. K. Johnson, editors, *Ecosystem Management for Parks and Wilderness* (Seattle: University of Washington Press, 1988): 41–61; Henry and Swan, "Reconstructing Forest History;" Oliver and Stephens, "Reconstruction of A Mixed Species Forest;" See also Oliver and Larson, *Forest Stand Dynamics.*

53 P. S. White, "Pattern, Process, and Natural Disturbance in Vegetation" *Botanical Review* 45 (1979): 229–299; C. D. Oliver, "Forest Development in North America Following Major Disturbances" *Forest Ecology and Management* 3 (1981): 153–168; H. H. West, H. H. Shugart, and D. B. Botkin, editors, *Forest Succession: Concepts and Application* (New York: Springer-Verlag, 1981); S. T. A. Pickett and P. S. White, editors, *The Ecology of Natural Disturbance and Patch Dynamics* (New York: Academic Press, 1985); Botkin, *Discordant Harmonies;* Oliver and Larson, *Forest Stand Dynamics;* D. G. Sprugel, "Disturbance, Equilibrium, and Environmental Variability: What is 'Natural' Vegetation in a Changing Environment?" *Biological Conservation* 58 (1991): 1–18.

54 W. K. Stevens, "New Eye on Nature: The Real Constant is Eternal Turmoil" *New York Times* July 31, 1990.

55 S. T. A. Pickett, review of *Forest Stand Dynamics,* by C. D. Oliver and B. C. Larson, *The Quarterly Review of Biology* 66 (1991); Botkin, *Discordant Harmonies;* Oliver and Larson, *Forest Stand Dynamics.*

56 J. W. Thomas, E. D. Forsman, J. B. Lint, E. C. Meslow, B. R. Noon, and J. Verner, *A Conservation Strategy for the Northern Spotted Owl. Interagency Scientific Committee to Address the Conservation of the Northern Spotted Owl* (Portland, Oregon: USDA Forest Service, USDI Bureau of Land Management, USDI Fish and Wildlife Service, and USDI National Park Service, 1990); K. N. Johnson, J. F. Franklin, J. W. Thomas, and J. Gordon, "Alternatives for Management of Late-Successional Forests of the Pacific Northwest," A Report to the Agriculture Committee and the Merchant Marine Committee of the U.S. House of Representatives, 1991; FEMAT, "Draft Supplemental Environmental Impact Statement on Management for

Late-Successional and Old-Growth Forest Related Species Within the Range of the Northern Spotted Owl" (Volume and Appendix A), USDA Forest Service, USDI Bureau of Land Management, 1993; R. Everett, *Eastside Forest Ecosystem Health Assessment* (Wenatchee, WA: USDA Forest Service, National Forest System, Forest Service Research. Wenatchee Forest Sciences Laboratory, 1993); R. N. Sampson, and D. L. Adams, editors, *Assessing Forest Ecosystem Health in the Inland West* (New York: The Haworth Press, Inc., 1994).

57 M. L. Hunter, Jr., *Wildlife, Forests, and Forestry*, (Englewood Cliffs, NJ: Regents/Prentice Hall, 1990); R. H. MacArthur and E. O. Wilson, *The Theory of Island Biogeography* (Princeton: Princeton University Press, 1967); L. D. Harris, *The Fragmented Forest, Island Biogeography Theory and the Preservation of Biotic Diversity* (Chicago: University of Chicago Press, 1984); R. T. T. Forman and M. Gordon, *Landscape Ecology* (New York: John Wiley and Sons, 1986); J. F. Franklin and R. T. T. Forman, "Creating Landscape Patterns by Forest Cutting: Ecological Consequences and Principles" *Landscape Ecology* 1 (1987): 5–18; M. R. Young, "Conserving Insect Communities in Mixed Woodlands" in M. G. R. Cannell, D. C. Malcolm, and P. A. Robertson, editors, *The Ecology of Mixed-species Stands of Trees* (London: Blackwood Scientific Publications, 1992): 277–296.

58 Hunter, *Wildlife, Forests, and Forestry*; Thomas et al., *A Conservation Strategy*; Johnson et al., "Alternatives for Management;" FEMAT, "Draft Supplemental;" USDA Forest Service, *Record of Decision for Amendments to Forest Service and Bureau of Land Management Planning Documents Within the Range of the Northern Spotted Owl; and Standards and Guidelines for Management of Habitat for Late-Successional and Old-Growth Forest Related Species Within the Range of the Northern Spotted Owl* (Washington, D.C.: USDA Forest Service and USDI Bureau of Land Management. GPO, 1994-589-111/00003 Region No. 10. April 13, 1994).

59 D. E. Lyon, F. C. Beall, and W. L. Galligan, "Concerns for the Technological Infrastructure of Wood Products Development" in D. P. Hanley, C. D. Oliver, D. A. Maguire, D. G. Briggs, and R. D. Fight, editors, *Forest Pruning and Wood Quality in Western North American Conifers,* Institute of Forest Resources Contribution No. 77, CINTRAFOR No. SP-19, (Seattle: College of Forest Resources, University of Washington, 1995): 164–168; Botkin et al., "Some Ecological Consequences;" D. R. Larsen, "Adaptable Stand Dynamics Model for Integrating Site-specific Growth and Innovative Silvicultural Prescriptions," *Forest Ecology and Management* 69 (1994): 245–257; K. N. Johnson, "Consideration of Watersheds in Long-term Planning Models: the Case for FORPLAN and its Use on the National Forests" in R. J. Naiman, editor, *Watershed Management: Balancing Sustainability and Environmental Change* (New York: Springer-Verlag, 1992).

60 Boyce, "Management of Eastern Hardwoods;" S. G. Boyce, "Forestry Decisions," USDA Forest Service General Technical Report SE-35, 1985; S. G. Boyce, and H. McNab, "Management of Forested Landscapes" *Journal of Forestry* 92(1) (1994): 27–32; S. G. Boyce, *Landscape Forestry* (New York: John Wiley and Sons, 1995); C.

D. Oliver, "A Landscape Approach: Achieving Biodiversity and Economic Productivity" *Journal of Forestry* 90(9) (1992): 20–25; Everett, *Eastside Forest Ecosystem*; Sampson and Adams, *Assessing Forest Ecosystem Health.*

61 M. J. Kelty, B. C. Larson, and C. D. Oliver, *The Ecology and Silviculture of Mixed-species Forests: a Festschrift for David M. Smith* (Boston: Kluwer Academic Publishers, 1992); T. J. Drew, and J. W. Flewelling, "Stand Density Management: an Alternative Approach and its Application to Douglas-fir Plantations" *Forest Science* 25 (1979): 518–532; J. B. McCarter and J. N. Long, "A Lodgepole Pine Density Management Diagram" *Western Journal of Applied Forestry* 1 (1986): 6–11; K. L. O'Hara, *Forest Pruning Bibliography* Institute of Forest Resources Contribution 67. (Seattle: College of Forest Resources, University of Washington, 1989); J. K. Agee, *Fire Ecology in the Pacific Northwest Forests* (Washington, D.C.: Island Press, 1993); Sampson and Adams, *Assessing Forest Ecosystem Health.*

62 Oliver "Forest Development"; J. N. Long, and F. W. Smith, "Relations Between Site and Density in Developing Stands: a Description and Possible Mechanism" *Forest Ecology and Management* 7 (1984): 191–206; T. W. Daniel, J. A. Helm, and F. S. Baker, *Principles of Silviculture.* Second Edition. (New York: McGraw-Hill, 1979); Spurr and Barnes, *Forest Ecology*, Third Edition; Smith, *The Practice of Silviculture*; Kimmins, *Forest Ecology*; Oliver and Larson, *Forest Stand Dynamics*; T. T. Kozlowski, P. J. Kramer, and S. G. Pallardy, *The Physiological Ecology of Woody Plants* (San Diego: Academic Press, 1991); Kelty et al., *Ecology and Silviculture*; Agee, *Fire Ecology.*

63 Forest Products Research Society, *A Technical Workshop: Juvenile Wood— What Does it Mean to Forest Management and Forest Products? Proceedings of a Symposium* (Madison, WI: Forest Products Research Society, 1986); Lyon, et. al., "Concerns for the Technological Infrastructure."

64 R. S. Seymour, D. G. Mott, S. M. Kleinschmidt, P. H. Triandifillou, and R. Keane, "Green Woods Model: A Forecasting Tool for Planning Timber Harvesting and Protection of Spruce-fir Forests Attacked by the Spruce Budworm" USDA Forest Service General Technical Report NE-91; C. D. Oliver, and M. D. Murray, "Stand Structure, Thinning Prescriptions, and Density Indexes in a Douglas-fir Thinning Study, Western Washington, U.S.A." *Canadian Journal of Forest Research* 13 (1983): 126–136.

65 For references, see Oliver and Larson, *Forest Stand Dynamics*; O'Hara, *Uneven-Aged Management.*

66 J. W. Barrett, *Regional Silviculture of the United States.* Second Edition. (New York: John Wiley and Sons, 1980); J. W. Barrett, *Regional Silviculture of the United States.* Third Edition. (New York: John Wiley and Sons, 1995).

67 Boyce, "Management of Eastern Hardwood Forests"; Boyce, "Forestry Decisions"; Boyce, *Landscape Forestry*; R. S. Seymour and M. L. Hunter, Jr., "New Forestry in Eastern Spruce-fir Forests: Principles and Applications to Maine" Maine Agricultural Experiment Station Miscellaneous Publication 716, 1992; B. Lippke and

C. D. Oliver, "How can Management for Wildlife Habitat, Biodiversity, and other Values be more Cost-effective?" *Journal of Forestry* 91(12) (1993): 14–18; Oliver "A Landscape Approach;" Boyce and McNab, "Management of Forested Landscapes."

68 G. Baskerville, "Adaptive Management: Wood Availability and Habitat Availability" *Forestry Chronicle* 61(2) (1985): 171–175; C. Walters, *Adaptive Management of Renewable Resources* (New York: Macmillan Publishing Company, 1986); Feigenbaum *Total Quality Control*; Deming, *Out of the Crisis*; Reich, *The Next American Frontier*.

AUTHOR AFFILIATION

STEPHEN D. BOYCE Professor of Forest Ecology, Duke University, retired.

DAVID BRUCE Biometrician, USDA Forest Service, deceased.

BONNIE CHRISTENSEN Doctoral Candidate in History; University of Washington.

MARION CLAWSON Senior Fellow Emeritus, Resources for the Future, deceased.

STANLEY P. GESSEL Professor of Forest Soils, University of Washington, deceased.

ROBERT B. HARRISON Associate Professor of Soil Chemistry, University of Washington.

DOUGLAS HELMS Senior Historian, National Resources Conservation Service.

CHADWICK D. OLIVER Professor of Silviculture and Forest Ecology, University of Washington.

GERMAINE M. REED Professor of History, Georgia Tech University, retired.

WILLIAM D. ROWLEY Professor of History, University of Nevada, Reno.

JERRY R. SPRAGUE Tree Improvement Specialist, North Carolina State University.

RICHARD A. SELLARS Historian, National Park Service.

RICHARD A. SKOK Dean Emeritus, University of Minnesota, Department of Forest Resources.

GEORGE R. STAEBLER Director of Research, Weyerhaeuser Company, deceased.

HAROLD K. STEEN Environmental Historian, Duke University.

KEIR STERLING Historian, U.S. Army.

GERALD W. WILLIAMS Regional Sociologist, USDA Forest Service.

ROBERT L. YOUNGS Professor Emeritus of Forest Products, Virginia Tech.

BRUCE J. ZOBEL Professor Emeritus of Forestry, North Carolina State University.